Recent Advances in GRAPES

The Authors

Dr. M.K. Sheikh born on 1st January 1952 at Bijapur, Karnataka, did his B. Sc. (Agri) M.Sc.(Agri) Ph. D. from UAS Bangalore & UAS Dharwad. He served for two years in Saudi Arabia in Ministry Of Commerce Jeddah. During 1982 he joined University of Agricultural Sciences Bangalore. He worked in the capacity of Research Assistant/ Instructor/Assistant Professor/Associate Professor. Now he is Professor & Head, Department of Horticulture College of Agriculture Bijapur. He participated in 1st International Symposium on Pomegranate & Minor Fruits at Cuckurova University, Adana, Turkey. He visited South Africa, Israel and Spain. He guided number of PG students. He published books on Grape and Pomegranate in Kannada and English. He published Research Papers and popular articles in National and International Journal. He organized 2nd International Symposium on Pomegranate and Minor Fruits during 2009 at UAS, Dharwad, Karnataka, India as Organising Secretary, for this he has been awarded gold medal ISHS Belgium. He has been awarded a fellowship of the year NESA-2008 and FAHS.

Dr. N. Manjula, completed B.Sc. (Ag. Maco), M.Sc. (Ag. Extn.) and Ph.D. from UAS, Bangalore. She has eighteen years of experience in Teaching, Research and Extension. She has made significant contribution in preparing number of laboratory manuals. She has more than 100 publications to her credit. She has also brought out two extension folders and six books both in English and Kannada. Guided PG students and served as Staff Adviser, also worked as RAWE Co-ordinator. In her distinguished service she served in various capacities as Extension Guide, Assistant Professor, Associate Professor and presently as Professor and Head, Department of Agril. Extension Education, College of Agriculture, Bijapur (Karnataka) under the UAS, Dharwad. She is also nominated as Member of Board of Studies (FA) of UAS, Dharwad. She has been awarded a fellowship of the year NESA-2009 and FAHS.

Dr. S.R. Mulla, completed B.Sc. (Agri) and M.Sc. (Agri.) from UAS, Dharwad. He has ten years of experience in Teaching and Research. He worked in the capacity of Jr. Merit Associate/Research Associate at Bijapur campus, UAS Dharwad. Now he is Assistant Professor, Department of Soil Science and Agricultural Chemistry, University of Horticultural Sciences, Bagalkot. He published books on *viz.*, Beekeeping and Aerophonics, also he published 13 research papers, 20 popular articles, 8 extension leaflets; He has made significant contribution in preparing of laboratory manuals and annual reports and he has been awarded Choudhary Devilal outstanding contribution in dryland agriculture during 2009-10 (By : ICAR, New Delhi). He is also Secretary to the Alumni Association, College of Agriculture, Bijapur, UAS, Dharwad. (www.alumniagribijapur.com)

Recent Advances in GRAPES

— *Authors* —

M.K. Sheikh

N. Manjula

S.R. Mulla

2014

Daya Publishing House®

A Division of

Astral International Pvt. Ltd.

New Delhi – 110 002

ISBN 9789351301240

Published by : **Daya Publishing House®**
A Division of
Astral International Pvt. Ltd.
– ISO 9001:2008 Certified Company –
4760-61/23, Ansari Road, Darya Ganj
New Delhi-110 002
Ph. 011-43549197, 23278134
E-mail: info@astralint.com
Website: www.astralint.com

Laser Typesetting : **Classic Computer Services**, Delhi - 110 035

Printed at : **Replika Press Pvt. Ltd.**

PRINTED IN INDIA

Preface

Grape cultivation is very lucrative farming in India. Grape is grown in Maharashtra, Karnataka Andhra Pradesh and Tamil Nadu. India produces 22 lakhs tones of grapes, which is 3.31% of the world production. India has the advantages of harvesting grape at such a time when no grapes are harvested anywhere in the world. In Northern Karnataka two pruning and one crop is the practice, where as two pruning and two crops is the practice in south Karnataka. In Karnataka there is a scope to establish a wine park and lot of demand for table grape, raisin locally and export quality grapes in foreign countries. At present, USA and Turkey together produces 80% of the total raisin in the world. UK is largest importer of raisin. In this book "Recent advances in grapes" all aspects like soil and climates, varieties, propagation, training and pruning, girdling and thinning, irrigation, nutrient management, use of growth regulators, physiological disorders in grapes, pest and disease control, processing, wine making, transport and marketing, scope for export along with related research done on grape in the world is covered.

We sincerely hope that this book will be useful to scientists, teachers, extension workers, students and farmers.

Dr. M.K. Sheikh

Dr. N. Manjula

Dr. S.R. Mulla

डा. एन. के. कृष्ण कुमार
उप महानिदेशक (बागवानी)
Dr. N. K. Krishna Kumar
Deputy Director General
(Horticulture)

भारतीय कृषि अनुसंधान संस्थान
कृषि अनुसंधान भवन - II
पूसा, नई दिल्ली 110 012
Indian Council of Agricultural Research
Krishi Anusandhan Bhavan-II,
Pusa, New Delhi-110 012

Foreword

Cultivation of grape is one the most remunerative farming enterprises in India. The medicinal properties of grapes were mentioned as early as 1356-1220 BC by famous Indian medicine scholars, *Sasruta* and *Charaka* in their medical treatises entitles '*Sasruta Samhita*' and '*Charaka Samhita*', respectively. Cultivated grapes are believes to have been introduced into the north India by the Persian invaders in 1300 AD, from where they were spread to the south. Indigenous varieties like '*Rangspay*'.'*Shonltu White*' and '*Shonltu Red*' are still grown in Himachal Pradesh. It is grown under a variety of soil and climate conditions in sub-tropical, hot tropical and mild tropical regions in India.

Although, India has the distinction of achieving the highest productivity in the world with an average yield of about 30 t/ha, potential for grape production development remains in sustaining productivity and minimizing risks, refining technologies to achieve high productivity with food safety and technologies to produce export quality grapes and raisings. The constraints such as heavy initial investment for establishing a vineyard, recurring coasts in vineyard management, narrow genetic base and high risk of losing the crop due to unprecedented changes in weather, soil and water salinity, drought, short period available for ripening in the north, wine not being a popular drink at present, and last but not the least marketing problems need to be addressed.

Since the highest productivity in grapes has been achieved, efforts are needed to extend grape cultivation to newer areas. Suitable rootstocks are to be identified since soil and water salinity and drought are serious impediments in this direction. There is a need to diversify the utilization of grapes. Further, frequent market glut due to inadequate cold storage facilities need infrastructural backup. Maintenance of quality of table grapes through crop regulation in a priority to increase exports. The scenario of invasive pests, pesticide residue, organic production, tissue culture propagation of newer root-stocks, water-nutrient use efficiency, greater focus on application/use of micronutrients, mechanization and varieties that are self thinning with good horticultural attributes are some of the points to be considered for future.

In this endeavour, the book written and compiled by Drs. M. K. Sheikh, Manjula N. and S. R. Mulla will be useful to farmers, researchers, students teachers and extension workers. I wish them all the best for bringing out such edition.

N. K. Kumar

(N. K. Krishna Kumar)

Contents

1

Introduction

In India, we are cultivating grape on an area of 1,50,000 acres and producing 12,00,000 tonnes of fruits out of which 70 per cent are sold as table grape and from remaining 30 per cent fruits resins are prepared.

On an average approximately around 60,000 tonnes of resins are produced in India and around 40,000 Tonnes are required for internal consumption. Looking to these figures grape cultivation looks to be not so profitable in India. Therefore, the farmers instead of producing resins they should concentrate on production of export quality grape, and little diversification is also required to cultivate wine varieties and cultivation of off-season crop. The following suggestions are needed to be adopted:

1. They should think of starting a wine park by cultivating 200 acres of grape.
2. The excise duty on wine should be removed and wine should be brought under food item on line with Maharashtra state.
3. The National Research Centre, Pune, should come forward to spare wine variety suitable to the farming community of Bijapur area.

Grape Cultivation

In horticulture, before starting the cultivation of agri horticulture crops, their marketing knowledge and cultivation aspects are essential. The availability of land and the water with good condition suitable for irrigation and selection of rootstocks and ways to take up early crop and sustaining the crop for longer duration in more economical ways are important aspects in order to make the grape cultivation more suitable in Karnataka.

Important Points about Grape Cultivation

1. Grape, after planting, gives crop in two year's time and if goes on stock cultivation takes upto 3 years.
2. Grape is an assured crop which harvests crop till 10 years and proves to be economical.
3. When you calculate the cost of cultivation and returns, the cost-benefit ratio or the returns are good and assured.
4. Alert mind, knowledge, work in time and correct decision will lead to progress provided the farmer will help in good cultivation of grape.
5. In grape cultivation and round-the-clock-work ought to be there and the labours can be utilized all the year round.
6. Available land, water, working willingness together helps in bringing about prosperity to the grape farmers.
7. Grape cultivation is a combination of art and science, otherwise the farmers will face problem if they show negligence in any stage of crop cultivation.

Programming

1. How Many Acres of Grape to be Cultivated

How much land you have and out of which how much is available of good land is important? While deciding about how much area you want to bring for grape cultivation, don't select 35 guntas, 43 guntas, 41 guntas etc.

2. Where to Plant Grapes

Where to take up grape cultivation for example, if you have a land in 2-3 places select only one place and for that purpose get the soil and water examined including roads, and neighbouring areas.

Farmer Category	*Beginning Area*	*IIrd Stage*	*IIIrd Stage*
Economically weak	1 AC	2 AC	3 AC
Medium farmers	4 AC	8 AC	12 AC
Good farmers	10 AC	20 AC	30 AC

3. When to Plant Grape

In Karnataka there is a practice of 'April pruning' for which the rooted cuttings are to be planted in July-August, whereas for October pruning rooted cuttings are planted in December-January.

In Karnataka, grape fruits are available around March. If December last week or January first week is the planting time then, next year in the month of March, first crop comes for harvesting *i.e.*, minimum of 15 month duration is required to get the first crop. That is why 4-6 months earlier planting is taken up otherwise there will be a wastage of money for maintaining upto 6 months and a loss will incure.

4. Grape Varieties Suitable for Cultivation

There are several varieties of grape but considering the scope of different grape varieties decide on which are in demand in the market and which are good potential yielders. After harvesting if they are going to stand good storage life/shelf life then or in the near future it may appear that only seedless grape may have good demand in the market.

For Example

1. Thompson seedless
2. Sonaka
3. Kishmish charni
4. Kishmish bele
5. Ruby seedless
6. Kishmish long beli
7. Monaka
8. Centinial seedless
9. Meribian seedless
10. Black munnaka
11. Beauty seedless
12. Perlette
13. Sharad seedless
14. Flame seedless
15. Red globe
16. Clone - II

5. Maintaining Distance in Grape Cultivation

Distance between row to row should be 8 feet or 10 feet and within the row 4' and 6'. The height of the pendal should be around 6' and 6½'. After deciding the distance, the soil layers, the varieties suitable for such soil, the spraying convenience of the crop surrounding the area should be taken into account.

Digging/Filling of Pits

For cultivation the following points should be born in mind. The slope of land, the north and south direction.

Type of Soil	*Trench Depth*	*Trench Width*
Medium	45 cm	45 cm
Heavy, black soil	75 cm.	45 cm.
Murrum light	45 cm	30 cm
Light soil	60 cm	60 cm

In the month of June, crops like greengram, horse gram, bean, redgram, soybean or sowing of green manure crop, viz sunhemp are advocated and in September-October after filling up the trench with good soil, the planting of rooted cuttings should be taken up in December-January. Examine the soil profile of one meter depth and get it examined for N.P.K., micronutrient, pH, EC, RAC etc.,

7. Rooted Cuttings

Observe a good variety, without any disease, and get it guaranteed also from a reputed garden as to when it is going to have fruits. While bringing the canes confirm that it should have an age of 5-6 years and after bringing the canes 2-3 cutting should be made from each cane. Decide the rooted cutting you want in that case; you should take double the quantity and treat them with fungicide, IBA-1000 ppm, then plant it in the bed or packets applying FYM and DAP. Afterwards, spray BM-1 per cent, two to three times and mix DAP 1 kg plus magnesium sulphate ½ kg in 100 litres and spray. The cuttings will be ready for planting in a couple of months time.

Filling of Trench with the following Material

Sl.No.	*Quantity of Fertilizer/Ac*	*Medium Soil*	*Heavy Soil*	*Light Soil*
1.	Green manure (sunhemp)	One acre	2½Acre	1½Acre
2.	Compost	3 Truck load	4Truck load	5 Truck load
3.	7.10.5 (Sterameal)	750 kg	750 kg	750 kg
4.	DAP	100 kg	100 kg	150 kg
5.	SOP	100 kg	100 kg	100 kg
6.	Single superphosphate	100 kg	100 kg	100 kg
7.	Neem cake	750 kg	750 kg	750 kg
8.	Magnesium sulphate	25 kg	20 kg	30 kg
9.	Calcium: magnesium	50 kg	100 kg	100 kg
10.	Ferrous sulphate	5 kg	15 kg	10 kg
11.	Zinc sulphate	5 kg	15 kg	12 kg
12.	Borax	1 kg	2 kg	5 kg
13.	Mixture of micro-organism	15 kg	30 kg	30 kg
14.	Vermicompost	20 kg	15 kg	15 kg
15.	Organic fertilizer	100 kg	500 kg	500 kg

Use of Root Stocks in Grape

Root stock means what:

1. In Santara, Mosambi-Jamberi Rangpur rootstocks are used.
2. Sapota is grafted on Khirni, Rayan. Therefore, Grape should be grafted and used with justifying background of root stocks requiring;
   (*a*) To take maximum yield/acre.
   (*b*) To increase grape quality,
   (*c*) To reduce cost of grape cultivation,
   (*d*) To overcome more akalinity/sodicity problem.

At present grape is cultivated by planting the rooted cuttings, from April and October pruning. Now grape area is expanding and people want to take up grape cultivation in problematic soils by using root stocks which can overcome the problems of high EC, high sodicity, akalinity, high pH etc.

Nowadays, our grape export is increasing and from 2004 free trade has begun.

Characteristics of Root Stocks

1. Resistance to phylloxera;
2. Resistance to salt;
3. Increase in yield;
4. Resistance to drought;

Types of Root Stocks

Dogridge, Ramsay, Salt Creek, 1619, 1616, 1613, SO_4 Champion, 49B, 1103P.

In situ Grafting

1. First planting of rootstocks in the month of January-February *viz.* Dogridge, 1619, Ramsay (anyone)
2. Wedge grafting in the month of September-October.
3. Selecting two good branches of root stocks that can be utilized for making cuttings of root stocks.

There are two methods of grafting:

1. *In situ* grafting;
2. Preparation of rooted cutting to go for wedge grafting and after 3 months of hardening the grafted plants are ready for sale/planting.

Layout of Garden

The planting distance needs to be decided by the farmer, after which he has to mark the points by using a rope. Dig the Trenches thereafter, so that planting is taken up by selecting disease and insect-free planting material. Drench with bordeaux mixture (1 per cent) and press the soil around the plant.

Care after Planting

During November-December after planting, keep an eye on the plant by observing whether proper irrigation is going on or not. The plant should grow uniformly in November but as cold temperature exists, the growth might not be fast.

Time/Month	*Nutrition*	*After Care*	*Watering*	*Other*
October	During planting time apply starter solutions.	After and before planting apply BM 1 per cent	After planting one light irrigation	Full plant
November 1-10	Foliar spray of Mahagreen	BM 1 per cent spray insecticide + wet sulphur spray	One irrigation	
November 11-20	Ammonium Sulphate-50 kg $Mg.SO_4$, 10 kg to soil/AC.	Foliar spray of insecticide	One-irrigation	

Contd...

Table–*Contd...*

Time/Month	*Nutrition*	*After Care*	*Watering*	*Other*
November 21-30	Ammonium Sulphate	Wet. Sulf. spray. 75 kg	One irrigation	—
December 1-10	$MgSO_4$ 12 kg/AC	BHC -10 per cent 5kg plus 300 mesh Wet. sulf. Powder.		—
December 11-20	DAP-50 kg 7: 10:5-100 kg Neemcake-200 kg	Fungicide + insecticide spray	Irrigation	—
December 21-31	Magnesium Sulphate spray 0.2 per cent	—	Irrigation	—

The phosphate might not be fully available. Taking all the points into consideration, decision has to be taken.

Aftercare of New Plantation

Give support to the plants with the help of bamboo and encourage proper and uniform growth, take bottom cut, keep 3 buds at the bottom when they sprout. By February work should be completed. Before the rains, keep the garden clean, encourage two T and each T should have 12-14 canes and each cane should be nipped at 15 leaves.

Time/Month	*Work to be Attended*	*Crop Care*	*Nutrition*	*Irrigation*
January 1-10	Bottom cut, bamboo support	BM 1 per cent drenching	—	Irrigation
Time/month	Work to be attended	Crop Care	Nutrition	Irrigation
Jan -11-20	Tying of vine to bamboo support	Spraying of insecticide	—	One irrigation
Jan 21-30	Tying of vine and erecting of pendal	Fungicide spraying	—	One irrigation
Feb 1-10	–do-	Insecticide + fungicide spraying	—	One irrigation
Feb 11-20	–do-	Lihocin 500 ppm spraying	DAP 50 kg or 7:10: 5 100 kg Neem cake 250 kg	One irrigation
Feb 21-28	—	—	Bone meal 100 kg	One irrigation
March 1-15	Encouraging main arm or T on both sides	Fungicide and insecticide spraying of BM-0.5 per cent	—	2 irrigation
March 1-15	Nipping on T tip at 8th leaf	BM spraying 0.5 per cent	Lihocin of 500 ppm	2 irrigation

Contd...

Table–*Contd...*

Time/Month	*Work to be Attended*	*Crop Care*	*Nutrition*	*Irrigation*
April - 1-15	Keeping 5-7 canes on each T	DAP 1 kg + $MgSO_4$ - 1.5 kg/100 lit spraying	—	2 irrigation
April - 16-30	Above work to be completed	BM spraying 0.5 per cent FYM -1 truck 7:10:5 100 kg Neem cake-100kg	—	2 irrigation
May 1-15	—	200 kg bio-fertilizer	—	3 irrigation
May - 16-15	—	BM-0.5 per cent	—	3 irrigation

By February end, the sprouting will be there and by keeping 5-6 leaves the auxiliary growth should be removed.

June to September

1. Before the rain starts, the number of canes should be encouraged properly.
2. Wherever the cane is kept, its excess growth should be stopped.
3. Nip the canes at the 15th leaf.
4. Double plants removed from the garden.
5. Keep the pest and disease in control.
6. Excess canes to be removed.
7. Encourage green manure cultivation.

To implement the above points, the future course of action will be as follows:

June

1. Maintaining the canes *i.e.*, for one square feet area and one cane should be retained and excess canes be removed immediately.
2. Stop growth of each cane at 15th leaf.
3. The canes usually possess excess and unwanted growth in the leaf axil, which should be removed. *i.e.*, 2-3 leaf.
4. Take up spraying of lihocin @ 500 ppm.
5. Take up foliar spraying of Bordeaux mixture (0.5 per cent).

July

1. Examine the condition of canes if they are not good, restrict them to 10 per cent.
2. To enrich the canes give micronutrient spray.
3. Don't allow excess rainwater to stand in the field.

August

If excess growth is observed then curb it and whatever the necessary steps required implement during September and October. Prepare to carry out necessary things such as, labour, dormex application, secatures which should be dipped in formalin to disinfect, observe to dieback and stem borer etc.

Erection of Pendal

The erection of pendal is necessary because it gives support to the plant/vine, helps the vine spread properly, helps in pruning. Then spread the canes on wires, facilitate easy application of fertilizer, spray insecticide and fungicide. Harvest sunlight etc. It helps to save cost of cultivation.

Types of Training System

There are several types of training system:

1. Pendal or bower system
2. Four-arm kniffin system
3. Y-Trellis
4. Telephone system
5. Extended Australian gable system.

Every system has got its own advantages and disadvantages. Lot of researches have been conducted on different training system, of which T and Y system are most ideal to the present situation.

The Advantages of T&Y System

1. The system will be tough and long-lasting.
2. Care of every cane is possible.
3. Exposure to direct sunlight can be avoided.
4. Spraying of insecticide, fungicide and hormone will be easy.
5. Drip irrigation system is most suitable.
6. Increases the grape bunch quality.
7. Per acre production of grape will be enhanced.

Requirement of One Acre 'T' Training System

Area-1 Acre, or 0.4 hectare, 43560 cm square ft *i.e.*, 212 x 205 = 43560 + 100 sq. ft, tank, filter fixing area, cultivable area = 200 feet long x 200 width.

= 40, 000 sq. ft.

Details of Requirement of Steel for 'T' System

Types	*Number*	*Size cm x mm*	*Per Acre 200 Lines*	
			Number	*Total*
Corner stone poles	2	8 x 2.5 x 2	40	320
Supporting poles	2	8 x 2.5 x 2	40	320
Types	Number	Size cm x mm	Per Acre 200 Lines	
			Number	Total
Inside poles	7	8 x 2.5 x 2	140	1120
T	9	4 x 2.5 x 2	180	720
T support	18	2 x 2.5 x 2	360	720
Hole pass bar	22	1 x 1 x 0	440	440

One line requirement **One acre requirement**

	Holes	*Nut and Bolt*	*1 Bolts*	*Holes*	*Nut and Bolt*	*1 Bolts*
Inners	7 x 10	7 x 3	–	1400	420	–
Outers	2 x 10	2 x 3	2 x 10	400	120	400
Support	2 x 3	1 x 2	–	120	40	–
Total	1920	580	400			
	2000	600	400			

Leave 10 feet space in the east and west sides and 25 feet for north south.

16 long steel (20 + 20 + 70 + 90 = 200 Nos of pieces)

Iron bar - 20 feet long = 22 in Nos.

GI bar - 12 gauge

One line (running foot) One acre requirement

200 x 5 = 1000 ft. = 20000 ft.

Cutting works = Poles 16 ft. long for every two units in two 220 = 440 pieces.

16 feet for every 4 unit x 180 and from 8 Nos.

360 Numbers have to be made.

Steel bar 5 feet, 440 in number, with a total of 2200.

Wires 200 ft. and, 100 pieces.

For Erection of 'T' Pendal and of Fix Up the Poles

For every line	For every acre
2 x 2 2.5 x 11 pits	11 x 20 = 220 pits
11 x 10 = 110 Ghan foot	2200 Ghan ft.

Details of Concrete Requirement (for one acre)

1. Sand = sieved = 4 brass
2. Khadi = ½″ = 3 brass
3. Khadi = ¼″ = 3 brass
4. Dambar or coal tar = 2 brass
5. Cement = - 40 bags.

After completion of concrete work, wire work needs to be taken up and to all steel poles apply oil paint or red oxide.

Bamboo Support

To each vine, bamboo support is necessary. About 1000 bamboos are required. Bottom of the bamboo should be applied with coaltar and on top apply oil paint or red oxide.

Details of Pendal Erection

1. Steel pole, approximately 2 tonnes
2. G.I. wire, approximately 1.5 tonnes
3. Pass bars, approximately 0.4 tonnes
4. Nut and bolt, approximately 0.4 tonnes
5. I bolt, approximately 0.2 tonne

Expenditure on Erection of Pendal (Rupees)

1. Sand and transport=	1000-00
2. Khadi and transport =	1000-00
3. Dambar =	500-00
4. Cement =	6500-00
5. Painting cutting/fitting/Erecting =	6000-00
6. 1000 bamboos Nos. =	5000-00
7. Painting, fixing in pits and tying =	2500-00
Total	***Rs. 16,000-00***

Nutritional Value per 100 g (3.5 oz)

Energy	288 kJ (69 kcal)
Carbohydrates	18.1 g
Sugars	15.48 g
Dietary fiber	0.9 g
Fat	0.0 g
Protein	0.72 g

Contd...

Table–*Contd...*

Thiamine (vit. B_1)	0.069 mg (6 per cent)
Riboflavin (vit. B_2)	0.07 mg (6 per cent)
Niacin (vit. B_3)	0.188 mg (1 per cent)
Pantothenic acid (B_5)	0.05 mg (1 per cent)
Vitamin B_6	0.086 mg (7 per cent)
Folate (vit. B_9)	2 μg (1 per cent)
Vitamin B_{12}	0 μg (0 per cent)
Vitamin C	10.8 mg (13 per cent)
Vitamin K	22 μg (21 per cent)
Calcium	10 mg (1 per cent)
Iron	0.36 mg (3 per cent)
Magnesium	7 mg (2 per cent)
Manganese	0.071 mg (3 per cent)
Phosphorus	20 mg (3 per cent)
Potassium	191 mg (4 per cent)
Sodium	3.02 mg (0 per cent)
Zinc	0.07 mg (1 per cent)

Percentages are related to US recommendations for adults.

Top Producers of Grapes for Wine Making, by Area Planted

Country	*Area (km²)*
Spain	11,750
France	8,640
Italy	8,270
Turkey	8,120
United States	4,150
Iran	2,860
Romania	2,480
Portugal	2,160
Argentina	2,080
Chile	1,840
Australia	1,642
Armenia	1,459
Lebanon	1,122

World Grape Production and some of the Important Producers

Country	Productionin 2009 (Tonnes)	‡	Productionin 2010 (Tonnes)	‡	Percentage of World Production in 2010
China	8,039,091		8,651,831		12.67 per cent
Italy	8,242,500		7,787,800		11.40 per cent
United States	6,629,160		6,220,360		9.11 per cent
Spain	5,573,400		6,107,200		8.94 per cent
France	6,104,340		5,848,960		8.56 per cent
Turkey	4,264,720		4,255,000		6.23 per cent
Chile	2,500,000	(F)	2,755,700	(I)	4.03 per cent
Argentina	2,181,570		2,616,610		3.83 per cent
India	1,878,000		2,263,100	(I)	3.31 per cent
Iran	2,255,670		2,255,670		3.30 per cent
World	67,901,744	(A)	68,311,466	(A)	100 per cent

‡ Footnotes:

No symbol = official figure

(A) = May include official, semi-official or estimated data

(F) = FAO estimate

(I) = FAO data based on imputation methodology

Note: The individual country figures do not add to the world totals because many countries have been omitted from the table. The listed countries accounted for approximately 71.38 per cent of world production in 2010.

Source: Food and Agricultural Organization of United Nations: Economic and Social Department: The Statistical Division.

Chadha, (2006) stated that the grape cultivation in India has been commercially taken up under a wide range of soil and climatic conditions. While there are three distinct regions, *e.g.*, temperate (J.K. and H.P.), sub-tropical (Punjab, Haryana, Rajasthan and Western U.P.) and tropical (Maharashtra, Karnataka, Tamil Nadu and Andhra Pradesh), nearly 94 per cent of the area falls in the tropical region. The current area and production under grape is estimated at 60,200 ha with an annual production of 1,546,300 tonnes. Maharashtra accounts for more than 80 per cent of the total production. Despite doubts of success of viticulture in tropical regions, this region has resulted in world's highest productivity in spite of numerous adversities for the normal growth and productivity of vines. There has been a continuous shift in grape varieties commercially grown in different regions. While in the sub-tropics, early ripening variety Perlette is grown, Thompson Seedless, Sharad Seedless among seedless and Anab-e-Shahi, Gulabi and Bangalore Blue among seeded varieties are cultivated in Maharashtra and southern States. A number of clones of Thompson Seedless namely Tas-A-Ganesh, Sonaka and Manik Chaman are also being grown commercially primarily in Maharashtra. While a single crop of grape can be taken in sub-tropics, it is possible to take more than 1 crop under tropical conditions. The constraints in grape production also vary from region to region. In India, while 78 per cent of grape produce is used for table purposes, nearly 17–20 per cent is dried for raisin production, while the remaining 2 per cent is used for manufacturing of juice and wine. Fresh grape exports from India are estimated at 35,936 tonnes valued at 1106.7 million rupees to markets in 23 countries during 2004–05. In the same year, India imported 1111.27 tonnes of fresh grapes valued at Rs 519.86 lakh. With high quality grapes meeting international quality standards being produced, India has a good future to expand its fresh grape, raisin and wine industry to meet both the ever increasing domestic demand and expanding export market particularly in the European Union and South Asian countries.

Possingham (2006) observed that in less than 50 years grape production in the tropics has changed from being somewhat of a horticultural curiosity to grapevines becoming the most widely grown temperate fruit in the tropics. Currently world production is over 3 million tonnes annually, with India producing more than a million tonnes and Brazil just under this figure each year. Other tropical producers include Yemen 163,000, Peru 136,000, Thailand 40,000, Colombia 19,000, Tanzania 14,000 and Venezuela 11,500 tonnes. Grapes are also produced in the tropical part of Mexico and Saudi Arabia, while many other tropical countries produce small amounts for local consumption. Many different climates exist within the tropics resulting from variations in temperature, humidity, elevation and in the amount and distribution of the rainfall. Those best suited to growing ever-green vines and producing grapes have average minimum temperatures above 15°C and a reliable dry period of at least 3 months during which bunch formation and ripening occurs. In both the dry and wet tropics, the growth and cropping cycle of the vine can be manipulated to extend from 5 to 12 months by a combination of pruning, modifying vine water status and the judicious use of chemical regulators. In the tropics overhead trellises, which separate bunches from the vine canopy and reduce bunch rotting, are the most common form of vine training. Some elevated areas within the tropics have cold periods that induce leaf fall and in these areas the growth cycle of the vine is climate controlled and not able to be manipulated. Initially grape vines were cultivated in the tropics for the production of table grapes and this is still their main end use. With their capacity to produce more than one crop per year grapevines can provide massive yields of nutritious fresh fruit at times of the year when supplies of tropical fruits are limited. New developments in this area include the growing of an increased range of seedless varieties and the selection and breeding of new clones and varieties better suited to the tropics. A further innovation has been the use of plastic shelters and plastic-houses with open sides to, shield the vines from rain, reduce the incidence of bunch rots and slightly modify ripening times. The adoption of post harvest techniques originally developed for grapes from temperate areas has been of major importance for tropical grapes. These involve fumigating the grapes with gaseous and slow release sulphur dioxide papers combined with storage at temperatures just above zero. These treatments greatly extend the shelf life of tropical grapes and allow them to be transported to distant markets. A relatively new development has been the use of tropical grapes for the production of both wine and dried raisins. Good wine is now produced in the slightly elevated, wet tropical areas of northern and eastern Thailand, from both the Shiraz and Chenin Blanc varieties. Grapes from the larger crop that ripens in the cooler low rainfall months are of better quality and have less disease than those from the monsoon crop. In a region with a somewhat similar climate, the Deccan Plateau of South India, a range of winemaking varieties are now being grown and cropped once per year, with the grapes being used exclusively for wine production. Good wine is also made from grapes, from vines grown with irrigation and cropped twice per year, in dry tropical areas of both Venezuela (Carora) and Brazil (San Francisco Valley). As with virtually all wine grapes grown in hot areas it is essential to supplement the ferments with tartaric acid to provide an appropriate acid balance to the wine. Raisin production in tropical areas of high humidity represents perhaps the ultimate challenge for tropical grapes; however, this problem has largely been solved in South India using techniques developed in Australia for producing raisins rather than drying them on paper on the ground. In this scheme of things, the Thomson Seedless (Sultana) grapes are dried on wire netting racks with roofs that protect them from the rain, after treating them with emulsions of fatty acids (mainly ethyl oleate) and potassium carbonate that hasten drying and lead to the production of light-brown coloured raisins. It is clear that tropical grape production is now a major world industry and the grapes are able to be utilized for fresh eating, wine and raisin production.

Camargo, *et al.* (2006) stated that the grape growing was introduced in Brazil during the colonial period, in the 16th century and has become a commercial activity in the temperate climate regions in the south of the country only in the 20th century. Grapes of American origin, especially *Vitis labrusca* and *Vitis bourquina* cultivars used to be predominant. Tropical viticulture has started in the 1960s, when fine table grapes of the cultivar 'Italia' were grown in the Submédio São Francisco Valley in northeast Brazil. It has spread to other regions in the following decades. Important technological advancements have been achieved with the refinement of the plant management techniques and the development of rootstocks and cultivars for table grapes, for wine and juices which are adapted to tropical climates. In this article, the main grape growing regions of Brazil are presented. It contains information on the cultivars used, the technology of production, the products and the markets to which they are destined.

A trial was conducted (Kumar and Rajan, 2006) on twelve grape varieties and two hybrids at the Institute's farm during 1995–2002 at a spacing of 3 x 3 m to find out suitable cultivars of grape for commercial cultivation under north Indian conditions. The bud sprouting and ripening were earliest in Beauty Seedless, *i.e.*, 28 Feb and 28 May, respectively, but it was a poor yielder with only 7 kg fruits/vine. Flame Seedless sprouted at almost the same time, on 1st March and ripened 9 days later on 6th June, but gave the highest yield of 27.5 kg fruits/vine. Pusa Navrang recorded the next highest yield of 21.5 kg fruits/vine, which ripened on 10th June. Other cultivars viz: BA × BS-71-50, Delight, Pusa Urvashi, BA × Per-75-32, Gold, New Perlette, Perlette, Pusa Seedless, Kishmish Beli, Cardinal, Kishmish Cherani, sprouted between 3–14th march and ripened between 9–16 June, but like Beauty Seedless, they yielded less than 9 kg/vine. It was also observed that fruit quality of Flame Seedless and Pusa Navrang was not much affected by rain, although, these cultivars were also exposed to rain during ripening time. On the basis of marketable yield, fruit quality and tolerance to rains cultivar Flame Seedless for table purpose and Pusa Navrang for juice purpose can be recommended for commercial cultivation under north Indian conditions.

Anirudh *et al.* (2006) reported that eighteen grape varieties planted in 1997 were evaluated for their performance at Regional Research Station, PAU Bathinda (Latitude 30°17′ N; longitude 74°58′ E; and elevation 211 m amsl) with respect to yield, bunch and berry characteristics, and fruit quality. The early maturing varieties required lesser heat units (890.1 to 1051.1 day °C) as compared to the late maturing varieties (1020.1–1491.7 days °C). The varieties Angur Early and Banqui Abyad consumed minimum heat units (890.10 day °C) to reach the ripening stage. The variety Arka Shyam consumed highest heat units (1491.7 day °C) from fruit setting to ripening and this was on par with Arka Hans and Arkavati. Perlette consumed 1051.10 day °C heat units. Banqui Abyad required minimum number of days from full bloom to ripening (DFFB; 51.20) that were on par with Angur Early, Khalili, Sindhu, Delight, Beauty Seedless and Perlette. The variety Perlette produced the highest fruit yield (25.07 kg/vine) and was followed by Beauty Seedless (19.90 kg/vine). In respect to the bunch breadth and length, Beauty seedless, Delight and Perlette were found to be good performers. The highest berry weight was found in Cardinal (4.80 g) and it was on par with Banqui Abyad. The highest TSS was found in the variety Delight (18.23 per cent) and it was on par with Beauty Seedless, Khalili, Unknown, Banqui Abyad, Black Muscat, Black Prince, Arkavati, Arka Shyam and Shadipur Local. The cultivars, Perlette and Beauty Seedless were on par with respect to the TSS and acidity. The cultivar Perlette outshined all the other cultivars as far as the fruit yield is considered.

Chanana and Gill (2006) reported that the grapes are grown on an area of 1400 hectares in Punjab state. The varieties under cultivation are Perlette, Flame Seedless and Beauty Seedless. Although these varieties are early ripening, they suffer from one or another drawback, like poor quality bunches. With advancement of research made in the department of Horticulture, P.A.U., the suitable technology has been developed to enhance the quality of grapes. The bunch quality can be significantly improved by adopting fruit quality improvement technology. It comprises of flower bud thinning + girdling + GA_3 dips. Flower bud thinning is done one week before flowering thereby leaving 100–120 flower buds per panicle. When berry size is 4 mm, girdling is done by removing 4 mm wide ring of bark from the main stem and dipping the cluster in 40 ppm GA_3. The field should have sufficient moisture during girdling and at least 3 weeks after girdling. In addition, to this practice, two foliar sprays of potassium sulphate one at one week after fruit set and again at veraison, significantly improves the quality of bunches, having amber coloured and bold berries.

Viticulture is a millennia-old activity, traditional in many countries in Asia Minor and in Mediterranean Europe. The number of used cultivars is very large. Usually every region possesses its specific cultivars, selected over the centuries, which gave rise to typical regional products. Beginning in the XIXth century, however, new wine growing countries appeared, and in the old, traditional ones, diseases and pests were introduced triggering the beginning of genetic breeding programs. In our days, genetic breeding of the grape is an important tool to make viticulture possible in traditional wine growing countries as well as in the new, emerging grape growers. New quality concepts in terms of human health and environmental protection, as well as the permanent changes of people's consumption habits show that technological innovation is paramount to ensure competitiveness of grape growers in traditional and emerging regions alike. Cultivars bear particular importance in this respect. This paper reports on general strategies of genetic grape breeding and presents new cultivar of table grapes and grapes for processing, with emphasis on the scenario in Brazil.

Song (2006) stated that the grapevine is one of the most important fruit crops in terms of crop production in Korea. From 1999 to 2003, the area planted to grapes decreased at an annual average rate of 3.8 per cent, mainly due to import of foreign grapes and tropical fruits. Among the varieties 'Campbell Early' cultivar occupies 74.4 per cent of the total area cultivated to grapes. 'Kyoho' cultivar occupies 12.9 per cent, 'M.B.A.' 5 per cent, and 'Sheridan' 4.9 per cent. Since the 1990s, eight new table grape cultivars developed by NHRI have been released. For 'Campbell Early' grapes, many farmers use the Wakeman's system for trellis system. They are modifying this system having one or two arms and alternative canes pointed at different directions (the height of the trunk is around 1.5 m) to save labor. Farmers use shelves for 'Kyoho' grapes, which produces big berries around 12 g. This system has narrow (1.8 m x 53.6 m) rows. Vines in alternate row are removed four to five years after planting. Researchers at NHRI are now testing the different trellis systems: improved rain shelter, improved H-shaped and natural one-branch for rain shelter. Physiological disorders caused by erratic weather are common. Simple plastic houses reduce berry cracking and berry shattering with protection from high rainfall. Efforts have been made to solve brown spot and leaf burn symptoms caused by poor air conditioning in plastic house under hot sunlight after heavy rain. Chemicals are applied about 10 times a year, but this frequency is reduced to less than three to four times by using plastic house. A number of farmers are increasingly using cover crops to control soil erosion and weeds and environment-friendly agricultural practices such as organic farming, integrated fruit production (IFP), good agricultural practices (GAP). This year NHRI is completing three-year IFP programs to be conducted in five regions in Korea. This will include GAP to be implemented by the Ministry of Agriculture and Fisheries. The Grape Research Center established on June 21, 2005 through farmers' requests from the Korean government is conducting researches extensively on grapevines including wine.

Patil *et al.* (2006) reported that four promising genotypes have been identified after evaluation and multi-location trials. These include one mutant (ARI-302), one intra-specific hybrid (ARI-27) and two inter-specific hybrids (ARI-144 and ARI-516). The ARI-302, a seedless mutant of Anab-e-Shahi has bumper yield and best fruit quality with high T.S.S. It may be used for table purpose and in processing industries for raisin making. The ARI-27 a hybrid of grape varieties: Diamond Jubilee × Rubi red of *V. vinifera* is tolerant to the mildews and anthracnose and can be used for making red port wine and juice. Likewise, ARI-144 (Cheemasahebi × Catawba) and ARI-516 (Catawba × Beauty seedless) are combinations of *vinifera* × *labrusca* and *labrusca* ×

vinifera, respectively. Both the hybrids are promising for their fruit quality and also tolerant to fungal diseases. ARI-144 shows best fruit quality, foxy flavour and tolerant to both the mildews, while ARI-516 shows early ripening of bunches with pleasant musky flavoured berries. Both the hybrids are useful for table grapes, otherwise for wine making. However, all these promising selections have special significance in Indian grape industries for their commercial cultivation.

Gill *et al.* (2006) opined that the work on introduction and evaluation of grape varieties suitable for agro-climatic conditions of Punjab is going on at P.A.U. Ludhiana. Under this project, a number of varieties and hybrids were introduced from different sources and evaluated for earliness, yield and quality characteristics. As a result of this programme, one intra-specific hybrid-27, two inter-specific hybrids-516 and were introduced from ARI Pune. Out of these three hybrids, H-516 has shown promise under N-Indian conditions w.r.t. fruit yield, quality, juice content and its processing potential. H-516 was relatively more tolerant to anthracnose than the commercial varieties grown in this Northern region. Being a coloured variety and having loose bunches, H-516 has an edge over Perlette. The number of clusters per vine and T.S.S. were recorded maximum in this hybrid as compared H-27, H-144 and the recommended cvs. Perlette and Flame Seedless. H-516 ripens in 2nd week of June. It is relatively more tolerant to pre-monsoon rains as compared to Perlette. Based on the experimental results, it has been found that the superior quality fruits and higher yield per unit area can be obtained from this variety by planting at a distance of 5 x 12 feet on Y-trellis system of training. This grape hybrid has the potential to widen the narrow varietal base of grapes under N-Indian conditions.

In Indian vineyards, some of the rootstocks have been introduced as mixtures. B2-56 (resembling to 110R) is one such rootstock, which has reached the vineyards as a mixture along with Dog Ridge rootstock, which is the ruling rootstock in the Indian vineyards. Earlier the growers and scientists in India used to refer to the B2-56 rootstock as 'Dog ridge A' for the sake of differentiation. At certain locations the vines grafted on these rootstocks outperformed those grafted on Dog Ridge rootstock in the growers' vineyards particularly those that suffered from moisture stress and received saline irrigation (Sharma and Upadhyay, 2006). To know the potential of B2-56 for use under adverse situations like saline soils, the petiole samples of Tas-A-Ganesh from own rooted *Vitis vinifera* L. vines grafted on Dog Ridge (*Vitis champini*) and on B2-56 rootstocks were analyzed for sodium and chloride concentrations over three years at six different stages of vine growth. The tissue age varied from one month to five months. The vines received saline irrigation through drip irrigation system. The studies indicated that the grafted vines accumulated significantly lower chlorides in the tissues than the own rooted vines. The own rooted vines exhibited leaf-burning symptoms. No significant differences were observed between the two rootstocks for chloride content at any stage of sampling. The vines raised on B2-56 rootstock accumulated significantly less sodium content than those raised on Dog Ridge and own rooted vines at all the stages and years of sampling. At two stages, the petioles from vines grafted on Dog Ridge rootstock had higher sodium content than own rooted. The vines grafted on Dog Ridge rootstock produced higher pruned biomass than the other two stock-scion combinations. The results indicated that Dog Ridge, the most preferred rootstock introduced by the growers for salinity and drought tolerance did not prevent Na from accumulating in the vines to a desirable level under saline irrigation. Further the B2-56 rootstock exhibited better exclusion for sodium than Dog Ridge.

Somkuwar *et al.* (2006) opined that with the increase in salinity in the soil and bad quality of irrigation water due to salt, the yield and quality of grape start declining. To avoid the adverse effects of salinity and chlorides in irrigation water, use of rootstock in grape cultivation has become mandatory as the vineyards on rootstocks give better yield and also quality as compared to the own rooted vines. The variation in yield and quality in grapes may be due to the differences in uptake of the nutrients and the dry matter partitioning in vines. The present study was conducted during 2004–05 on seven-year-old Thompson Seedless grapes grafted on different rootstocks, *viz.* Dogridge, Salt Creek, 1613-C and St. George. The individual plant parts were analyzed for per cent dry matter distribution and the nutrients content. Significant differences were recorded for per cent dry matter distribution in different rootstocks. Among the different parts, the roots of Dogridge had the highest dry matter distribution of 56.19 per cent followed by St. George (45.11 per cent) and Salt Creek (40.11 per cent), respectively over the own rooted vines. The rootstock 1613-C had higher dry matter distributed in stem (40.92 per cent) compared to other rootstocks. The maximum total dry matter was recorded in secondary cordon of Salt Creek, however, 1613-C showed higher dry matter accumulation in canes as compared to other rootstocks. Among the nutrients, significantly higher N was recorded in canes as compared to other parts (trunk, cordons and roots) in all the rootstocks. Salt Creek had higher N (1.96 per cent) content in canes while lowest N (1.06 per cent) was estimated in 1613-C rootstock. Phosphorous content was less in the trunk than all other parts. Higher per cent of potash was recorded in canes as compared to other parts studied.

Statish and Adsule (2006) reported that the grape rootstocks are playing a major role in Indian viticulture to combat adverse effects of water scarcity, soil and water salinity. Apart from Dog ridge, which is the leading rootstock in grape growing regions of India, some other rootstocks are also gaining importance for both table and wine grape varieties. Rootstocks vary in their ability to root, based on biochemical constituents of mother vines and also IBA (Indole butyric acid) concentration used for treatment of cuttings. An experiment to standardize IBA concentration for better rooting of grape rootstocks revealed maximum success per cent in *Vitis longii*, SO-4 and 99 R using 1000 ppm of IBA, while it was maximum using 750–1000 ppm of IBA in Dog ridge, Salt Creek, 110 R and 1103 P rootstocks and Thompson Seedless recorded highest success per cent using 500 ppm. Early sprouting was noticed in *Vitis longii*, SO-4, 99 R when treated with 2000 ppm; Dog ridge, Salt Creek, 110 R and 1103 P treated with 1000 ppm and Thompson Seedless with 500 ppm. Cuttings taken from mother vines with moderate C/N ratio and less phenol content recorded maximum success per cent with maximum root length. This suggests rooting ability of different rootstock depends on biochemical content of mother vines and also concentration of IBA used for treatment of cuttings.

Barreto *et al.* (2006) stated that the response of grapevines to tissue culture has been reported to be genotype dependent, hence it becomes imperative to optimize culture conditions for newly developed varieties or cultivars, needing a large scale planting but availability of sufficient planting stock is a constraint. In the present study, conditions for in vitro propagation of six popular table grape cultivars have been standardized. Single node stem segments of cultivars 2A-Clone of Thompson Seedless, Red Globe, Crimson Seedless, Thompson Seedless, Flame Seedless and Italia cultured on ten different basal media showed varied percentages of bud break and morphogenetic responses. The percentage of bud break varied among the six cultivars. Direct rooting at basal ends of single node segments was observed in all the cultivars except in Italia. Woody plant medium induced the highest response and Eriksson medium the least in all the cultivars tested. Rooted nodal segments with shoots in axils could be established into whole plants on potting. Six cultivars showed different optimum concentrations of growth regulators for induction of maximum number of multiple shoots in both primary as well as secondary nodal segments. Shoot proliferation could be enhanced by several fold in the majority of cultivars on culture of initial shoot clumps to glass bottles instead of culture tubes. Six cultivars showed significant differences in optimum requirement of nutrients and growth regulators for shoot elongation, in vitro and ex vitro rooting and survival rate. Tissue culture plants of all the six cultivars could establish successfully in a greenhouse.

Murthy and Smita (2006) reported that the genotypic influence on in vitro rooting and further plantlet establishment was investigated employing single nodal cuttings of seven different grape hybrids and their respective parents. All the genotypes produced adventitious roots in response to the growth regulator stimuli, however, pronounced differences were observed with respect to their rooting potential for applied growth regulators. Black Champa, Arka Neelamani, Anab-e-Shahi, Shweta Seedless, Thompson Seedless and Arka Krishna gave better rooting response in terms of both: number of roots and length of roots on MS medium supplemented with IBA compared to those cultured on NAA supplemented medium. Arka Thrishna, Bangalore Blue, Arka Chitra, Angoor Kalan and Arka Majestic produced less number of roots which are shorter in length. The various indicators/parameters indicate that the rooting potential of the progenies were in line with the rooting potential of their respective parents more so conforming with their female parents.

Tamhankar *et al*. (2006) stated that the study of genetic relationships in grapes is not only important for breeding but also for evolutionary studies and germplasm conservation. In the present study, Inter Simple Sequence Repeat (ISSR) markers were used for the DNA profiling and characterization of total 143 grape varieties and rootstocks. A set of selected polymorphic primers was used to generate DNA amplification profiles. The size of amplification products ranged from 300 to 2000 bp and the number of polymorphic bands from 3–22. Cluster analysis resulted into separation of varieties belonging to *vinifera* and *labrusca*. Separate groups based on the berry colour were generally observed. Varieties belonging to *V. vinifera* appeared more diverse and formed distinct sub-clusters while varieties belonging to *V. labrusca* showed homogeneity. Among 43 seedless varieties, a high degree of similarity was observed. Seeded grape varieties showed more polymorphism than seedless ones. Among the 26 rootstocks analyzed, three main clusters were observed which were mostly according to the original species from which the rootstocks were derived.

Thus, the ISSR markers appear to be reliable and efficient for the assessment of genetic relationships among grape varieties.

Upadhyay *et al.* (2006) opined that downy mildew is one of the most important diseases of grape, which affects the quality and productivity of grape production in India to a great extent. It is caused by the obligate fungal pathogen *Plasmopara viticola* and requires preventive and remedial spray of fungicides for management, which pose environment and health hazards besides increasing the cost of production. Development of downy mildew resistant commercial varieties is the top priority of grape breeders. The perennial and heterozygous nature of grapevines makes conventional approaches of resistance breeding laborious and time consuming. Marker assisted selection (MAS) offers an opportunity to these limitation. In MAS, a molecular marker closely linked to the trait is identified and used for the selection of the progenies. Microsatellites or SSR are highly reliable class of molecular markers. In this study, 15 Vitis specific SSR primers were used to amplify DNA from several resistant and susceptible grape accessions. One primer was monomorphic and 14 were polymorphic and detected 135 alleles with an average of 9.64. The number of alleles per primer ranged from 5 (VVMD28 and VVS16) to 16 (VVMD7). These accessions were also screened against the pathogen and were rated on a scale of 1–9. The disease rating was 1–3 for resistant accessions and 7–9 for susceptible accessions. Association mapping approach was applied for the analysis of marker and disease data. Eleven promising SSR markers showing significant association with disease resistance/susceptibility were identified. Six SSR markers showed promising association with resistance whereas five SSR marker bands were found to be associated with susceptibility. This work demonstrated the feasibility of association mapping for identification of linked marker in a perennial crop.

Chougule *et al.* (2006) reported that the present investigation was carried out in order to study the effect of canopy management and cycocel application on yield and quality attributes, viz: weight of bunch, length of berry, diameter of berry and weight of berry. The investigation was laid out on a four-year-old orchard of Thompson Seedless grape with two replications in Factorial Randomized Block Design. The 7-5 sub-cane pruning was done after April pruning. The treatment consist of: three cane densities, *i.e.*, 30, 35, 40 per vine; three leaf densities, *i.e.*, 14, 16 and 18 per cane after October pruning with application of cycocel at five concentrations viz: control, 250, 500, 750 and 1000 ppm at five leaf stage after October pruning. Significantly highest yield (15.96 kg/vine; 35.44 MT/ha), TSS acid ratio (45.7) with attractive grayish yellow colour of berries and significantly minimum acidity (0.49 per cent) was recorded in cane density 35 per vine, leaf density 16 per cane with application of cycocel 500 ppm at 5 leaf stage after October pruning; however, the highest weight of bunch (439.99), length of berry (k.69 mm), diameter of berry (19.78 mm), weight of berry (3.51 g) was recorded in cane density 30 per vine, leaf density 16 per cane with application of cycocel 500 ppm at five leaf stage after October pruning.

At this centre, the National Repository of Grape Germplasm presently contains 112 indigenous and 288 exotic accessions of grapes (Mundankar and Karibasappa, 2006). Information on different aspects of these accessions like notification details, distinguishing morphological characters, area of adaptation, yield potential, bunch and berry characteristics, resistance to different biotic and abiotic stresses are required by scientist, managers, administrators, extension personnel and farmers. A user-friendly Information System for Grape Germplasm in India has been developed, which makes storage and modification of data easy and enables quick and accurate retrieval of data. The software provides the facility to store and edit data on 113 characteristics as per the UPOV, IPGR and NBPGR descriptors for grapevines. Presently, the database contains data for 138 accessions on 42 characters that have been evaluated under tropical Indian agro climatic conditions.

Adsule *et al.* (2006) observed that the paper begins by dealing with the world raisin production scenario in the context of Indian raisin production. The necessity of diversification to raisin product in India is further elaborated giving characteristics of good raisins and varieties of grapes used for raisin production. Further, cultural practices used for raisin grapes, technology used for raisin production and locations of raisin production in Maharashtra and Karnataka is further elaborated giving more stress on drying structures. Quality of Indian raisins was also discussed in detail keeping in view the Codex standards and deficiencies in Indian raisins were highlighted for improvement and compliance to Codex standards.

Gill and Arora (2009) reported that the performance of six hybrid grape varieties *viz.*, Chasan-B, Portan, Muscat Hamburg, H-27, H-144 and Punjab Purple (H-516) on Y-system of training was compared with commercial grape variety Perlette. The grape variety Perlette was earliest (end May) to ripen, followed by Punjab Purple which ripened in 1st week of June, while Muscat Hamburg, hybrids H-27 and H-144 ripened in 1st week of July. The pooled data of 2007 and 2008 showed that the variety Punjab Purple had significantly higher average number (89.1) of clusters/vine than all the other varieties and hybrids. The least number of clusters/vine (40.4) were obtained in hybrid H-27. The average weight of cluster was also maximum (350.5 g) in variety H-27. The average yield/vine was maximum (17.5 kg) in Punjab Purple, however its yield was at par with commercial grape variety Perlette. The significantly lowest (10.4) yield was obtained in variety Portan. The yield in Muscat Hamburg was at par with grape hybrid H-144. The quality parameters like TSS, Acidity and TSS/Acid ratio also varied among different grape varieties and hybrids. The grape variety Punjab Purple had significantly higher (17.7 per cent) T.S.S, followed by the hybrid H-144 (16.7 per cent). The per cent acidity was lowest (0.48 per cent) in variety Portan. The TS.S/Acid ratio was maximum (33.8) in Portan, which was significantly at par with Punjab Purple. Amongst the hybrids evaluated in the present study, the hybrid H-516 (Punjab Purple) recorded higher yield and better fruit quality.

2
Soil and Climate

Soil colour and texture are important in grape cultivation. Soil depth in relation to the thickness of surface soil and depth of the soil upto watertable is important for grape cultivation.

Soil pH is important in plant nutrition in view of two reasons; where pH is linked with the composition of exchangeable cation of the soil, acid soils have low cation-exchange capacity with the cation largely dominated by hydrogen, whereas alkaline soils have relatively high CEC value with clay particles dominated by calcium, magnesium, potassium and sodium ions.

The acid soils tend to limit the availability of phosphorus and calcium, while the alkaline soil limits the availability of iron, manganese, boron and zinc. In acid soil acid is fixed as iron and aluminum phosphate that are not available to vines. Phosphorus fixation was as high as 90 per cent in Indian soil and soil required 3.8 to 13.6 fertilizer P equivalent for affecting one unit increase in available P by soil test. Correction of soil acidity by liming the soil. pH is generally associated with high content of soluble salts in the soils of India. Correction of soil alkalinity is due to presence of excess sodium mostly in the form of carbonates, tend to reduce the availability of calcium, which renders insoluble and is precipitated as calcium carbonate in non-saline alkaline soil, in reclaiming such soils this process is needed to be reversed by replacing excess sodium in the exchange complex of soil by calcium. the North Karnataka soils are having pH 7.8 to 9.0, EC (mmhos/cm) 0.8 to 6.2, ESP (per cent) 10-40, organic carbon (per cent) 0.3 to 0.46 and CEC (me/100g) 13 to 50.

The grapes were grown successfully with ease between 340 and 390 latitude on either side of the equator. In India it is grown successfully around 15° N latitude. It requires hot and dry climate of summer places with humid summers are not suitable.

The rainfall is having very impact on grape development, during fruit bud formation from April-May it requires 10-100 mm rainfall. For bloom and fruitset it requires 5-50 mm rainfall in November to December, for berry growth it requires 0-50 mm rainfall in December-February and for ripening 0-10 mm from February to March.

In India where, the areas receiving rainfall not exceeding 900 mm are good grape cultivation. Rain during flowering and ripening are not favourable for grape. Humidity with rainfall during flowering favours downy mildew, where as drizzling for 2 days continuously at flowering can devastate the crop, at ripening 10 mm rainfall causes berry crack and rotting. Rainfall is association with clouds may the initiation of floral primordium. Rain association with cold weather may reduce the bud break and consequently the yield.

Low temperature in association with high humidity during shoot growth and bloom (19-20 leaf stage) after October pruning can be affected by downy mildew. Low temperature less than 15°C in association with dry weather during berry growth stage prior to berry softening (Veraison) is favourable for incidence of powdery mildew on leaves and berry.

Mild temperature during berry ripening helps in colouration of grapes. At low light intensity grapevine leaves can carry out photosynthesis, they absorbs 90-95 per cent of solar radiation. Thompson seedless buds at 35°C could be favourable for fruitfulness.

Palaniappan and Varalakshmi (2005) reported that the adverse situations affecting the productivity of grape may be soil, water or climate; factors pertaining to climate such as temperature, rainfall and relief affect the physical and chemical characteristics of soil and water. Soils rich in free calcium carbonate, exchangeable sodium and chloride affect the growth of grape. High clay content along with sodium interferes with root formation and aeration. Excess sodium also raises the soil pH and there by affect the availability of nutrients and especially micronutrients. Similarly excess chlorides both in soil and water causes toxicity symptoms in plants restricting it's growth and quality of crop.

Maharashtra soils having high free calcium carbonate more than 8 per cent is quite common. Reactions of these soils are alkali making availability of zinc and iron more critical. In Andhra Pradesh, grape-growing soils are coarse textured and gravelly in nature with poor organic matter status. This situation leads to deficiency of boron, magnesium and zinc. In northern Karnataka and other places where ESP values are more than 12 in soil, K/Na ratios in the plant have been distorted leading to imbalanced cationic nutrition. In few cases, where undecomposed organic matter has been added regularly, zinc deficiency has been noticed.

Crux of the problem can be efficiently tackled by proper soil and water management methods. Rootstocks may help to overcome higher limits of sodium and chloride to a certain extent. Integrated understanding of soil, water, plant and atmospheric continuum can help the grape growers to optimize various inputs in order to sustain the productivity of quality grape.

Schaefer (2006) reported that the paper deals with commercial wine making under tropical conditions. It is not only the heat of 25–35°C during the harvesting and fermentation period that makes this a special challenge but also the general conditions of grapes grown under tropical conditions without a dormant period of vines and permanent short day conditions. These factors result in an irregular sprouting and consequently irregular ripening of bunches and with some cultivars even extreme irregular ripening of berries within one bunch.

Narendra Babu and Prakash (1996) conducted a pot trial experiment on six grape genotypes (Dogridge, Saltcreek, 1613, St. George, *Vitis champini* clone and cultivar Thompson Seedless) with three levels of moisture stress for a period of 14 days. Of the six grape genotypes studied, the rootstocks, 1613, St. Geroge, *Vitis champini* clone along with cultivar Thompson Seedless did not survive the stress under 100 per cent moisture stress condition. However, all of them survived under 50 per cent stress condition. Under 100 per cent stress, all the genotypes showed general reduction in Relative Water Content (RWC), Osmotic Potential (OP), Leaf Water Potential LWP),Transpiration rate (AE), Net Photosynthetic Rate (Pn) and Stomatal Conductance (Gs).Under 50 per cent moisture stress RWC, AE, Pn, Gs, reduced in all the genotypes. Whereas, they were on par with the OP and LWP. Leaf temperature (LT) increased in all the genotypes at both the stress levels. The N, P and K content reduced in all the genotypes as a result of moisture stress. However, in Dogridge the N content in the leaves of stressed plants was almost on par with the well watered plants. Another experiment conducted on the diurnal variation of leaf WC, OP and LWP were maximum in the morning in all the three genotypes (Dogridge, saltcreek and Thompson Seedless) studied and reduced considerably by noon and slightly covered in the afternoon under complete moisture stress. Soil moisture content was reduced in all the genotypes due to moisture stress. In the root development studies with six grape genotypes the highest root dry weight among all the categories of roots was found in Dogridge. Similarly, the dry weights of the shoot as well as the plant height were maximum in the rootstock Dogridge.

Myburgh *et al.* (1996) stated that there is uncertainty about the optimum soil preparation depth for vineyards and to what extent irrigation can compensate for insufficient soil depth. Pinot noir was planted under non-irrigated conditions to soil preparation depths of 200 mm, 400 mm, 600 mm, 800 mm, 1000 mm and 1200 mm. Field lysimeters of 400 mm, 800 mm and 1 200 mm depths were constructed to study the effect of irrigation and to facilitate accurate measurement of water consumption. Vegetative growth and physiological activities responded positively to increasing soil depth under dryland conditions as well as under irrigation. Irrigation improved vegetative growth compared to non-irrigated treatments of corresponding depths. Under dryland conditions optimum vegetative growth and physiological response were achieved during the third year with 800 mm and 1 000 mm soil depths. Loosening the soil to 1200 mm induced excessive growth on a lengthened Perold trellising system. Irrigation stimulated vigour and consequently sufficient vegetative growth was achieved at the end of the second season for 800 mm and 1 200 mm deep soil. This situation was only attained at the end of the third season for 400 mm deep irrigated soil. At this stage irrigation stimulated excessive vigour for 800 mm and

1200 mm deep soils. Water consumption of irrigated vines increased during the first three years after planting. During the third year, water consumption of the 800 mm irrigated treatment was comparable to the 500 mm generally regarded as the norm for full-bearing vines in the coastal region of the Western Cape; Increased vegetative growth induced by increased soil depth reflected clearly in evapotranspiration. Three-year-old vines, which grew too vigorously on 1 200 mm deep soil, consumed 614 mm of water. On the other hand, where 400 mm soil depth limited growth, water consumption amounted to only 387 mm. During the first and second seasons irrigation should be reduced to 44.6 per cent and 65.5 per cent respectively of the irrigation generally required by bearing grapevines.

Costantini *et al.* (2009) stated that the adoption of precision agriculture in viticulture could be greatly enhanced by the diffusion of straightforward and easy to be applied hydropedological models, able to predict the spatial variability of available soil water. The Lin's and Host hydropedological models were applied to standard soil series descriptions and hillslope position, to predict the distribution of hydrological functional units in two vineyards and their relevance for grape yield and wine quality. A three-year trial was carried out in Chianti (Central Italy) on Sangiovese. The soils of the vineyards differentiated in structure, porosity and related hydropedological characteristics, as well as in salinity. Soil spatial variability was deeply affected by earth movement carried out before vine plantation. Six plots were selected in the different hydrological functional units of the two vineyards, that is, at summit, backslope and footslope morphological positions, to monitor soil hydrology, grape production and wine quality. Plot selection was based upon a cluster analysis of local slope, topographic wetness index (TWI), and cumulative moisture up to the root limiting layer, appreciated by means of a detailed combined geophysical survey. Water content, redox processes and temperature were monitored, as well as yield, phenological phases, and chemical analysis of grapes. The isotopic ratio ^{13}C was measured in the wine ethanol upon harvesting to evaluate the degree of stress suffered by vines. The grapes in each plot were collected for wine making in small barrels. The wines obtained were analysed and submitted to a blind organoleptic testing.

The results demonstrated that the combined application of the two hydropedological models can be used for the preservation of the moisture status of soils cultivated with grape during summertime in Mediterranean climate. As correctly foreseen by the models, the amount of mean daily transpirable soil water (TSW) during the growing season differed considerably between the vineyards and increased significantly along the three positions on slope in both vineyards. The water accumulation along slope occurred in every year, even during the very dry 2006. The installation of indicators of reduction in soils (IRIS) tubes allowed confirmation of the occurrence of reductive processes in the shallowest soil.

Both Sangiovese grape yield and quality of wine were influenced by the interaction between TSW content and salinity, sometimes contrary to expectations. Therefore, the studied hydropedological models were not relevant to predict grape yield and wine quality in all the hydrological functional units. The diffusion of hydropedological models in precision viticulture could be boosted considering salinity along with topography and soil hydrological characteristics.

The paper presents the results obtained from a methodological interdisciplinary study realised with the aim of setting the viticultural vocation of the "Vino Nobile di Montepulciano" territory and zoning the vineyards. The study provided for climate monitoring, soil biological, physical, chemical, hydrological characterisation, as well as soil survey and cartography. Moreover, a large set of winegrape characteristics was measured during the vegetation of vines and at the harvest. Finally, the wines obtained by the standard techniques for small-lot winemaking winery, have been sensory analysed.

The trial showed the importance of environmental characteristics to define the quanti-qualitative answer of the winegrapes; in particular it specified the fertility conditions of different lots and the relative oenological answer of the wines. So, the most fertile soils, those lacking permanent limitations to crop cultivation, owing to excessive productivity, provided the worst viticultural and oenological results. On the other hand, the best results could be obtained in quite fertile soils, but with some pedological limitations. The least fertile soils, those which have been more or less eroded, produced always less than the better preserved ones, but with very variable oenological results according to the influence of the year.

Jackson and Lombard (1993) stated that the world production of wine has steadily risen over recent years and consumption has not kept pace with this increase, thus many countries have surpluses of wine which pose problems in international trade. Despite these problems, there is not, generally, a surplus of high quality wines. Quality is not easy to define, but ideally, it should be related to intrinsic visual, taste, or aroma characters which are perceived as above average for that type of wine. Usually this is reflected in the price paid for that wine – although price is not necessarily a reliable indicator since it can be influenced also by fashion, tradition, availability and personal preferences. Unfortunately, despite the many references to quality and the amount of work which directly or indirectly refers to it, there is still confusion over what contribution climates, sites, and viticultural practices really make. This paper is a review of the effects of these environmental and management practices which may change grape composition and wine quality. Its scope is limited to table wines rather than fortified wines.

Cristina *et al.* (2013) reported that in rain-fed vineyards, water availability strongly influences vegetative and reproductive growth and, thus, berry quality. Narrow-band spectral indices might provide information on vine structure and physiological status and might be a useful tool for predicting berry quality. The present study tested the feasibility of using the normalized difference vegetation index (NDVI) and the photochemical reflectance index (PRI) to assess berry quality in five commercial vineyards (cv. Xarel·lo) in 2009 and 2010. Measurements of predawn water potential (ψ_{pd}), canopy to air temperature difference, fractional intercepted photosynthetic active radiation (fIPAR), exposed leaf area, and canopy reflectance were taken at veraison. Berry weight, total soluble solids (TSS), and titratable acidity (TA) were determined at harvest. Values of ψ_{pd} indicated mild to moderate water deficits. NDVI characterized the effects of water availability on vine vigor (fIPAR), while PRI was related to water status (ψ_{pd}). The extent of water deficits was a key factor in determining the aptitude of spectral indices at estimating berry quality attributes. Thus, consistent with the influence of vine vigor and water status on berry quality attributes, NDVI was related to both TA ($r^2 = 0.46$) and IMAD (the ratio of TSS to TA; $r^2 = 0.27$), whereas PRI was related to both TSS and IMAD ($r^2 = 0.23$ and $r^2 = 0.34$, respectively). In addition, PRI was positively related to berry weight ($r^2 = 0.68$). The results suggest the potential of reflectance indices of canopy vigor (*i.e.*, NDVI) and photosynthetic functioning (*i.e.*, PRI) at estimating berry quality attributes in vineyards experiencing mild to moderate water deficits.

Anthony *et al.* (2012) stated that the current study explores the relationship between sensory characteristics and wine composition of Cabernet Sauvignon wines in relation to Australian geographical indications (GIs). Descriptive sensory analysis was conducted to characterize the sensory attributes of commercially produced Cabernet Sauvignon wines from the Barossa Valley, Clare Valley, Coonawarra, Frankland River, Langhorne Creek, Mount Barker, Margaret River, McLaren Vale, Padthaway, and Wrattonbully GIs. Canonical variate analysis using the significant sensory attributes demonstrated that each GI could be distinguished from the others. A recently developed analytical method was used to analyze over 350 volatile compounds in the wines assessed, and measures of the major nonvolatile components were also determined. Compositional results were analyzed using partial least squares discriminant analysis to identify candidate components that were unique to certain GIs, including 2-isobutyl-3-methoxypyrazine, menthone, isomenthone, carvacrol, δ-octalactone, *p*-methylacetophenone, *m*-dimethoxybenzene, protein-precipitable tannin, and monomeric anthocyanins. Results demonstrate that Australian Cabernet Sauvignon wines have common sensory attributes related to geographic origin. The work also identifies a number of candidate components that are related to individual GIs which warrant further investigation. The study is the first to explore the concept of regionality in Cabernet Sauvignon wines from Australia.

King *et al.* (2012) opined that the manipulation of vine balance is widely practiced to enhance grape and wine quality. Reported benefits from crop thinning are ambiguous. Studies were undertaken over two growing seasons (2006–2008) in Hawke's Bay, a cool-climate region of New Zealand, to investigate the influence of basal leaf and crop removal on fruit and wine composition and the indices of vine balance associated with optimum quality. Three levels of basal leaf removal in the fruiting zone were used. Crop removal achieved average reductions of 15 per cent with moderate treatment and 35 per cent with severe treatment below the average nonthinned yield of 17.5 t/ha. Treatments were undertaken at the preveraison growth stage on Merlot vines. Leaf removal had no effect on fruit composition, but levels of total anthocyanins and the flavonol quercetin-3-glucoside were enhanced in the wines. In contrast, crop removal increased sugar concentration and decreased titratable acidity of the must. While crop removal had no effect on wine anthocyanins, the proportions of anthocyanins as malvidin-3-glucosides and total phenolics in the wines were significantly increased. Relative to accepted indices of vine balance, the study vines were unbalanced with excessive shoot growth and dense canopies. A crop reduction of ~6.0 t/ha brought the capacity of the exposed leaf area into balance with that required to ripen fruit to optimum maturity and produce high-quality wines.

Justin *et al*. (2012) conducted the study to determine the key environmental and viticultural variables affecting the concentration of 3-isobutyl-2-methoxypyrazine (IBMP) in Cabernet franc grapes. Berries were sampled from individual vines at 30 days after anthesis (DAA), 50 DAA, and harvest from 10 and 8 commercial New York State vineyards in 2008 and 2009, respectively. IBMP concentrations at 50 DAA were significantly higher in the warmer 2008 growing season (2008, 103 to 239 pg/g; 2009, 12 to 87 pg/g). However, in the cooler 2009 growing season a smaller per cent decrease in IBMP from 50 DAA to harvest was observed, so that IBMP at harvest was not significantly different between years (2008, 1 to 13 pg/g; 2009, 5 to 14 pg/g). IBMP accumulation up to 50 DAA and log-fold decrease of IBMP from 50 DAA to harvest was modeled as a function of >120 viticultural and environmental variables (122 in 2008 and 140 in 2009). Important variables identified for modeling IBMP at 50 DAA were those associated with vine vigor, which was positively correlated with IBMP accumulation. Cluster light exposure did not explain differences in IBMP accumulation across sites, but it was important for modeling smaller differences within some sites. IBMP decrease could not be satisfactorily modeled across multiple sites, but within sites the decrease was most consistently correlated with classic fruit maturity indices (total soluble solids [TSS], $TSS*pH^2$). The intensity of herbaceous aromas in wines produced from each site was not correlated with IBMP concentration, but multivariate models indicated that lower vine water status was the best predictor of increased herbaceousness.

3

Varieties

Grapes belong to the genus *Vitis* of the family Vitaceae. The genus *Vitis* is divided into 2 sub-genera, namely, *Euvitis* and *Muscadinia*. There are nearly 57 species under *Euvitis* and 3 species, namely, *rotundifolia, munsoniana* and *popenodi* under *Muscadinia*. Besides differing in other characteristics, the *Euvitis* has a somatic chromosome complement of 38, while the *Muscadinia* has 40. Among all the species, *Vitis vinifera*, the Old World Grape, of the subgenus *Euvitis* is the most important as more than 90 per cent of the grape varieties grown in the world today are derived directly or indirectly from this species. Other important species of *Euvitis* are *V. labrusca, V. cineria, V. riparia* and *V. rupestris*. *V. rotundifolia* is the most important species of the sub-genus *Muscadinia*. This is practically immune to diseases and pests including phylloxera and nematodes.

There are 3 main commercial uses of grapes, namely, table, raisin and wine. Grapes are also used for juice purpose which is unfermented. For breeding, definite objective should be kept in mind. The objectives vary from place to place, *e.g.*, for North India, earliness is very important, *i.e.* the varieties should ripen before the onset of monsoon rains in June. For south India, disease resistance is very important. This is in addition to high yield and good quality. The general objective of grape breeding is to develop high quality, high yielding and disease resistant varieties for specific purposes like table, raisin, wine and juice. For table varieties, seedlessness, large berry size, loose cluster and good colour are important characters. Raisin grapes should be seedless with good texture. Wine grapes should be high yielding and disease and insect resistant. In foreign countries like U.S.A., France and Germany, where there are phylloxera and nematode problems, breeding for rootstocks is also important. The rootstocks should be vigorous and resistant to nematodes and phylloxera. In our country, breeding rootstocks for resistant to salinity and drought is important.

Grape as Breeding Material and Breeding System

The *vinifera* grape has originated in the region between Black and Caspian seas in Middle Asia. To being with, the native vine was dioecious with the male and female vines occurring in about equal numbers. Rarely a male vine, at least in some species, may produce a few hermaphroditic flowers and some fruits, so that vines may in this sense be considered subdioecious. Female vines, on the other hand, have not been known to revert to the other sexual types. Breeding results based on many varieties and species, that the simple genetic explanation for the 3 sexual types (0, 0 and 0) of grapevine can be formulated on a multiple hypothesis at a single locus, a follows:

Male vine	:	Su F Sum
Devied males	:	Su F Su F, Su F Su +
Female	:	Su m, Sun
Homozygous from a phordite	:	Su + Su+
Heterozygous from a phrodite	:	Su+ Su m

The dominance relationship of the 3 alleles is thus: SuF> Su+ > Sum.

The native vine was originally cross-pollinating, but selection for hermophrodite or female types) the only ones to produce fruits) must have occurred very early. As larger vineyards and single varieties were selected and established, in more isolated areas, even the female varieties were really discarded because of partial or complete failure to be cross-pollinated.

The grapes are highly heterozygous. The heterozygosity at first was maintained sexually at a higher level by enforced cross-pollination to produce new varieties. Added to this are the long periods of asexual propagation and an accumulating mutational load to further increase heterozygosity. In the world collection of varieties, there are no varieties of any importance that can trace their origin to selfing of a parental variety. In breeding results in rapid loss of vine vigour and fertility, even a in the first generation.

The most useful breeding method is crossing of unrelated parents with good combining ability followed by raising a large number of hybrid seedlings in each combination and rigorous selection in F1. Test crossing with a common parent is also useful in exploring the genetic potential of a large number of varieties. The aim is to produce the maximum state of useful heterozygosity since many selections can be quickly and easily reproduced by asexual means.

Breeding Methods

1. Introduction and Selection

In breeding, introduction of germplasm from various sources is the first step. Selection of introductions best suited to local condition is one of the methods of breeding. In India, the first extensive collection of *vinifera* varieties was assembled at Lyallpur in the Punjab before partition of the country. They found 7 varieties out of 116 as promising. These varieties were, however, lost to Pakistan after partition. Collection were then made at Ganeshkhind Fruit Research Station, Pune; Horticultural Research Institute, Saharanpur; Indian Agricultural Research Institute, New Delhi and Indian Institute of Horticultural Research, Hessaraghatta, Bangalore. Perlette is the most outstanding selection made for North India. It ripens before the onset of Monsoon. Other promising selections are Beauty Seedless, Himrod and Delight.

At I.I.H.R., Bangalore, nearly 1000 grape varieties and 20 *vitis* species were collected. They were evaluated thoroughly for various purposes like table, raisin, wine and juice. Ten varieties were found to be promising for fresh table purpose, 8 varieties for dry white table wine, 12 varieties for dry red table wine, 2 varieties for red dessert wine and 2 varieties for raisin making. The varieties are:

Fresh table purpose: Angur Kalan, Athens, Black Champa, Coarna Rosie, Convent Large Black, Medeleine Celine, Red Prince, Katta Kurgan, Queen of the Vineyrads and Taifi Rozovi.

Dry white table wine: Bayan Shirei, Chenin blanc, Excelsior, Fateasca Alba, Ova Pagola, S.V. 5276, Sauvignon and White Riesling.

Dry red table wine: Black Champa, Black Cornichon, Black Muscat, Convent, Large Black, Gamsa, Fateasca Neagra, Muscat, Red Prince, Rubired, Saperavi, Tievoasa and Zinfandel.

Red dessert wine: Black Champa and Convent Large Black.

Raisin purpose: Thompson Seedless and Black Corinth.

2. Intervarietal Hybridization

A large number of hybrids were raised and evaluated at IIHR, Bangalore and IARI, New Delhi. At I.I.H.R., Bangalore, 4 very promising hybrids were named and released for cultivation. They are: 1. Arkavati, 2. Arka Kanchan, 3. Arka Shyam and 4. Arka Hans.

Arkavati: It is a hybrid between Black Champa and Thompson Seedless. It is seedless with good quality (T.S.S. 22-25 per cent, acidity 0.6-0.7 per cent). it is excellent for making raisins and good for making champagne, dry white table wine and also for fresh table purpose. It has yielded up to 30 tonnes per acre as against 12 tonnes per acre in Thompson Seedless.

Arka Kanchan It is from a cross of Anab-e-Shahi and Queen of the Vineyards. Seeded with good muscat flavour. T.S.S. 19-22 per cent and acidity 0.5-0.6 per cent. It is good for making dry and sweet table and dessert (Muscatel) wines.

Arka Shyam: This hybrid is from a cross of Bangalore Blue and Black Champa. Seeded with good quality (T.S.S. 22-25 per cent, acidity 0.5-0.6 per cent). Resistant to anthracnose. Two crops can be taken in a year in southern and western India. It is good for making juice and for making dry and sweet table wines.

Arka Hans: It is a hybrid between Bangalore Blue and Anab-e-Shahi. Seeded with good quality (T.S.S. 18-21 per cent, acidity 0.4-0.5 per cent). Resistant to anthracnose Double cropping possible in peninsular India. It is good for making wines.

All the above 4 varieties produce good cross on bower as well as head system of training.

At IARI, New Delhi, 3 early ripening hybrids have been developed, which are under testing.

Dr. G.S. Cheema from Ganeshkhind F.R.S., Poona, planted naturally cross-pollinated seeds of Pandhari Sahebi (0 sterile) and obtained 2 new varieties, Selection-7 and S-94. S-7 is now called Cheema Sahebi. These varieties have had some success in grape industry, especially around Poona. Their major defect is poor keeping quality.

In U.S.A.: Some of the important varieties developed by Prof. H.P. Olmo and released by Davis, California are as follows.

1. Perlettee: Queen of the Vineyards x Sultaniana Marble
2. Delight : Queen of the Vineyards x Sultanina Marble
3. Deauty Seedless: Queen of the Vineyards x Black Kishmish
4. Ruby Cabernet : Carignane x Cabernet Sauvignon
5. Emerald Riesling ; Muscadelle x White Riesling
6. Rubired : Alicante Ganzin x Tinto Cao
7. Royalty ; Alicante Ganzin x Trousseau
8. Scarlet : Golden Muscat x Teinturier
9. Niabell : Niagara x Campbell Early
(IV) (IV)

Varieties Released by U.S.D.A, Fresno

1. Cardinal: Flame Tokay x Alponse Lavellee
2. Calmeria: resulted from seeds of open-pollinated Ohanez (Almeria)
3. Black Rose: (Damas Rose x Black Momukka) x Alphonse Lavallee.

In U.S.S.R.: late Prof. A.M. Negrual developed 16 varieties, which are given below:

1. Tarnau (Seedless): Nimrang x Kishmish Chiorni
2. Volgo-Don: Katta Kurgan x Zabalkaski
3. Djura Izum: Nimrang x Taifi Rozovi
4. Pabeda: Zbalkanski x Muscat Hamburg
5. Rani VIR a: Chauch x Kishmish Chiorni
6. Muscat VIRa: Muskat Beli x Muskat Hamburg
7. Oktyabraski: Nimrang x Kabarnai
8. Maskat Uzbekistanski: Charas x Muskat Hamburg
9. Charas Muskatski: Charas x Muskat Hamburg

10. Pozdni VIR a: Nimrang x dobrelyabi
11. Guzel Kara: Katta Kurgan x Dobrelyabi
12. VIR Adin: Takveri x Tibushe
13. Prima: (Muskat Alexandriski x Pufliakovski x *V. rupestris*)
14. Salava: Tagobi x Amuski
15. Uzbekistan: (Tagobi x Amurski) x Tagobi
16. Perri May: Tinturye x Aleatiko

3. Interspecific Hybridization

All the species within the subgenus *Euvitis* cross readily. In North America, the spontaneosu and intentional hybridization of *V. vinifera* with *V. labrusca* resulted in evolution of *American Hybrids* such as Concord, Catawba and Isabella. These hybrids are hardy and more resistant to diseases than *vinifera* varieties. These varieties are grown commercially in Easter U.S., Japan and also in China.

When the rootlouse, phylloxera, decimated the vineyards in France and other countries of Europe, American species, namely *V. rupestris, V. lincecumii* and *V. riparia* were introduced to France. The private French grape breeders attempted crosses involving these species and *V. vinifera* to create a "direct producer", a vine having phylloxera resistance and also good wine quality. This quest was such as Seibel 1000, Seyve-Villard 5276 emerged. These hybrids are more resistant to diseases and are capable of higher production than *vinifera.* These hybrids are easily identified by the breeders name. These hybrids occupy one fifth of a total vineyard acreage of France and produce average quality wine. These are outlawed in other parts of the world as their wine quality is poor.

An Oriental *Vitis* species, *V. amurensis* of Siberia has been used by Russians to impart cold resistance to the *vinifera* grape.

Native species of *Euvitis* are confined almost entirely to the North Temperate zone. However, one species called *Vitis caribea* has been found growing in abundance in the tropical Caribbean area. In Venezuella, this vine has hybridized spontaneously with the poorly adapted *vinifera* to produce a new race of very vigorous and disease resistance vines locally called "Criollas". These may be very useful for improving tropical viticulture.

So in *Euvitis,* there is no serious bar to gene exchange. In wider or intergeneric cross between *rotundifolia* of the subgenus *Muscadinia* and *vinifera,* thus exists a unilateral type of incompatibility. Varieties of *Vitis vinifera* can be successfully and easily crossed with the pollen from various clones of *rotundifolia,* but all reciprocal combinations fail. This failure is not due to lack of pollen tube growth but must reside in the cytoplasm of the embryo sac or maternal tissue bordering the embryo sac. The F1 hybrid of *vinifera-rotundifolia* crosses are uniform and vigorous, but they are with few exceptions completely pollen and ovule sterile. Some hybrids have shown to carry high resistance of immunity to phylloxera and could be used rootstocks. The particular *vinifera* varieties used as female parent is very important it determines to a large extent the number of viable seeds obtained. Success have then been achieved in obtaining family fertile forms. Then diploid *vinifera x rotundifolia* hybrids are expected to be of great value in the transfer of diseases and insect resitance into the *vinifera* group as *V. rotundifolia* is almost immune to all diseases and pests.

4. Tetraploidy

It is of interest in table grape production as the chief benefit from tetraploidy would be increase in berry size and reduction in bushiness of the vine (fewer shoots). Autotetraphoidy are found to be considerably sterile and are less productive than the parents. Allotetraploids, however, have been more desirable as commercial varieties. Examples are production of tetraploid Niabell and Early Niabell from cross between tetraploid Niagara and tetraploid campbell Early.

5. Induction of Mutations

Irradiation with gamma and X rays have been used to induce mutations in grapes. The most significant achievement is the production of "loss clustered" variety of perlette by irradiating the dormant cuttings with X ray at 2500 r.

6. Biotechnology

Embryo rescue technique: Now two seedless (stenospermocarpic) parents can be crossed and with the aid of embryo rescue technique *in vitro,* there is possibility of getting a high proportion of seedless progeny for selection. Also, there is a possibility of increasing the number of hybrids from *vinifera x rotundifolia* crosses.

Hybridization Techniques

1. Floral Biology

Knowledge of floral biology is very essential for carrying out successful hybridization work. In grapes, the peak of anthesis and dehiscenc is between 8 and 9 A.M. and the dehiscence of pollen generally takes place immediately after anthesis. The stigmas are receptive one day before to one day after anthesis. So, the clusters of male and female parents to be used in crosses should be bagged 3-4 days before anthesis. A clothe cage should be constructed around the vines, particularly around the female parent, to prevent contamination caused by wind. The emasculation with forceps cleared regularly with 70 per cent ethyl alcohol should be done a day or two before anthesis and then clusters bagged immediately. Pollen from the clusters of male parent should be collected between 8 and 10 A.M. by shaking the clusters against the bottom of a petridish. Controlled cross-pollination should be done 1 day before to 1 day after anthesis with a camel hair brush used for transferring pollen.

Choice of Parents

This is most important aspect in perennial crop breeding.

(a) *Use of female (male-sterile) vines:* Certain varieties like Pandhari Sahebi and Angur Kalan have female flowers characterised by reflexed stamens containing pollen without germpores. As mentioned earlier, these pollen do not germinate and sterility is complete. Such varieties with desirable characters should be used as female parents. In such cases, the tedious emasculation by forceps is not necessary.

(b) *Varieties with low seed viability as male parents:* They viability of seeds is largely a characteristic expressed by maternal parent. If varieties with low seed viability are needed as parents, they should be used as male (pollen) rather than seed parent.

(c) *Seedless varieties as male parents:* Seedless varieties should always be used as male parents for producing vigorous, high yielding and disease resistant new varieties.

3. Seed Germination and Transplanting

Grape seeds should be ripened for about 3 months at 40-45°F to break the rest period. The stratified seeds should be sown in seed pans containing well-sterilized soil mixture to avoid damping off of the seedlings. When the seedlings are 3-4" high, they should be transplanted to 5-9" pots.

4. Field Planting

When the transplanted seedlings are about 12" high, they can be planted in the field in rows. When transplanting growing seedlings all the leaves may be stripped off by hand. Normally, 5 feet distance is kept between the rows and between the seedlings. If any machine be run, the distance between the rows is increased.

The seedlings are permitted to grow without training in the first season. They are then cut back in the season and the most vigorous shoot is trained up the stake to a height of 4-5 feet then pinched back to form a head. Once the shoot is vigorous enough, fruit can be forced at any time by pruning back to a prim primary eye before it enters the rest period. However, fruit production should only be encouraged in the season considered best in the locality.

5. Evaluation of Hybrid Seedlings

The breeder has to be ruthless in rejecting the bad seedlings. Data on each seedling to be collected for 2-3 seasons on quality and yield potential. Promising hybrids should be selected. Normally, there is one promising hybrid out of 1000 seedlings.

6. Testing of Promising Hybrids

Promising hybrids are given numbers and then they are tested in comparison to existing commercial varieties (standards) by adopting proper statistical design for yield, quality and other characteristics. The best ones should be tested at different centres in the country under the All India Coordinated Fruit Improvement Project and also at farmers' field. Testing can also be done at different centres within a state.

7. Release of Variety

Based on three years performance data, the best hybrids are released for commercial cultivation through State Variety Release Committee/Central Variety Releases Committee.

(*Courtesy*: Dr. S.S. Negi, *Ex. Director* CIHNP Lucknow)

1. Anab-e-shahi

Trunk large, to very large, cane large and thick, dark brown. Leaves cordate moderally pubescent, emerging leaves light green with pink tinge, inflorescence fairly compact bunch very large, slightly shoulderd or conical, uniform ripening, medium to high shattering.

2. Bangalore Blue

It is a vinifera and labrusca hybrid vines vigrous, bark very loose, canes brown long, thick, tendril greyish green mostly bi-fid, some time tri fid. Fully grown leaves tri lobed to 5 lobed. Ripening fairly uniform in South.

3. Thompson Seedless

Berry are elongish it ripens uniformly it does well on Kniffin system, telephone system, it is succeptible to downy disease flower hermophrodite.

4. Arkavati

A hybrid, crossed between Black champa x Thompson seedless, vines vigrous, cluster medium size, well filled long conical to cylindrical, berries medium. TSS–25 per cent

5. Arka Hans

A cross between Bangalore blue x Anab-e-Shahi are moderatly vigrous cluster medium size, TSS–18-21 per cent. It is suitable to head system.

6. Arka Kanchan

A cross between Anab-e-Shahi x Queen of vine yard medium to large. TSS–19-22 per cent. It is late ripening with poor keeping quality.

7. Arka Shym

A cross between Bangalore blue x Black champa vines are moderatly vigrous, cluster medium size, berries medium size. Yield potential very good. It is very good for dry red table wine.

8. Maroo seedless, superior seedless, Flame seedless, Red globe are newly introduced.

9. Among the wine varieties Shiraz, Merlotte, Cabernet savignon, Zinfenda, Pinot Noir are promising.

10. Sonaka

A mutant of Thompson seedless, bunch medium to large, berries long, fetch good price in market.

11. Sharad Seedless

A black varitey, medium size bunch. Alternate bearing is observed. More preferred at Christmas festival.

12. Flame Seedless

Uneven ripening is a problem, otherwise a good variety.

13. Redglobe

It is getting premium price in market.

14. Fantacy

It is also a very good variety in grape, medium size bunch.

15. Tas-A-Ganesh

A mutant of Thompson Seedless

16. Clone II

A mutant of Thompson Seedless yield 10 per cent more than Thompson Seedless.

Aulakh *et al.* (2003) observed the Arid irrigated region of Punjab is located at latitude of 30.08° N and Longitude of 74 12°E having 300 mm annual precipitation with minimum and maximum temperatures ranging from0.6 to 4.8°C and 39.85 to 44.2°C respectively most of the grape cultivars ripen during June coinciding with the monsoon rains. Cultivation of Perlette grapes has already become very popular due to earliness in ripening. But this cultivar has certain problems keeping this in view; the present studies were carried out on the three years old vines of seven early ripening cultivars *viz.* Beauty Seedless. Bharat Early, Early Muscat, Khalili Loose Perlette, Pearl of Csaba including Perlette at Regional Fruit Research Station, Bathinda during the year 1993-94. Three vines of each cultivar planted 3x3 m apart and trained on two wire kniffin system were realized for this study. The results showed that Pearl of Csaba was earliest to sprout (19th Feb.) flower (30th March) and ripen (last week of May). the maximum fruit yield (4.6kg/vine), bunch weight (163 g), bunch size (19.0 x 12.2 cm) TSS (22 per cent) and lower acid content (0.76 per cent) were also recorded in Pearl of Csaba as compared to Perlette and other early ripening grape cultivars. Maximum berry weight (2.60 g) and berry size (1.80 x1.64 cm) were observed in Perlette grapes. Khalili although superior in quality was found to be a shy bearer. Based on these studies, Pearl of Csaba seems to be a promising variety for cultivation in the arid irrigated region of Punjab.

Ravikumar *et al.* (2006) conducted an experiment in grapes to standardize the optimum pH and TSS levels for quality wine production. Grape bunches of cultivars Muscat, Bangalore Blue, Sharad Seedless, Thompson Seedless and Tas-A-Ganesh were harvested during March–April 2004 for this study. Results on wine quality parameters revealed that the wine from Muscat was ranked as 'excellent', Bangalore Blue as 'Commercially acceptable', Sharad Seedless as 'Deficient wine', Thompson Seedless as 'Good' and 'Tas-A-Ganesh' as 'Poor and objectionable'. Browning, a problem in white wine was noticed in both the cultivar 'Thompson Seedless' and 'Tas-A-Ganesh', but that was controlled in 'Thompson Seedless' at pH 3.0 and TSS 22 °Brix but not in 'Tas-A-Ganesh'.

The yield and quality of 14 early-ripening grape cultivars, planted during 1995 in Rehmankhera, Lucknow, Uttar Pradesh, India, were evaluated during 1997-2000 to identify the most suitable cultivars for North Indian plains, where pre-monsoon showers is a limiting factor for grape cultivation (Ram Kumar *et al.*, 2002). The period of bud sprouting and ripening varied with the year and cultivar. Bud sprouting (20 February-5 March) and fruit ripening (16 May-5 June) were earliest in Beauty Seedless. BA x Per-75-32, Gold, Delight and Kishmish Beli exhibited late fruit ripening. Flame Seedless and Pusa Navrang were high-yielding, and the mean annual yield of these cultivars was approximately 20 kg per vine. Both cultivars, which were resilient to damage by heavy rains, exhibited phenotypic yield stability even under unfavourable conditions. Cardinal (5.40-8.61 kg per vine), Kishmish Charani (4.69-12.66 kg per vine), Beauty Seedless (1.13-22.63 kg per vine), Pusa Seedless (5.02-7.13 kg per vine), Gold (1.36-8.88 kg per vine) and New Perlette (2.69-9.84 kg per vine) had moderate yields, whereas Pusa Urvashi (1.20-2.89 kg per vine) and Kishmish Beli (1.15-3.04 kf per vine) had low yields. Hybrid BA x BS-71-50 did not produce fruits. The highest bunch weight was recorded for Pusa Seedless (279.75 g). Cardinal had the highest berry weight (3.01 g). The fruits of Kishmish Beli were characterized by the lowest acidity (0.44 per cent). The results suggest that Flame Seedless, which had high fruit yield, number of bunches and bunch weight, was the most suitable for cultivation under Lucknow conditions.

Erfani *et al.* (2008) stated that the seedlessness is the most important characteristic of fruit quality for raisin and table grapes. Grape breeding at University of Tehran, Karaj branch, Iran started in 1999 with crossing some selected seedless and seeded cultivars at the end of evaluation of 90 cultivars in grapevine collection of College of Agriculture in Karaj. Out of 1400 progeny obtained from 26 different crosses, 381 of them which produced fruits were evaluated during growing seasons of 2006 and 2007. Progenies were divided into four groups of completely seedless, semi-seedless, semi-seeded and completely seeded, according to seed trace or seed weight. Results classified progeny to 11 per cent completely seedless, 13.6 per cent semi-seedless, 24 per cent semi seeded and 51.2 per cent completely seeded. Percentages of seedless progeny for four male parents of Askary, Yaghooti, Bidane Sefid and Bidane Ghermez were 15.4 per cent, 10.8 per cent, 9.3 per cent and 10.6 per cent, respectively. Meanwhile, percentages of seedless progeny for female parents of Muscut of Hamburg, Ghezel Uzum, Dizmary, Rajabi Sefied, Ali Baba, Alhaghi Ghermez and Tabariz were 5.4 per cent, 5 per cent, 17.5 per cent, 13.2 per cent, 1.4 per cent, zero and 36 per cent, respectively. Results also showed that among male parents, Yaghooti and Bidane Ghermez cultivars and among female parents, Tabarze cultivar had better backgrounds to transmit stenospermocarpic seedlessness characteristics.

Leandro *et al.* (2008) studied the effect of different types of packages and, the action of sulfur dioxide (SO_2), in post harvest of fine grapes var. Crimson Seedless and Italia. The fruits, harvested in a farm located in Boa Vista, State of Roraima (Lat. 2° 50' 06" N and Long. 60° 40'28"W), showed, soluble solids (SS) medium of 16.50 and 14.80°Brix, for the varieties Crimson Seedless and Italia, respectively. Prior to treatment, bunches were cleaned in sodium hypochlorite (NaOCl) solution at 100 $mg.L^{-1}$, previously acidulated, for 10 minutes. For the modified passive atmosphere were used bags of low density polyethylene (LDPE), not perforated, with 0.010, 0.015 and 0.020mm of thickness and, arranged in corrugated cardboard (4kg) and wood

(7.5kg) boxes. To produce SO_2 it was used Kraft papers of quickly liberation, with 3 and 8g of sodium metabysulfite ($Na_2S_2O_5$). After the treatment, the fruits were storage in cold chamber at 4 ± 1ºC and 95 ± 3 per cent of relative humidity (R.H.) The evaluation were carried out in the moment of harvest and, at 7, 21, 35, 42 and 56 days of cold chamber, for cluster weight loss, percentage of detached and deteriorated berries, quality of stem and, content of SS of fruits. After eight weeks it was carried out a sensorial analysis for the two varieties. It was identified in both varieties, that grapes submitted to generate SO_2, containing 3g of sodium metabysulfite and, wrapped in packages of LDPE of 0.020mm of thickness, regardless type of secondary package showed smaller weight loss, smaller percentage of detached and deteriorated berries and, best quality of stem. The results in the sensorial analysis had agreed to the physical-chemical results. There is no significant difference in the contents of SS between the treatments, in both varieties.

Claude Dubois *et al.* (1996) stated that the effect of variety, year, and grape maturity on fermentation kinetics was studied. The differences were quantified by comparing the values of the maximum CO_2 production rates ($(dCO_2/dt)_{max}$) and the trends of the (dCO_2/dt) curves at the end of fermentation. Fermentation kinetics were very different among grape varieties, but also changed significantly with year and maturity. Year above all influenced the maximum CO_2 production rate $(dCO_2/dt)_{max}$ when maturity dramatically changed the kinetics at the end of fermentation. Kinetic variations were also related to must composition changes. The maximum CO_2 production rate was correlated to the must assimilable nitrogen concentration while the trend of the dCO_2/dt curve at the end of fermentation (expressed as DM80) was above all a function of the sugar concentration.

Lopez *et al.* (1997) reported that the free and bound aroma compounds of eight Spanish grape cultivars (Albarino, Treixadura, Listan, Viura, Xarel.lo, Parellada, Garnacha, and Tempranillo) used in winemaking were studied. With a single-extraction procedure it was possible to provide information about the potential flavor of these varieties: bound aroma was determined directly by TFA glycoside analysis and indirectly by enzyme hydrolysis release of aglycon. An aroma data base of these varieties and a comparison between the grape flavor compositions of each variety were established. Grape characteristics (sugars, acidity, polyphenols, juice yield), mainly due to climate and cultivar factors, were related with the aroma composition. Non-terpenyl compounds such as 2-phenylethanol and benzyl alcohol were the main glycosidically bound compounds of Spanish varieties. Direct determination of glycosides is better than hydrolysis of aglycon as an approach to the potential aroma of grapes.

Hossein *et al.* (2012) opined that powdery mildew is one of the most serious and destructive diseases of grapes around the world, including Iran. Although the application of chemical pesticides is the most common control method, the use of resistant varieties may be the most effective and environmentally sound strategy for managing the disease. Fourteen grape varieties were evaluated for testing powdery mildew (*Uncinula necator*) resistance under natural infection conditions and artificial inoculation methods were used between 2007 and 2009 in Ardabil province of Iran. Disease severity on leaves and fruit (cluster) were evaluated by the following procedures: 0: immune, 1: highly resistant, 2: resistant, 3: tolerant, 4: susceptible, and 5: highly susceptible. Results of the experiments showed that Shahani, Yagothi and Tabrizkishmishi cultivars were immune, Sahibi cultivar was highly resistant, Agshilig and Tokoulgan were resistant, Kishmishi, Khalili and Tabarzeh cultivars were tolerant, Aldarag and Koupakbogan were susceptible and Garashilig, Seyrakpousteh and Rasmi cultivars were highly susceptible to the disease. In general, results of this study were promising and some immune, highly resistant and resistant cultivars to *U. necator* were identified and they may be used as a resistance genetic source for management of the disease in national and international programs.

An investigation (Gatta *et al.*, 2007) concerning phenolic fraction of some varieties of grapes and relative wines, from Southern Italy was carried out. For this purpose, the low molecular weight phenols of 'Aglianico', 'Negroamaro', 'Bombino Nero', 'Uva di Troia', 'Gaglioppo' and 'Primitivo' were analyzed by RP-HPLC and monitored at 280 nm (catechins and proanthocyanins), 350 nm (benzoic and cinnamic acids, flavonol compounds), and 520 nm (anthocyanins). The chromatographic profiles of the investigated samples were rather different. 'Negroamaro' and 'Primitivo' wines showed a high content of catechin and proanthocyanidin B2. In these wines the most abundant phenolic acid was the trans-caffeoyl-tartaric acid. 'Aglianico', 'Bombino Nero' and 'Uva di Troia' grapes and wines had the highest anthocyanins content. These results supply some important information about the most suitable oenological technologies in order to exalt the potentialities of the considered grapes.

4

Propagation

Grape is grown on an area of 45000 hectare with an annual production of 1,20,000 Tonnes in Maharashtra and in Karnataka it is grown on an area of 12000 hectare with a production of 1,65,000 tonnes. With reports of declining yield is the states viz Maharashtra and Northern Karnataka due to salinity of soil and water and severe drought the farmers have realised the importance of root stock. In California the root stocks are used against phylloxera a root aphid, salt resistant and drought resistant, and the root stocks are the answer for this problem. Grape vine when grown on their own root, production of export quality grape because a difficult task and when they grown on root stock looking to the soil and water reports, different root stock are available and they can be utilized for different situation. The soil is alkaline saline.

In Northern Karnataka chloride content in irrigation water is a major problem because of excess level of sodium and calcium which are freely available root stock species *vitis champini* can grow very well because it is deep rooted and it can absorbs the filter water from deeper zone. The root stock can be propagated by seeds, cuttings, layers and to get the 100 per cent purely they can also be clonaly propagated. The grafting on these root stock is easy by wedge grafting by planting the root stock in January - February and grafting in October same year. To overcome the problem of salinity the Horticulture practices like salt leaching amendment of soil, application of sawdust, opening of trenches to leach out the soluble salts like sodium etc. green manuring like growing of sunhemp, and incorporating in soil when it flowers, mulching with straw, plastic etc also help to overcome these problems, but the results were not encouraging. Hence growing of root stocks like Dog ridge (All situation) 1103 R (Drought tolerant), 99 Ritcher (Drought tolerant) 110 Ritcher (Drought tolerant) So4 (1 lime tolerant) Ramsey (tolerant to saline soil) St. George (Drought tolerant) The root stocks which are Drought resistant and good to solve the problem of salinity because they are having effective mechanism of ion from the soil, as they will not absorb the Cl^- and Na^+ and take up only beneficial ions which are not harmful to plants.

Why to Use Root Stock

Try to find out the problem of grape cultivation because the ideal soil is the one where pH less than 8.5, EC less than 2-2.5 mmhos/cm^2 Exchangeable sodium percentage less than 15 and free calcium 10-12 per cent. The water for grape cultivation the Electric conductivity should be less than 1 mmhos/cm^2, SAR less than 8, meq/ litre, all the above soil situations are difficult to available. If the above said situation of soil and water are available then there is no need to employ the root stock otherwise which leads to fast growth, more production of leaf, naturally the cost of cultivation may go up and to suppress the growth use of CCC may cost more because it is ant gibberlin. Root stock also helps in nutrient uptake but zinc deficiency is more predominant. By cultivation on root stock the quality and productivity will also be more. Under saline soil situation if EC > 4 mm hose, ESP <15, pH < 8.5 EC is directly related to chlorine which causes leaf margin burning, leaf becomes rubbery when crumbled in hand palm, the plant show stunted growth, pre-mature leaf drop, internodal length is reduced, less yield Dogridge, 110R, 1103P. Salt Greek are good in overcoming the problem of salinity, with Bangalore. Dogridge root stock the conc. of soluble salts EC 3.8 dsm^{-1}, safe limit of chlorine is 3 mequivalent (me) per litre and with root stock upto 18 meq/litre. In sodic soils which is a problem in sub-tropical regions. The soils are said to be alkaline when the EC of saturation extract is less than 2 dS/m, ESP > 15 per cent, pH 8-10, the growth under such situation will be affected with slender shoot, smaller size leaves, short internodal length and such soils are not fit for cultivation.

In sub-Tropical regions the demand for water will be more as the maximum evapotranspiration loss in Northern Karnataka is 9-10 mm/day, specially at Bijapur, the water is available at a very deeper level, they go for rigging upto 650 feet, the quality of water is poor because less per colation, the rain water go as run off and forms gulleys but do not go in to the soil because these region soil are Black soil or clay soil where the potash content is high (KCl) and water per colation is very less and the problem get aggravated in summer or wherever drought situation are observed not necessary in summer only but other part of the year also, under such situation the berry size get reduced, less number of canes, sun burn of grape bunches, which further leads to reduction in quality for export because the export quality bunch should be milky green.

As the root stock roots go very deep and away from stem, hence water and nutrient availabity of nutrient also will be good and it can also be minimized. Dogridged root stock is an efficient user of available potassium and available nitrate. Root stocks have an important role to change the shoot vigour and bring equilibrium in vegetative and productive growth, quality and yield, they also help in fruit bud differentiation and they are also capable of Cytokinin synthesization in their roots. 1613 root stock also synthesize more cytokinin than own root Thompson seedless roots, high cytokenin. Gibberlin is known for good fruit bud differentiation, because more bud break more yield, when Thompson seedless was grafted on 1613 root stock the yield were more in farmers field.

Research conducted in India and other parts of the world have identified that root stock are resistant to salinity, alkinity, drought, root stock helps in increased fruit bud differentiation resulting in more productivity.

The root stock 1103 R, 140 R, 110 R, 450 A, 99R St. George are suitable to drought situation or water shortage problem, the root stock like Dogridge, 140 U, 99R. 110 R, Ramsey are suitable to the situation of soil EC>2 mmhos/cm^2, water EC > mmhos/cm^2, the root stocks Dogridge, 1613, Ramsey to a situation of soil ESP more than, 15 or water SAR > 8, the root stock Fercal, SO_4, 420 A are suitable where free content of the water is > 4 meq/litre than the root stocks *viz.* Dogridge, Ramsey Teleki for increased Nitrogen potassium content Dogridge is suitable. The Morphological character of Dogridge *viz.*, leaf shape is cordate (Heart shape) petiole colour purplish.

Propagation of root stock is mostly done by hard wood cuttings they should be of last season growth. The plant you select should grow vigorously and free from any disease *viz.*, Viral, Bacterial and Fungal.

The cuttings should be medium in size with optimum internodal length they should not be flat or angular when you observe the section, the cutting should be selected from the middle portion and preferably of 6-9 inches in length with 4-5 buds, the cuttings should be soaked in water for 24 hours then in IBA solution with a conc. of 2000 to 2600 ppm for 5-10 seconds helps in better rooting or some time applicaiton of *Trichoderma herzianum*, also helps in rooting. The poly bags should have a size of 9" x 6" with 200 -250 gauge thickness, the planting media is good Red soil with vermicompost : sand : Farm Yard Manure in 1:1:1:1 proportion and insert two buds into the packet and 2 to 3 buds above the media. The packet should have holes at sides and at bottom and apply few granules of phorate or Furadon, or Carbofuron etc. to overcome the problem of Termite attack or apply chloropyriphos 4 ml/litre of water.

There is also a practice of grafting by machine where selected scion and root stock pieces get interlocked by pressing the machine, they are dipped in liquid paraffin, then they are kept in controlled atmosphere chamber where relative humidity and temperature are controlled, in few days scion and root stock sprouts without any soil media such plants are available at Rs. 20-30/each some time Ramsey root stock which is most suitable to wine varieties like Shiraz pinot Noir, Merlot, Zinfendal, Cabernet Savignon are available at Narayan Gaon, Nasik.

Sometime after opening the trenches and application of super phosphate 500 gm per running meter, FYM micronutrient, Green manure/5 kg per plant, 2-3 cuttings of Thompson seedless or desired variety can be planted directly after treatment with IBA solution having a conc. of 2000-2500 ppm after sprouting of the cutting the vigorous one will be retained and grafting is done during October. Extra care should be taken against the termite attack by application with Neem cake/Chlorophyriphos 4 ml/litre of water every 15 days once. For control of powdery mildew spray carbendenzim 1 gm/litre of water, Anthracnose spray Thiophenate methyl 1 gm/lit,

Downy mildew 2 gm (metalexil + Mancozeb)/litre of water. Some time mist chambers are utilized for proper rootings in grape cuttings. The grape cuttings can be selected from the September-October pruning because the temperature and humidity will be suitable for sprouting rootings the cuttings also get 3-4 months time to develop till the planting starts by February.

Some time the beds are prepared having a size of 5' x 2' x 6" a good red soil, FYM are applied after treating with IBA 2000 to 2600 ppm for 5 seconds, they are planted. But cuttings should be planted in polythene bags only for long distance transport.

The farmers can also construct the zero energy chambers at their field. They are double wall tanks with15-20 cm gap between two walls, the inner diameter of the tank should be 180 cm in width, 150 cm in height, and 600 cm in length outside wall should have a length of 650 cm, height 150 cm and width 230 cm, sand is filled inside the inner tank upto a height of 50 cm, the sand is filled in between two walls and soaked till it reaches saturation or a dripper is installed in between the two walls. Half matured cuttings are selected with 4-5 nodes the cuttings are dipped in 1100 ppm IBA for 60 seconds and planted in sand; the tank is covered with 750 gauge covers. It takes about one month to start rooting then rooted cuttings are transferred to poly bags filled with vermicompost, sand, FYM, Red soil in 1:1:1:1 proportion and kept outside to sustain the outside heat.

Once the cuttings attains the age of 6 months they are transplanted in the field for establishment. The land should be prepared by levelling with tractor, the weeds like *Cynadon dactylon* and *Cyprus rotundus* should be completely controlled by application of Glyphosate @ 5-6 ml/lit for Cyanadon and 10-12 ml/lit. for *Cyprus rotundus* or manual labours can be engaged. some time two to three sprays may also be used to control the weeds otherwise they compete for nutrient and water, if there is a water shortage problem, then diffuser can also be used which minimises the water requirement of the crop. A 2' x 2½' trench with convenient length is opened preferably in south-north direction and keep the top soil at one side, after sufficient exposure to sunlight the top soil is filled or some time if it is murum soil then red soil should be filled grape is also cultivated in Medium black size in Belgaum Distt. in Karnataka with good drainage system with 1' width x 2-2½′ deep Trench. So that the water table is brought down, then the root stocks are planted based on soil and water report, the quantity of water discharged by a bore well, the quality of water during summer also matters because simply mulching doesn't help in solving the shortage of water problem because grape requires 10,000 to 15000 litres of water per day per acre. While filling the trenches the top soil should be filled at bottom upto a height of 2' depth, FYM 20 kg/plant, 500 gm super phosphate, 100 gm Urea, SOP 100 gm, 200 g magnesium sulphate, 10 mg micronutrient per running foot, the remain trench is filled with down surface soil and leveled, sometime neem leaf, or castor cake Glycerdia leaves are also applied.

Pour the water after filling the soil in the trench, this helps in easy permeable of water, soil becomes loose, good aeration the root stock penetrate better in soil because they produce good mass of roots, further they can apply with humic acid for root growth. The direction of the trench are not important if vines are proposed to train on pendal system (Bower system). But if the vines are proposed to train on GDC (Y) (Geneva double curtain) then North-South directions are important, this can also minimise the sun burn of the bunches under sub-tropical regions. E-W planting should be avoided. JCB are utilized to open the trench in rocky soil with 2½ bucket.

The best season of planting is February to March, when the temperature starts rising. The pits of 15x15x15 cm are opened in the trench at a spacing of 5' (10 x 5'), 4' (8' x 4'), 6' (12' x 6') 4' (6' x 4') etc. After planting the root stocks press the soil and watered immediately. If the polybags raised plants are there then the packets with sharp blades without damaging the ball of earth and put in the pit and press the soil and apply chloropyriphos to control the termites.

After planting light frequent irrigation rather than heavy application since the feeder roots are at the top only, then apply water based on evapotranspiration loss with 4 litre/hour discharge if Electricity availability is better if not, then it is given only for 4-6 hours then 8 litre discharge/Hour may be utilized or diffuser may also be utilized, and the drippers are placed at 2' distance.

After 15 days (16-30 days) of planting give urea 1kg/AC/once in two days for 15 days 19:19:19 from 30-45 days 2 kg/AC/once in two days, urea 1kg/Ac/once in two days (46-60 days), 19:19:19 2 kg/AC/day once in two days (61-75 days) Urea 2 kg/Ac/once in two days (76-90 days), 19:19:19 2 kg/AC/once in two days (91-105 days) Urea 2 kg/AC/once, 13:0:46 or 1:47:35, 2 kg/AC/Once in two days (121-150 days), 0:0:50, 2kg/AC/once in two days (151-180 days) otherwise after 30 days of planting apply 25 kg urea, 15 kg DAP per acre and after 60 days of planting apply 25 kg urea per acre, after 90 days of planting apply 20 kg urea + 25 kg of SOP/AC and after 120 days of planting apply 25 kg of SOP/AC.

The roots stocks not susceptible to diseases like anthracnose, powdery mildew, down mildew however, root are not resistant to rust and alternaria disease, therefore, to control rust spray Kavach (Chlorothalonil 2gm/lit., Macozeb 2gm/lit, Tridemefone 1gm/lit, Alternaria can be controlled by spraying copper oxy chloride (Blitox) copper hydroxide (Kocide) 2 gm/lit. The chaffer beetles are serious in attacking the root stock which can be controlled by spraying Lanate (Methomyl) 1gm/lit. or Carbaryl 2gm/lit of water.

The root stocks are allowed to grow for 2 months which help in conserving the food, after 2 month weak thin shoots are removed, by maintaining 2-3 good shoots, generally two shoots are maintained for grafting Bamboo sticks are erected in each basin and they should be applied with tar (Dambar) to avoid attack of termites, in due course of time the leaves are removed at 15 days interval. The graftable diameter of the shoot should be 6-8 mm by September October and by October sap flow will be good for formation of dead parenchyma cells (Callus) and success of the graft. A temperature of 35°C and relative humidity of 75 per cent has been found the best for good grafting for successful grafting trained malis (gardeners) should be engaged for grafting ? usually they charge one rupee each for grafting. By virtue of some problem if grafting fails immature green shoots are utilized for grafting during next June month, sometime if two-tier system (4-arm Kniffin system) is to be followed, then at bottom layer wine varieties can be utilized (*viz.* shiraz, pinot Noir, Zinfendal, Cabernet savignon, merlot etc.) and on top any table variety viz Thompson seedless, Flame seedless, Sharad seedless, Sonaka, Manik chaman, Tas-A-Ganesh Red globe, Crimson red are used.

The scion of any commercial table variety Thompson seedless, Manik Chaman, Tas-A-Ganesh, Sonaka, Sharad seedless, Red globe may be selected which is free from any disease, the scion stick should have short internodes with 9-12 months of age and 3-4 nodes and drenched in 0.1 per cent carbendazim for 2-3 hours to avoid Botrydiplodia disease which subsequently kills the plants after grafting.

In *in situ* grafting wedge shape scion inserted into the slit open vertically on root stock and tie with polythene film, at a height of 1½′ above the ground level, the grafting should not be done close to the ground level otherwise because of contact with irrigation water rotting of grafting union takes place and sometime laboures while weeding also damage the grafted plant. The scion and root stock selected may be same thickness for good union. Some time it is difficult to get both mature scion and root stock. It is always better to select both mature scion and root stock.

There is lot of demand for grafted plants in grape which are available in polybags, by which the time required to raise the root stock and *in situ* grafting may be reduced. The lacuna in grafting in polybag are the root stock thickness may not attain 8 mm and it will not be possible to graft at 1½′ above the ground level, and there is a possibility that roots may get damaged after transplanting it is costly. So farmers prefers to go for *in situ* grafting which is cheap and best some time in grape because of in compatibility between scion and root stock inverted bottle neck are formed this is due to uneven growth in both scion and root stock. This condition results in low yield after 4-5 years of grafting.

At certain occasion October grafting fails under such situation green wood grafting is done in May-June but the scion should have good food reserve. The other method of grafting is by chip budding, this is more easier in summer because of better sap flow and slippery bark and easy bud availability.

Better management and care is required to irrigate at proper time to see that the buds should not dry, equally better care is required to avoid damage near the joint or graft union. The grafting taken at a time but formation of union may take different time, so after February bottom cut is taken above the graft union, this helps in uniform growth. Based on the training system selected say Bower system. Telephone system, 4-arm Kniffin system, Single stake system, Y-Trelliss, Geneva double curtain system, develop the permanent frame work. Selection of training system is very important because in Thompson seedless there is apical dominance problem exists which is more predominant in bower system, the cane develops horizontally and due to over lappig and shade, sprouting of buds is very difficult, such situations are easily overcome by California System (Y-Trellis) which is a split canopy and low head system, where as Bower system or pendal system is high head and close centre.

Harvesting of sunlight from Eastern and Western side is very easy in Y Trellis system of training.

After grafting proper irrigation, pest and disease control is very important and loosening of polythene strips tied around the graft region to avoid girding of the stem.

As soon as the plant reaches a height of 120-140 cm head back the plant &allow apical shoots to grow in the E-W directions. These will form primaries and allow to grow to a length of 30-45 cm and nip the apical portion allow two apical buds to grow from each primary area in N-S direction, this will form 4 cordons, when they attain 90 cm length again nip or pinch off, develop canes on cordon and develop 5-6 canes/9 sq.ft. this takes 12 months to develop the frame work of the plant. Afterwards allow one cane/sq.ft. after back pruning and restrict the shoot length to 4 feet after July till October pruning or fore pruning, during July-August take up cycocel spray at 250-500 ppm conc. to reduce the vigour of shoots to increase maturity otherwise spray KH_2PO_4 10gm/litre of water if April pruning delayed.

After October pruning keep two shoots/cane and arrange the shoots at 40-50 angle to the ground for better harvesting and keep only healthy canes, because the thin canes grow vertically and have more apical dominance. Maintain single stem rather than double stem, because single stem helps in reducing the diseases it also advisable initially to show cordon or primary arms, double cordons immediately in first year is not advisable because it reduces food reserve resulting in poor growth also it reduces the thickness of main stem. So by 2nd year of foundation pruning, the shoot at the basel node of the first year cordon may be turned into other side as another cordon thus this way we can develop double cordon and increase yield can be obtained. The length of cordon should be 2.5 to 3 feet, if the length increase apical dominance increases resulting in poor growth.

It is also advisable to maintain 2' in less vigrous varieties and 2½′ in vigrous varieties between the codon, during summer burning of primary is observed if spacing is more than 2½′ between two cordons and build micro climate and aeration will be affected.

In Tropical and sub-tropical region bud break is not a major problem during back pruning but it as a major problem in fore pruning the reasons are lack of chilling buds, poor cane maturity, low temperature after pruning and higher rate of apical dominance. Early cane bending, twisting used to follow which is cumbersom but now a days application of Hydrogen Cynide 1.5 per cent, Thio urea 4.5 per cent potassium Nitrate 4.5 per cent, pactobutrazol 1000 to 1500 ppm by swabing to increase the fruit fullness of basal buds or lateral buds and to increase cluster cane ratio one has to go for sub-cane. After April pruning the buds starts sprouting and shoots develops into cane and there will be different growth rate sometime vigrous, moderately vigrous. At 5th leaf stage spray cycocel 1 ml/lit. to 2 ml/lit of water to check the growth. When the vigrous shoots attains 10-11 leaves pinch at 9th leaf stage with 7-8 mm thickness. The moderately vigrous shoots. When attain 8-9 leaf stage pinch 7th leaf, when less vigrous shoots of 4-5 mm thickness attain 6-7 leaf stage pinch at 5th leaf stage. Later new shoots appears from the apical buds which are designated as sub-canes. After maturity the sub-canes and apical canes are pruned at 3rd - 2nd, 1st leaf stage on vigrous, moderately vigrous, less vigrous canes. This kind of sub-cane development helps in overcoming the bud-break after October pruning.

Application of nutrient helps in increased yield and quality, the requirement of nutrient also vary according to growth stage based on the growth select the fertilizer.

Root stock plays an important role in uptake of nutrient as compared to on their own root.

Before the production starts apply more of nitrogen and low quantity of phosphorus and potash which also vary depending on soil type. So apply ½ kg super phosphate 100 g urea, 100 g SOP, once in 3 months with ½ kg Neem cake and 10-15 kg FYM.

After attaining the fruit bearing stage apply the NPK in following manner. As Grape requires 660:880:660 kg/year and apply the fertilizer in three split doses both at April–October pruning. At April pruning apply 20 per cent N, 60 per cent P_2O_5 after 30 days 20 per cent N, 40 per cent P_2O_5, 20 per cent K_2O at 75 days after october pruning for Thompson seedless variety and its mutants *viz.* Tas-A-Ganesh Sonaka, Manikchaman, Clone -III.

Application of fertilizer *viz.* urea, phosphoric acid, sulphate of potash can also be applied through drip. After April pruning NPK/kg/Ac/day is as follows. For one month 1.50 kg N, 1.4 kg P_2O_5 2nd month, 0.50 kg N, 2.0 kg P_2O_5, 0.45 kg K_2O/AC/Day, 3rd month 1.40 kg N, 1.40 kg K_2O/Ac/day and at 4th month apply 0.40 kg K_2O/ Day/AC.

The recommended dose of 275, 368, 275 kg of NPK/AC/year, 30-40 per cent can be saved by Fertigation. The dose to be given through Fertigation is 180:250:200 kg/ AC/Year.

Based on petiole analysis exact nutrient recommendation on grape is possible. Research done at various research stations recommends critical limits of the various nutrients elements for grape cultivation. If petiole analysis reveals more concentration of the nutrients than the critical level then fertilizer applied is more and if it is less then it is deficiently applied.

Root stocks shows deficiency symptoms of micronutrients like Mg, Mn and Zn because of more vigorus growth, so apply Zinc sulphate after 35 days of October pruning as foliar spray. Apply 50 gm $MgSO_4$ per plant as soil application. At monthly inerval after blooming apply 0.15 to 0.25 per cent.

Grape irrigation can be based on evapotranspiration of that area and farmers think that root stock do not require much irrigation, thinking that they have deep root system and absorb thin film of water also. Some time Tensiometer can be installed to see the requirement of water to see proper growth, fruit bud differentiation and yield. As in maize cultivation sunflower plants are used as indicator plants, in Grape also there is a need to develop indicator plants. It is better to irrigate before it show the stress symptoms like wilting of leaves in afternoon, proper irrigation after back pruning to get good growth is very important so better to apply light irrigations at frequent intervals than applying more water, some time mulching, spraying PMA, Kaoline, anti-stress also help to over come water demand.

Application of more water affect the soil health because it contain salts and in due course of time soil becomes saline which will leads to poor aeration less infiltration, water logging etc. Subsequently root rotting takes place. By drip irrigation 25 per cent of irrigation water can be saved and yield can be increased by 30-45 per cent. The wetting of soil depend on its texture, the sandy soil will have more infiltration rate and wetting area is deeper where as in case of clay/Black soil the rate of infiltration is lower and wetting area is shallower broader.

If the canopy area is 4m^2 the emitters should be 40 cm away from stem < the wetting area with the 50 cm and keep 2 emitters/vine. If the canopy area is 2m^2 then put emitters 30 cm away from the stem put 2 emitters/vines.

Depending on evapotranspiration rate in Northern Karnataka the water inputs by drip irrigation is as follows.

Complete harvest before 10 days of pruning and apply 18000 to 20,000 litre of water/AC/Day, 1-40 days after April pruning apply 20,000 to 24,000 lit./Day/AC, 40-70 days, 10,000 to 12,000 lit/Day/AC. From 71 day till October pruning 12,000 to 14,000 litrs/Day/AC. At October pruning from the day of October pruning to 40 days 16000 to 18000 litre/Day/AC at 41-75 days 10,000 to 12,000 litre/Day/AC, then 76 to 110 days of October pruning 18000 to 20,000 litres/Day/AC after 110 days till harvest 14000 to 16,000 litres/Day/AC.

In Grape major diseases are Anthracnose Downy mildew, powdery mildew. The root stocks *viz. Vitis champini, Vitis rupestris, vitis berlandieri, Vitis riparia* are resistant to diseases and these are used in breeding of root stocks like salt, Greek, St. George 1613, Dogridge etc. However necessary precaution are necessary when diseases are observed.

Rust is noticed in Bangalore blue variety where as root stocks are susceptible to rust.

To control Downy mildew spray Bordeux Mixture @ 0.5 per cent copper oxy chloride 2.5 gm to 3 gm/lit. of water. Metalaxyl 2gm/lit. of water Ridomi MZ 2-2.5 gm/lit.

To control powdery mildew, spray Bayletan 1gm/lit, Mancozeb 2.5 gm/lit, Karathane 0.25 ml/lit. of water, wettable sulphur 2gm/lit, miclobutanol 10WC, 0.5-1gm/lit or Hexaconozol + captafol 1.5 gm/lit.

To control Anthracnose spray Hexaconozol/1ml/lit, Thiophenate methyl 1 gm/lit of water, 1 per cent Bordeux mixture, copper oxy chloride 2.5 to 3 gm/lit - copper Hydroxide 2gm/lit.

Salt Creek root stock helps in improving sugar content in Thompson seedless. To maintain leaf and fruit ratio, one has to maintain a leaf area of 2500 cm^2 *i.e.* 12-15 leaves per bunch, other wise quality of the bunch will be deteriorates.

Girdling at 60 days after October pruning helps in fruit development by hindering the translocation of carbohydrate above the girdling zone. After October pruning following schedule helps in improving the bunch quality

1. At April pruning at 5-7 leaf stage spray 250-500 ppm-CCC
2. Spray GA - 15 ppm at pre-bloom
3. Spray GA - 20 ppm at 25 per cent cap fall
4. Spray GA - 25 ppm at 25 per cent cap fall

5. Spray GA - 50 ppm at 100 per cent cap fall
6. Spray 6BA - 1—15 ppm at berry softening
7. Spray NAA - 20 -25 ppm 25-20 days before harvesting to reduce berry shattering.

The present investigation was undertaken to study the effect of IBA and number of internodes on rooting of stem cuttings of grape root stocks. It was found that the rootstock 1613-C did not require any chemical treatment for stimulating the root formation while Dogridge and Salt creek cuttings with five internodes were treated with 1500 ppm IAB for better sprouting percentage of 86.60 and 80.00 respectively. The significant results for root growth parameter of rootstocks were obtained at higher concentration of IBA.

Sumkuwar *et al.* (2006) stated that in recent years, many grape gardens have been raised on rootstocks particularly in peninsular India due to the shortage of irrigation water and salinity in the soil. There is still apprehension in the grape growers mind that which rootstock has more beneficial effects with respect to sustainable yield and quality and also the longevity of the vine. Thompson seedless was grafted on Dog Ridge, 110R, 99R, 1613C, St. George and Salt Creek and was compared with own rooted Thompson seedless for bud fruitfulness. The results revealed the significant differences with respect to rootstock and bud position. Among the different rootstocks, general performance of the and also the microscopic studies showed that the rootstock 110R can be the alternative to the present Dog Ridge rootstock which is popular among the grape growers of Maharashtra.

McKenry and Safdar (2006) reported that sixteen cultivars of grape were screened over a two-year period in the presence or absence of 10 different nematode populations. Populations of *Meloidogyne* spp., *Xiphinema index*, and *Mesocriconema xenoplax* developed more rapidly and caused greater damage than populations of *X. americanum* and *Tylenchulus semipenetrans*. Populations of mixed *Meloidogyne* spp. having a history of feeding on grape were among the fastest developing populations. Tolerance to nematode parasitism appeared to be based on different mechanisms. Slow developing, less pathogenic nematode populations often stimulated vine growth, thus vines appeared to possess tolerance. Likewise, cultivars selected for nematode resistance often stimulated vine growth when fed upon by the nematode. However, tolerance sources that resulted from nematode resistance are vulnerable due to the occurrence of populations that break resistance mechanisms. Growth of cultivars with phylloxera (*Daktalospharia vitifoliae*) resistance was unchanged by the presence of nematodes, indicating that phylloxera resistance may provide a useful source of nematode relief. These and several additional sources of specific tolerance are discussed.

Somukuwar *et al.* (2011) stated that four grape rootstocks belonging to different *Vitis* species were planted in September. Rooting behaviour, polyphenol oxidase (PPO; EC 1.14.18.1) activity and biochemical parameters were studied at different growth stages after planting the cuttings in polythene bags. Significant differences were recorded for rooting success among the different rootstocks, with the maximum sprouting percentage determined in the Freedom rootstock. Th e highest variation in

PPO activity was also recorded in Freedom. High PPO activity was recorded in rootstock 140 Ru, while the highest root length was recorded in Dog Ridge. In the rootstock 110 R, the PPO activity was the lowest during the initial stage, though it increased up to 60 days after planting DAP) and was then reduced up to 90 DAP. Increases in the number of rooting primordials were recorded in all the different rootstocks at different stages of growth. A higher number of rooting primordials was recorded in Freedom, and it was followed by 110 R. The present study suggests the differences in the rooting behaviour of the different rootstocks are based on PPO activity at regular time intervals up to 90 DAP.

Mahdi and Singh (2009) reported that the micropropagated plantlets derived from three different grape rootstock genotypes namely, Dogridge (*Vitis champini*), SO4 (*V. berlandieri*×*V. rupestris*) and ARI-H-144 (*V. vinifera*×*V. labrusca*) were subjected to randomly amplified polymorphic DNA (RAPD) and inter simple sequence repeats (ISSR) analyses in order to evaluate their genetic stability and/or detect likely existing variations among *in vitro* derived plantlets. A dozen RAPD (10-mer) and ten ISSR (dinucleotide contained repeats) primers were used for PCR and reproducible band profiles were obtained. The 84 and 81 distinct and scorable band classes (a total of 1,914 and 1,980 scorable bands) with an average of 7.0 and 8.1 bands per primer were obtained by RAPD and ISSR, respectively. Although higher numbers of bands were obtained by ISSR rather than RAPD, but none of the primers showed polymorphism among micropropagated plantlets and their respective mother plants. The profiles generated based on the two marker systems were found to be highly uniform and monomorphic. Cluster analysis further confirmed genetic stability of micropropagated plantlets. Jaccard's similarity coefficients obtained for both markers in mother plants and there *in vitro* regenerants were estimated to be 1.00 but three sets of genotypes were grouped into two major clusters with similarity coefficients of 0.53 (RAPD) and 0.63 (ISSR). The molecular analyses precisely proved the production of genetically stable grape plantlets and certified the application of micropropagation protocol to be developed on a commercial scale.

Chandrasekhara Reddy and Chadha (1998) carried out an on the effect of roostocks on Anab-e-Shahi grape employing the rootstocks St. George, Teleki 5-A, Dogridge, 1616, 1613 and Gulabi in comparison with own-rooted vines to identify the most suitable rootstock. Rootstocks Dogridge and 1616 were found to impart vigour to Anab-e-Shahi grape but adversely affected the yield and yield attributes. Rootstock Gulabi lowered most of the vigour attributes and resulted in the highest yield. St. George and 1613 similar vigour attributes as own-rooted vines, but influenced the yield favourably. Rootstocks Gulabi, 1613 and St. George increased the yields through increased fruitfulness of canes. Fruit quality parameters were not influenced by any of the root stocks. Higher levels of total and NO_3-N were associated with lower yields in the rootstocks Dogridge and 1616 and it was reverse in Gulabi. St. George appeared to be an ideal rootstock for economizing N inputs. Higher petiole K contents during June were associated with higher yields. Lower Mg and higher Zn contents were in general associated with higher yields. Rootstocks Dogridge, 1616 and St. George had low levels of chloride content and may be suitable for areas with high chloride in irrigation water. Among the petiole nutrients, NO_3-N, Mn, K and Zn contents in June and total N, Mg and Cl in December were found to be useful in yield prediction. Leaf sampling in June may be useful to work out relationship between petiole nutrient contents and yields under double pruning and single cropping conditions. Rootstock Gulabi may be employed to increase productivity of Anab-e-Shahi grape. St. George and 1613 are useful on sodic soils and in locations where irrigation water has high chloride content.

Three experiments were carried out at Indian Institute of Horticultural Research (IIHR), Hessaraghatta, Bangalore during the year 2006-07 to study the effect of various substrates and microbial combinations on rooting and graft success of grapevine rootstocks (Husameddin and Prakash, 2008). The experiments consisted of three grapevine rootstocks *viz.*, Dogridge, St.George and 1613 and three substrates namely plant sap, humic acid, vermiwash and a control and seven microbial combinations (M1: *Glomus mosseae* + *Trichoderma harzianum*, M2: *Glomus mosseae* + *Bacillus subtilis*, M3: *Glomus mosseae* + *Pseudomonas fluorescens*, M4: *Glomus mosseae* + *Trichoderma harzianum* + *Bacillus subtilis*, M5: *Glomus mosseae* + *Trichoderma harzianum* + *Bacillus subtilis* + *Pseudomonas fluorescens*, M6: IBA and M7: control). Among the substrates used, humic acid significantly influenced almost all the root and shoot parameters in hardwood grapevine cuttings of the three rootstocks. Next best treatments were vermiwash and plant sap. Among various microbial combinations tried, treatment M5 was the most effective for almost all the root and shoot parameters studied as compared to control. Other microbial combinations (M1, M2, M3 and M4) gave similar results as IBA treatment. Among different grape rootstocks tested, the root development was significantly superior in 1613, while St. George rootstock showed the best shoot development. Early budbreak was recorded in the cuttings treated with plant sap which was on par with the cuttings treated with vermiwash. Among the different microbial combinations M2, M3 and M5 gave early budbreak as compared to control. The percentage of graft success and root colonization was significantly increased by humic acid, whereas other substrate treatments were not significant. The per cent graft success was significantly increased by M4 and M5, while the percentage of root colonization was significantly increased by all the microbial combinations (M1, M2, M3, M4 and M5). The rootstock 1613 gave the highest per cent graft success, while St. George showed the highest percentage of root colonization.

Kanamadi and Singh (1987) carried out an experiment with an objective of identifying the superior grape hybrids in respect of different qualitative and quantitative parameters, studies were carried out on the comparison of parents comprising of twenty promising hybrids with ten parental varieties in a replicated trial in a completely randomized design at the Indian Institute of Horticultural Research, Hessaraghatta, Bangalore, during 11986-87. The ten varieties used either as female or male parents were; (1) Anab-e-Shahi, (2) Angur Kalan, (3) Bangalore Blue, (4) Bangalore Purple, (5) Black Champa, (6) Cheema Sahebi, (7) Convent Large Black, (8) Gulabi, (9) Queen of the Vineyards and (10) Thompson Seedless. Hybrids were evaluated against their parents in respect of vine growth parameters, yield attributes and quality components. Salient findings of the study are summarized below; 1. Significantly differences were observed among the parents and hybrids in respect of stem girth in June and October, and October pruning weight. Significantly differences were also observed in respect of the number of days taken for bud burst, shoot length, leaf number/shoot specific weight, leaf area during the fruiting season and leaf dry matter content, Chlorophyll 'b' and total chlorophyll (a+b) content during the fruiting season, yield attributes such as number of bunches/vine, mean bunch weight, bunch length and breadth, number of berries/bunch, 50-berry weight and the yield/vine. Quality components such as juice, total soluble solids and seed (number) content also varied significantly among the parents and the hybrids. Shoot length at 60, 75 and 90 days after bud break during the growth season was negatively correlated with yield/vine. Reduction in the yield on account of increased shoot length seems to have mediated through the number of berries/bunch. Dry matter content, area and specific weight of leaf during the growth season were positively correlated with yield/vine. Anab-e-Shahi hybrids 4/15 (Anab-e-Shahi x Bangalore Purple) and 5/17 (Anab-e-Shahi x Black Champa) which had less shoot length after 60, 75 and 90 days of April pruning than Anab-e-Shahi seemed to be promising for yield. The same Anab-e-Shahi hybrids *viz.*, 4/15 and 5/17 had more leaf dry matter content during the growth season, adding to their promise in respect of yield. 5/17 also holds promise for yield over Black Champa in view of higher leaf dry matter content. Anab-e-Shahi hybrids, 4/15 (Anab-e-Shahi x Bangalore Purple), 5/13 (Anab-e-Shahi x Black Champa), 5/17 (Anab-e-Shahi x Black Champa), F-13/20 (Anab-e-Shahi x Gulabi) and F-12/29 (Anab-e-Shahi x Gulabi); Thompson Seedless hybrids 29/6 (Black Champa x Thompson Seedless) and 30/16 (Black Champa x Thompson Seedless); Gulabi hybrids F-13/20 (Anab-e-Shahi x Gulabi) and F-12/29 (Anab-e-Shahi x Gulabi) and Bangalore Purple hybrid 4/15 (Anab-e-Shahi x Bangalore Purple) excelled their respective parents in yield/vine. Based on the observations made on the vine growth parameters, yield and quality, it can be concluded that Anab-e-Shahi hybrid 5/17 (Anabe- Shahi x Black Champa), Thompson Seedless hybrids 29/6 (Black Champa x Thompson Seedless) were improvements over their respective parents for yield and quality, Anab-e-Shahi hybrids, 4/15 (Anab-e-Shahi x Bangalore Purple) was improvement over Anab-e-Shahi for yield.

The present investigation was carried out during 1986 and 1987 at IIHR, Bangalore, to evaluate the hybrids of the crosses Anab-e-Shahi, Angurb Kalan, Bangalore Blue, Black Champa, Convent Large Black, Thompson Seedless and Queen of the Vineyard, with respect to growth, fruitfulness, yield, quality, petiole nutrient contents and resistance to diseases (Venkataram and Singh, 1989). The salient findings of these investigations were seen as considerable variations in stem girth observed among the hybrids and parents. The hybrids 5/12 and 4/30 from the cross Anab-e-Shahi x Black Champa, hybrids 9/3 from the cross Anab-e-Shahi x Queen of the Vineyard and hybrids 29/4 and 29/5 from the cross Black Champa x Thompson Seedless had more trunk girth than rest of the hybrids. The hybrids 8/29, 29/4 and 29/5 appeared to be highly vigorous, while rest of the hybrids had low vigour. The hybrids 26/8 registered the minimum pruning weight. The percentage of increase in pruning weight of the hybrids was from 88.68 per cent to 153.26 per cent over the best parent, indicating for the tendency of the hybrids vigour. The parent Black Champa as a seed parent and Queen of the Vineyard as a pollen parent appeared to be best combiner for imparting high vigour to the progeny. The hybrids 4/30 had highest number of shoots. The hybrids 7/12 and 8/24 had minimum number of shoots. The hybrids which recorded the lowest angle of the shoot was 7/12 and 8/24. The hybrid 4/30 recorded the lowest shoot length in the both season, while the hybrid 29/5 recorded the highest shoot length in growth season and the hybrid 5/12 in fruiting season. Cane diameter increased invariably in hybrids compared to their parents. The genotypes with high number of shoots had less percentage of cane maturity. The hybrids 29/4 and 29/5 from the crosses Black Champa x Thompson Seedless had invariably number of leaves the better parent Bangalore Blue had the highest tolerance to all the diseases. While Anab-e-Shahi and Thompson Seedless appeared to be highly susceptible. The hybrid 30/40 had the least incident of anthracnose, while the hybrid 29/5 was highly susceptible. The hybrid 21/28 had higher downy mildew disease tolerance among all the progenies, while the hybrid 5/12, 7/12 and 26/8 were moderately tolerant. Hybrids 4/30, 8/24, 29/3 and 29/4 were highly susceptible.

Pollamma and Yadav (1997) stated that the micropropagation is a technique for rapid multiplication of plant *in vitro*. The present investigation was carried out to develop a protocol for micropropagation of salt tolerant grape rootstock variety 'Ramsey'. The experiments were conducted at the plant Tissue Culture Laboratory – I of the Division of Biotechnology, Indian Institute of Horticultural Research, Hessaraghatta, Bangalore. Various experiments were conducted and result were analysed and discussed. Decontamination of 'Ramsey' nodal cutting were best achieved when they were pretreated for 1 hour or 4 hour and surface sterilized for 10 minutes. With this treatment the per cent survival was higher when compared to other treatments *viz.*, the contamination rate was lower in the above treatments. Softwood nodal cuttings gave lower contamination rate. Induction of multiple shoots from nodal explants was best achieved with modified MS medium supplemented with 3 ppm BA. At this rate of BA maximum number of leaves and nodes also were obtained. At a concentration of 5 ppm BA in MS basal medium callus with pink colour pigmentation was obtained. Further, it was found that half strength MS was a better basal medium with 3 ppm BA for induction multiple shoot. Half strength MS gave rise to a sturdy plantlet with a good shoot and root system which could be directly acclimatized. Hardening was done using the closed sachet method with a very high survival per cent.

Varghese and Murthy (2004) opined that among the six grape varieties Black Champa, Arka Neelamani, Anab-e-Shahi, Shweta Seedless and Arka Krishna taken up in the study Black Champa variety of grape responded the best to *in vitro* culture while Arka Krishna was the least responsive on MS medium supplemented with various levels of growth regulators; IBA and NAA. IBA supplemented to medium was found to be better for *in vitro* culture establishment compared to the one supplemented with NAA. Rooting, sprouting and growth response obtained varied with the variety. All the six varieties showed better rooting response in MS medium supplemented with IBA compared to those cultured on MS medium supplemented with NAA. Black Champa, Arka Neelamani, Anab-e-Shahi, Shweta Seedless and Thompson Seedless gave higher root number and root length at higher concentrations of IBA tried (3 mM) compared to the lower concentrations. Arka Krishna showed the lowest root number and root length, which was observed at a lower concentration of

IBA. All the varieties except Arka Krishna showed the maximum root weight at higher concentrations of IBA. Black Champa variety showed maximum rooting, sprouting and growth response at higher level of IBA (3 mM). In general it gave the highest root number, root length, sprout height, root weight, node number and leaf number when compared to the other varieties. This shows that *in vitro* culture can be taken up successfully for Black Champa variety of grapes. It was found that in Thompson Seedless higher IBA levels favoured rooting response, growth response and sprout weight while lower IBA concentration favoured better sprout height. Arka Neelamani resembled its paternal parent Thompson Seedless in rooting requirements (root number, root length and root weight) while shooting was better at a lower concentration of IBA. Anab-e-Shahi showed variable response at different levels of IBA. It gave the highest root number, root length, and sprouts height and root weight at higher concentration of IBA while the sprout weight was observed to be maximum at lower level of IBA. The growth variables leaf number and node number were maximum at higher levels of IBA. It was observed that Shweta Seedless variety resembled its parent Anab-e-Shahi in its rooting response and sprout height; while on the other hand, for higher sprout weight it needed a higher concentration of IBA than Anab-e-Shahi. Further, it was observed that in Shweta Seedless and its parents (Anab-e-Shahi and Thompson Seedless), the growth response (leaf number, node number) and sprout height were better on full-strength MS media compared to half MS medium. On the other hand the rooting response and sprout height was better on full MS medium. In vitro culture of Shweta Seedless single nodal cuttings on a combined 2 mM BA and 2 mM IBA supplemented media gave the best shoot and root system without any callusing. Nearly 80 per cent survival was obtained when *in vitro* derived plantlets of representative samples transplanted into polybags using the closed sachet method.

5

Training and Pruning

April Pruning: Why and How to Do it

The necessary information; Grape is cultivated in the Mediterranean region where the climate will be just like a coastal region. Under such a situation the grapevine grows continuously. Warm and dry weather is ideal for growth. It is successfully grown in southern parts of the country where the temperature ranges from 15° to 40°C and annual rainfall amounts to over 50 cm. The amount of rainfall is not the criterion but the distribution is important for its successful growing. The rainfall should not coincide with the fresh growth after pruning or during fruit ripening, cloudy weather, high humidity, low temperature and precipitation. During flowering and berry development it is detrimental as they are highly infectious diseases.

After April pruning, to get 16-20 branches or canes, it requires 30-40 days and to establish the flower primordial it requires another 25-30 days. It means 50-60 days before the onset of monsoon, April pruning should be done; if the pruning gets delayed, the period available from April pruning to the onset of monsoon gets delayed which results in the immature cane growth and coincides with the rains and ultimately gets affected by anthracnose. The farmers are advised to take up KH_2PO_4, 10 g/litres of water to bring immediate maturity to the canes. In Karnataka the South West monsoon starts from 15th June and if the farmers take the pruning in the first week of April, the time available to develop a cane will be 10 weeks, by which time the canes get copper coloured and the wood gets matured and which results in minimizing the anthracnose problems. Further if pruning gets delayed by one week then the time available will be 9 weeks. Under water stress situation, they are advised to go in for anti-stress spray in the summer to overcome water shortage problem.

It has been experimentally proved that the most fruitful part of Thompson seedless variety is during the middle portion *i.e.*, 6th to 9th node and accordingly the spur pruning or cane pruning is to be done.

Soil Management

To start with the soil, where vineyards have been established, it should be fertile and productive. Soil should be of good depth with good drainage having minimum lime content. The vineyard soil should be uniform. Any shortcoming should be avoided before planting, keeping the soil in good health. The fertility level should be checked every year. During the productive season the soil should have the following characters:

1. pH 6.7 to 7.7
2. Organic carbon– around one per cent
3. Electric conductivity – less than 20 mm hos/cm^2
4. Calcium carbonate – Less than 8 per cent.
5. Water-holding capacity – around 50 per cent.

Spacing, Primaries and Number of canes/vine

Spacing	*Spacing in Sq. Foot*	*Number of Primary*	*Canes on each Primary*	*Total Cane*
6' x 4'	24	2	8-10	16-20
8' x 6'	48	3	10-12	30-40
10' x 5'	50	3	10-15	40-50
10' x 6'	60	3	15-20	50-60
10' x 10'	100	4	20-25	80-100
12' x 8'	96	4	20-22	80-90
12' x 6'	72	4	15-20	60-70
12' x 4'	48	4	12-15	40-50
10' x 4'	40	4	8-10	35-40
8' x 8'	64	4	10-12	60-65

Knowledge about various year yield; based on that the future course of action depends upon the quality of the grape bunch existing in the previous years also reflected from the soil and water quality, the nutrition management and the pest and disease management, the proper and timely use of Hormones. On an average 10-12 tonnes per acre yield is expected if the yield is reduced by 10 per cent for various reasons as listed below:

1. After last April pruning new growth was not proper.
2. After October pruning flower production was not proper.
3. Proper attention was not given regarding flower formation and precautionary measures like spraying of 6BA -10 ppm, 1AA - 10 ppm, NAA - 10 ppm was not taken.

4. Last October pruning canes had not matured.
5. There might have been disease and insect problem for which precautionary sprays were not taken up.
6. Natural calamities also caused loss of crop viz, hail storm, continuous rainfall and cyclone etc.
7. The yield loss due to various reasons needs to be examined for which precaution measures were taken.

Water and Soil Testing

After the previous crop harvest by soil testing, one has to take two soil samples by digging one cubic feet pit and by scratching the soil from the inner surface of the pit it should be examined whether it is feasible for various nutrients and micronutrients. This is how the farmers will examine the crop yield to ascertain and implement the recommendation after analysis.

Watering

After the harvesters of the previous crop give heavy watering or flooding. the previously fertilizer and excess sodium alkalinity could have get accumulated in the plot, thereby leaching was made possible by flooding. After that soil testing should be taken up.

Inter-cultivation

Once after heavy flooding, if the soil is ready for ploughing, take up deep ploughing, then collect the weeds especially hariyali and Jek *Cynodon dactylon, Cyprus rotundus.*

Application of Fertilizer

Based on the soil report, application of fertilizer should be taken up usually by applying fertilizer as mentioned in the table for ready reference to the farmer.

Sl.No.	*Name of the Fertilizer*	*Fertilizer Requirement per Acre*
1.	Compost	12 tonnes
2.	7:10:5	300 kg
3.	DAP (18:46)	50 kg
4.	Bio-fertilizer	15 kg
5.	Micronutrient	15 kg
6.	Bombaymeal	1 tonne
7.	Magnesium sulphate	15 kg

Management of April Pruning

By keeping the main stem and primaries-pest and disease -affected branches and weak branches from last season should be removed keeping them at a proper distance by keeping 1-2 buds and wherever the number of canes are more on the primaries then throw them out. If there are dried swollen/barren region at the internodes region, then cut them from the bottom without leaving any stub. In 6' x 4', keep two primaries and 7-8 canes be maintained, with maximum of 12 canes on each primary.

Timely Pruning

Before the rain starts and after completion of harvesting of the crop the April pruning is so adjusted that at the time of the beginning of the monsoon, the canes should have 14-15 leaves. If harvesting gets completed by the end of March then pruning is possible by mid-April. Then onwards 1½ to 2 months time is available for cane maturity. During April the temperature will be of 40°C and humidity would be very low. Hot wind will be blowing and the soil would also be having less nutrient after the crop harvest.

Therefore, based on soil analysis report apply only fertilizer to the soil to get proper growth otherwise the canes will be unproductive. One should not compromise with fertilizer application at the time of cane growth, as the loading of fertilizer gets accumulated in the stem before the new growth starts. If there is going to be any mistake, it will reflect on the yield. In case you want to go for sub-cane, allow the cane to grow up to the 6th bud and pinch the buds from the 6th bud and allow another 7 buds to grow. In another method allowing the cane to grow up to 8th bud, pinch the 8th bud and retain 7 buds. Then allow another 5 buds to grow and in both the cases the total number of leaves should be twelve.

So by June first fortnight complete the operation.

Management of Hormonal Practices to be adopted after April Pruning

April Pruning	*CCC*	*6 BA*	*Uracil*
Pruning before 10th of April	500 ppm. at 5th leaf	–	–
April pruning between 10th-25th	1) 500 ppm at 5th leaf stage	20 ppm after 45 days of April pruning.	50 ppm. 45 days after April pruning.
April Pruning	CCC	6 BA	Uracil
	2) 750 ppm 7+5 leaf stage.		
April Pruning after 25th	1) 500 ppm at 5th leaf stage. 2) 750 PPM 7+4 leaf stage 3) 750 ppm at 7+4+4 leaf stage.	20 ppm. after 5 days.	1) 50 ppm. after 45 days. 2) 50 ppm. after 55 days.

Drenching of Bordeaux Mixture

After hard pruning, drenching of Bordeaux mixture is necessary; one has to spray on the primaries and secondly one has to apply Bordeaux paste.

Important Works to be Taken Up at April Pruning

1. Take up April pruning in time
2. Before pruning take up inter-cultivation and apply fertilizer and water.
3. After April pruning, pasting of Bordeaux and application of lime to the main stem is necessary.
4. Disease affected, stem borer should be eradicated as its precautionary measures.
5. Testing of soil and adopting the recommendations.
6. After harvest vine should not be disturbed for 15 days. Undertake April pruning by 10th.
7. Apply 20 tonnes of farm yard manures, one tonne bone meal, 20 kg nitrogen and 60 kg phosphorus.
8. About 4-20 days after April pruning, flea beetle control is necessary by spraying Alphamethrin 1 ml/litre.
9. Keep one cane for every one square foot area at the 4-6 leaf stage and remove the thin canes.
10. At 5th leaf stage, spray 500 ppm Lihocin (1ml./litre)
11. 25 to 30 days after April pruning, apply 30 kg nitrogen and 40 kg phosphorus.
12. At 9th leaf stage, keep 7 leaves and pinch the small two leaves.
13. Spray 500 ppm, Lihocin (2nd spray) at 12 leaves (7+5 stage). Means when the cane is pinched at 7th leaf, it attains another 5 leaves.
14. After April pruning at 45 to 50th day, apply 50 kg potash.
15. About 40 -50 days after pruning, reduce irrigation.
16. Grow green manure crops *viz.*, sunhemp in the month of July. If the soil is black then grow Dhaincha.
17. In September (1st week), cut the sunhemp and apply to soil up to ½′ depth and cover it.

Control of Pest and Disease

After April pruning, due to hot climate, the occurrence of disease will be less. But to control mealy bug, flea beetle, leaf eating caterpillar, and mites one or two sprays are required.

1. Carbaryl (3g). or Malathion (2 ml); In the month of June when the monsoon starts, the leaves and canes get matured, that is, within 6-8 weeks diseases like Anthracnose, powdery mildew begins in the, first and third weeks. In that case Bordeaux mixture of one per cent foliar spray is necessary. Then in the months of July, August and September with 12-18 days gap, follow the following spraying schedule:

First spray - Carbendizm of 1 g/litre quantity.

Seconds spray - Captaf of 2 g/litre quantity.

Third spray - Blitox of 2 g + streptocyclin ½g/litre quantity.

Fourth spray - Ziram 4ml/litre.

Use Apasa 80 0.25 ml/litre or any other sticker and if drizzling starts burn sulphur and apply its smoke.

Anti-Stress and its use in Grape Cultivation

If there is a problem of irrigation due to water shortage then watering is important because at the flower primordial stage sufficient water quantity is a must. If there is a shortage of irrigation water, use of anti-stress is very important. When Anti-stress is mixed with water and sprayed, it will not be visible to the naked eye because it forms a very thin layer and from this layer water loss from the leaf will be minimized and the vine will not face water shortage.

The stomata on leaves stops loss of water to some extent, because of which CO_2 and oxygen exchangeability is not affected; the mites and jassides are controlled because of thin layer of anti-stress. Germination also gets minimized but new bunch and small berries are affected because anti-stress are not protected. The bunch, when gets soft, anti-stress should not be used, that is, when we use more water for irrigation the vine from roots absorb more water which gets filled in berries and because of less transpiration loss the berries get cracks and bunch starts rotting.

Schedule of Anti-Stress After April Pruning

Sl.No.	Stage Bunch	Anti-Stress/litre
1.	After April Pruning at 30th day	2 m/litre
2.	After April Pruning at 55th day	3 m/litre
3.	After April Pruning at 80th day	3 ml/litre
4.	After April Pruning at 105th day	4 ml/litre
5 .	After October Pruning at 25th day	2 ml/litre
6.	After October Pruning at 55th day	3 ml/litre

Micronutrient Application After April Pruning

Time	Fertilizer Grade	Quantity
First spray June 15th	Micnelt - MS 32	500 gm in 100 lit (250 litre/AC)
Second spray July 15th	Fertilon combi - F	100 gm in 100 litre of water (250 litre/AC)
Third spray	Multiplex	250 ml in 100 lit of water (250 litre/AC)
Fourth spray	DAP + magnesium sulphate	(250 litre per acre) 1.5 kg + 1/2 kg/500 litre

Water Management After April Pruning (Litre/AC/Day)

State	*On Own Root*	*With Mulching*	*Mulching+ Anti-stress*	*Root-Stock*	*Anti-stress+ Rootstock*
10-20 day before April pruning	0	0	0	0	0
Between 3-10 days of bud sprouting	19,200	17,600	17,600	13,000	12,000
Cane development stage of 11-35	24,000	22,000	14,400	16,000	12,000
Establishment of flower primordia 36-70 days	12,800	11,800	7,800	9,000	6,500
Bud development stage 71-100 days	15,000	14,000	10,500	10,000	8,000
Cane maturity stage 101-150 days	9600	9000	6800	8000	6,600

Water Management

There is a wrong assumption among the farmers that rootstock gardens do not require irrigation at normal and regular intervals. In fact rootstock can use the applied water more efficiently penetrating their roots into deeper layers, and they even absorb the film of water around the soil particle and save water to some extent.

Hence irrigation is most necessary for enhancing vine growth, fruit bud differentiation and production.

Points to be considered for efficient water management:

1. Irrigate the vines before they start showing symptoms of moisture stress like temporary wilting of leaves in the afternoon.
2. During summer irrigate the vines adequately to meet the evapotranspiration requirement.
3. Plan the irrigation schedule meticulously to obtain rapid shoot growth in the initial stage; after April pruning it slows down its growth in the later stage to attain cane maturity and fruit bud differentiation.
4. Light and frequent irrigation are better than heavy and less frequent irrigation depending on soil type and availability of water.
5. Practice moisture conservation, like mulching anti-transpirants, spray and addition of sufficient organic matter into the soil to increase water use efficiency.

Excess Irrigation and Effects

Excess irrigation than the required amount has adverse effect on soil characteristics which in turn affects plant growth and productivity. Since water contains one or other dissolved salts, excess irrigation leads to build-up of salts which in later years converts normal soil into saline soil. Also excess irrigation leads to poor infiltration resulting in water logging which affects soil aeration and roots start rotting leading to ultimately death of vines.

Drip Irrigation

This is the most effective way of irrigating vine yards where the vines are irrigated frequently and slowly through emitters or various discharges. It also saves about 25 per cent irrigation water and increases yield by 35-40 per cent than in basin irrigation.

Placement of Emitters

The number of emitters to be fixed around each vine depends on the distance at which the emitters are fixed and the diameter of wetted area around each emitter (20-30 cm deep layer) in the soil. The wetting of soil depends on its texture and water holding capacity. Rate of infiltration is faster in sand soils and in wetted area it is deeper. Whereas in clay soil rate of infiltration is slower and wetting area is shallower and broader. The number of emitters and its placement for seedless cultivars of grape is shown below:

Canopy Area (m^2)	*Location of Emitters from Stem (cm)*	*Diameters of Wetted Area (cm)*	*No. and Location of Emitters/Vine*
4-5	45	60	2 on line
2-3	30	60	2 on line

Irrigation Schedule

Grapevine need water in different quantities during different growth stages Accordingly one should irrigate the vineyards. The quantity of water to be irrigated depends on the evaporation and transpiration rate. Based on the replenishment of evo-transpiration rate, water input by drip irrigation is as follows:

Sl.No.	*Stage*	*Water Kilo Litre/AC/Day*	
		South Interior Karnataka	*North Karnataka*
1.	From harvest to 10 days before April pruning	16-18	18-20
2.	10 days before pruning to pruning	0	0
3.	1-35 days after April pruning	18-20	20-24
4.	36-70 days after April pruning	10-12	10-12
5.	71 day after April pruning till October pruning	12-14	12-14
6.	0-45 days after October pruning	14-16	16-18
7.	46-75 days after October pruning	10-12	10-12
8.	76-110 days after October pruning	16-18	18-20
9.	After 111 days till harvest.	12-14	14-16

Importance of October Pruning

1. October pruning with prior planning
2. The plot where October pruning is going to take place should be divided into different blocks.
3. Note down the work to be attended before October pruning
4. Note down the work to be attended after October pruning
5. Keep developed canes and buds before going in for October pruning
6. Go for bud testing in the laboratory.
7. Pasting of buds should be done.

Note:

1. How to prepare black Bordeaux paste: Take half kilogram of copper sulphate plus lime 250 gm + 125 gm of Thiourea in 5 litres of water and smear the front 3 buds.
2. Take Dormax/Romax of upto 30 ml/litre of water and swab it on the canes leaving the backside 3 buds for uniform bud sprouting.
3. By application of black Bordeaux or Dormex the apical dominance is controlled.

Importance of October Pruning When and How to Do it

Before and after October pruning the details of work to be done should be understood thoroughly.

In Karnataka, the October pruning is done by the middle of October and under changing climatic condition, sometime early sometimes late pruning is advocated. By doing early or late pruning what effect it is going to have on the crop is also important and should be kept under check.

October pruning is very important in grape cultivation. In Karnataka grape harvest starts as early as January and completes by the first week of April. The grape harvested in January month is not sweet and for export the sugar-acid ratio should be between 35:1 to 40:1 which is a must and that is possible in the month of February and March and for this purpose October pruning has to be done in the first fortnight of October.

At present before undertaking October pruning two points are very important which one has to take into consideration. First is to get the highest yield possible and second is to get good market response in different parts of the country. To realize these things this pruning is advised between 15th September to 15th November. Sometimes if one gets higher yield the prices will be less. So, we can conclude that it doesn't make much sense but it makes difference in the total expenses and returns. Therefore to tap the moderate price in this market, early or late pruning in October doesn't make much of a difference.

In our places, from 15th November to the last week of December, cold winter temperature prevails. In severe cold temperature conditions, the bud sprouting is difficult, therefore, by 15th of October pruning should be initiated.

Following Points to be Considered

1. Too Early Pruning

For European markets, export of grape has a good scope, when Christmas and New Year are the two occasions when demand is more and one may get good returns. For this purpose if grapes are to be made available then in the 3rd week of December and in order to make harvest possible, October pruning has to be done by the 15th of August. There are heavy rains in August, causes more pest and diseases affections.

Early pruned grape coincides with low temperature and with shortened day it results in less sugar development and a greenish colour means good harvest.

2. Early Pruning for Suitable Varieties

In our country the old varieties are grown at the same time *viz.*, Kishmish Charani, Sonak, Manik Chaman are harvested 1-2 weeks earlier under this condition. TSS will be around 22 to 24 per cent but for export TSS of 18 per cent is required which is possible in these varieties.

3. Necessity of Hormones before October Pruning

When you decide to go for early pruning watering in time is necessary. If the leaves stalk/pedicel are good, then defoliation becomes difficult since the sap flow is continuous and good towards fruiting buds, which will have an adverse effect on new sprouting. Therefore before pruning when you are not removing leaves at that time hormonal spray use is advisable. Later on the sap flow gets slower and because of formation of abscission layer leaf fall takes place. For early pruning such sprays are advantageous, with manual defoliation getting reduced. Among the chemicals Thiourea, Potassium Nitrate upto 3 to 5 per cent and hormones like Ethrel upto 39 per cent Sl 1000 ppm. to 2500 ppm. *i.e.*, 2.5 ml/litres to 6 ml/litres and foliar spray are required. For coloured varieties Ethrel and others like Thiourea and Potassium Nitrate is necessary with use of Ethrel in white varieties which increases pink berry problem– so avoid using Ethrel in Thompson varieties etc. Before you take up spraying of chemicals for defoliation if there is a problem of diseases, then take up fungicide spray and defoliate. Before pruning affected branches affected by insects shortage of water and alkalinity may shoot up defoliation; that means the defoliation is not natural; one should know the function of leaves, necessity of removing the leaves; the purpose of October pruning etc.

Water Management and October Pruning

Early Pruning brings early fruits on the vine usually in the month of February when water requirement is more. If there is less availability of water then less watering is given to the vines. For the production of grape, watering should not be reduced but if reduced then the weight of the bunches get reduced. Under such situations by doing early pruning early harvesting is possible but when water filling starts in the berries, and comes across with water stress, the quality of the berries gets impaired. In the month of April and May the demand for grape will increase in North India, and to take advantage of this, pruning has to be done early to bring grape early in the market with good weight and quality. During April-May the temperature will be high and hence water loss from leaves also will be more, hence it is suggested to use anti-stress (5 ml/litre) or anti-transparent Kaolin (4 g/litre of water) and as the availability of water comes down naturally the rise in the context of pH intrudes to tackle this problem.

If late pruning is done in November, due to of cold the sprouting also comes late in November and the required berry swells and gets reduced and to overcome this problem use of hormones and nutrients is advised.

If pruning is done in the beginning of October then the sprouting will come into force, therefore keep the canes with good bunch and remove the others.

Early Marketing Area

For export of grape how much quantity you can produce that you have to decide for yourself, thereafter the area should be earmarked. At present 25-30 tonnes per hectare yields is expected. Therefore, one area you can earmark, and the corner area of thorny bushes should be separated; then make two divisions of the area with ½ an acre of blocks and while taking up October pruning proper planning is important.

Planning of October Pruning

Keeping the future line of work, pruning is planned as follows:

1. Pre-pruning preparation
2. October pruning
3. Post-pruning preparation

12-Point Programme

1. Applications of bio-gas starter solution
2. Inter-cultivation
3. Light irrigation
4. Application of compost
5. Removal of leaves
6. Application of fungicide/insecticides to control pest and diseases.
7. The farmers are advised to follow phytosantitation.
8. Taking up pruning work
9. Application of chemicals for uniform bud burst
10. Application of fungicides.
11. Light inter-cultivation.
12. Light irrigation.

About 1-7 pre-pruning work, 8-Actual October pruning, and 9-12 Post-pruning work, leading to a total of 12 points programme.

A) Pre-October Pruning Programme

Before you take up actual October pruning, the points mentioned above (1-7) should be taken into account. After deciding the dates of October pruning 22 to 25 days works should start before. If October pruning is to be taken in the first fortnight of October then 1 or 2 dates you have to fix up and if rains have occurred approximately 25 days early and are continuous, it will be congenial. If the garden soil is not good or the soil contains 8 per cent lime nodules, then give heavy flooding to wash out the salts that are accumulated near the root zone, which is the main purpose of this programme of pre-October pruning.

Fertilizer Application

Based on soil test the fertilizer should be applied as follows:

Sl.No.	*Name of the Fertilizer*	*Quantity/AC*
1.	Compost	10 tonnes
2.	Diammonium phosphate (DAP)	75 kg
3.	Micronutrient mixture	20 kg
4.	Magnesium sulphate	15 kg
5.	Borocal	10-15 kg
6.	Bio-inoculants	15 kg

Give irrigation, when the date of actual October pruning is over 4-6 days before and remove leaves from vine, as also weak canes, insect, disease–affected and thick canes. If the spacing of the garden is 8' x 6' then keep 40 canes *i.e.*, 75 per cent of the total canes.

Cleanliness

Collect all the fallen leaves and canes from the garden and burn them.

Foliar Spray

Disease and insect-affected branches should be given a spray of insecticide and fungicide. *viz.*, sulphur dusting on plants and ground should be done.

B) Actual October Pruning

Keep the required buds on every cane and go for October pruning. Try to understand the sub-cane methodology which means if 5+7 buds system is allowed in April, during October pruning it should be pruned leaving about 7 buds. Whereas in case of 7+5, the pruning is done leaving 7 buds, which is referred to as zero pruning. In long canes we find 10 to 11 buds that are close to each other, which is a genetic make-up in Thompson seedless under this condition, the October pruning is done at the 11th bud. In general the 6th bud is most potential in Thompson seedless.

In case of 8' x 6' spacing, every grapevine should have 40 canes and on every cane 5th, 6th and 7th buds are most productive, then you will get 120 productive buds per vine. Out of which 50 per cent buds sprout then you will get 60 bunches per vine. Usually you get 40 per cent flowering then 45 bunches you are sure to get.

Spacing and Number of Buds/Vine

Spacing	*No. of Canes/ Vine*	*Pruning at Buds*	*Total No. of Buds/Vine*	*Good Buds/ Vine*
6' x 4'	20	6th bud	120	20
	18	7th bud	116	30
8' x 6'	40	6th bud	240	40
	36	7th bud	245	54
8' x 4'	30	6th bud	180	30
	24	7th bud	175	36
10' x 5'	40	6th bud	240	40
	36	7th bud	245	54
10' x 10'	80	6th bud	480	80
	60	7th bud	420	90

As per the schedule given above, approximately 12 tonnes per acre of yield are expected. If the number of buds are more, then upto15 tonnes/AC yield also can be expected.

C) Post October Pruning

For getting bunches in the same shoot the pasting of black Bordeaux is required. This paste is applied leaving the last 3 buds on the canes. Take half kg copper sulphate, 250 gm lime and 125 gm Thiourea in 5 litres of water and paste it on the canes. Secondly, Dormex (Hydrogen Cyanide) 30 ml/litre also can be swabbed by lint with handgloves.

After application of paste take up insecticide dusting on soil. Pruning work, pasting, and insecticide dusting should be done on the same day and after completion of all this on 2nd and 3rd day, light soil stirring is required and then irrigate.

After October pruning, or when the pruning work starts take to irrigation and after completion or pruning once again watering is given, so that this you can say October pruning work is over.

Usually from the 15th of September, first work should start and complete by 15th October this way in one month all operations will be competed of pre-pruning operation and other related works cannot be carried out simultaneously. But if the October pruning is done properly, it will have an impact on the percentage of flowering. If pruning is done, don't think work is over. Then good grape production is not possible. One acre pruning is done in one day, but later on we will come to know that out of one acre only ½ acre flowering is good which means the sprouting is uniform and flowering percentage is also good but in the remaining ½ acre area, sprouting was less hence flowerings are also less. Under such a situation why such a problem has arisen one has to think it over. Everytime one will think why it is so, inspite of October pruning done in time taken all precautions.

This point, if you take it to your mind in a proper way, you will understand pre-pruning operation were not carried out properly hence good post–pruning results were not got. Only pre-pruning operation are not the reasons, the work carried out after April pruning one may have missed out something.

Development of cane, nipping of canes, application of fertilizers and watering work, all this work you divide into different sections from April pruning to October pruning and you keep in mind that every bunch should weigh around 500 gm and each berry should weigh around 4 gm. In case of 8' x 6' spacing, every vine should give 5 kg boxfull of production and the capability of each vine that you can see after October pruning, force it to flower and for that purpose October pruning management is very important.

Testing of buds before October Pruning

Testing is it Necessary? Yes

Just to see how many flower primordias have got established and got confirmed.

When to test it?–10-15 days before October pruning.

Which buds to be tested?– Usually 6-7 buds.

Method of bud Testing?–Through microscope.

Without cane Pruning bud Testing: Remove all the leaves from the vine by doing pruning on matured canes and apply thiourea/Dormex after which 4 buds testing of cane is possible.

After October Pruning Pasting is necessary?

Yes it is necessary to raise such question at the time of pruning.

1. **Apical dominance:** At apical region the dominance is more to overcome that, application of Thiourea or Dormex is necessary.
2. During October pruning, reduction in temperature during winters may cause slow sprouting of buds and uniform growth is not possible, by 15th October if there is heavy rainfall coinciding with pruning, delay in sprouting is caused and with soil affected by lime nodules the moisture holding capacity may increase. Late pruning means last week of October or early November, that is, Ist week of November. Where there is reduction in temperature and less sunshine, delay in sprouting of buds may be caused.
3. If number of canes on the number of bud–load are more than expected then also the sprouting after October pruning gets delayed and it will not be uniform too.
4. The bud burst may vary from variety to variety. The variety Tas-e-Ganesh as compared to Thompson seedless, Sonaka and Manik Chaman is more vigorous. In Kishmish Charani variety there is no apical dominance.

5. On canes, the lower 3-4 buds have lower establishment of flower primordia, which results in less flowering. It may be of any variety and in order to have sure shot sprouting it is compulsory that one has to go for pasting.

Which Paste One has to Select?

Application of Thiourea or black Brodeaux paste. And Dormex has shown not much difference in overcoming the apical dominance. But Thiourea percentage is controversial and before use of Dormex testing is also necessary. In some places, after the preparation of paste, they have to be applied immediately, sometime the application at the earliest possible may give good result. It is necessary that applied paste should be there for at least 5-6 hours on buds. After the application of paste, increase in temperature and the day length duration have a deciding effect on bud sprouting.

Post-paste Application and Other Points to Remember

1. As per the programme to have uniform bud sprouting, application of paste is necessary.
2. In Black Bordeaux mixture urea percentage should not be more than 3 per cent, and the Dormex percentage should not exceed 3 per cent.
3. Once the paste is prepared, immediately it should be used.
4. Application of paste should start from afternoon 3 pm to 5 pm.
5. The paste should be applied on apical 2-3 buds only.
6. Use gloves while applying paste and use cloth brush.

(*Courtesy* Dr. B.G. Bhujabal, Drakshi Virut, Pune)

Grape (*Vitis vinifera* L.) is an important commercial fruit crop of Maharashtra having vast export potential (Salunke *et al.*, 2006). The pruning is an important practice to increase the yield and quality of the grapes. Hence, the experiment was carried out to study the effect of pruning and various chemicals on pigmentation of Thompson Seedless grapes at Ahmednagar (M.S.). The experiment was comprised of treatments having three dates of pruning, *viz.*: 16th September, 3rd and 22nd October and seven chemicals, *viz.*: control (recommended schedule of plant growth regulators), 6-BA at 10 ppm at 3 to 4 mm berry diameter stage, CPPU at 2 ppm at 3–4 mm berry diameter stage, ascorbic acid at 500 ppm at version stage, SDDTC at 100 ppm at 10 days before and after veraison stage and at veraison stage which was laid out in a factorial completely randomized design (FRBD). Altogether there were twenty one treatments replicated by four times and each treatment comprised of five vines. The results of time of pruning revealed that significantly maximum number of bunches per vine (34.05), maximum bunch weight (454.33 g), 100 berry weight (294.28 g), length of berry (19.80 mm), diameter of berry (16.50 mm), yield (34.01 MT/ha) and TSS (22.7°C) were observed in vines pruned on 22nd October with application of CPPU at 2 ppm at 3–4 mm berry diameter stage with recommended schedule of plant growth regulators. The highest total number of bunches affected due to pink berry per vine (20.16), percentage of infected berries per bunch (40.52 per cent) and total anthocyanin content

(0.317 mg/100 g), total sugars (21.58 per cent) and reducing sugar (19.51 per cent) were recorded in the vines pruned on 16th September, while lowest in the vines pruned on 22nd October. The study also showed the negative correlation of pigmentation with maximum and minimum temperature while positive correlation with morning as well as evening relative humidity. Overall, the pruning for fruiting should be done in third week of October with application of CPPU at 2 ppm at 3–4 mm berry diameter along with application of recommended plant growth regulators.

Novello and de Palma (2006) stated that the horticultural market is characterized by an increasing demand of fresh extra-seasonal products and of early- or late-ripening fruits, both achieving a better economic income. In table grapes, the ripening period can be extended by selecting new varieties or by applying cultural techniques apt to force the plant growth under unfavorable climatic conditions or to protect the foliage and fruits from meteorological adversities. The harvest date may be advanced by inducing a praecocious budbreak. This is obtained by managing the vineyard as a protected cultivation, that is, by covering the roof and the lateral belts of vineyard with transparent plastic films, which allow a high fraction of solar radiation to pass through and then are able to retain a good portion of energy by limiting the convective and radiative thermal dispersions. Hence, the air temperature increases and induces a faster accumulation of growing degree days which, in turn, stimulate an earlier vine budbreak. The plastic covering is affixed on the vineyard at the end of winter; the earlier the covering, the earlier the budbreak. This latter can be advanced by 10 to 40 days depending on the covering date, the grape cultivar and the climatic characteristics of growing environment. The maximum effect can be obtained by covering the vineyard 50 days before natural vine budbreak in open field; however, the advancing effect could be partially lost during the following phenological phases, when the internal thermal regime is too high respect to the requirements of the reproductive and photosynthetic processes. The plastic film optical properties greatly influence the canopy development and the qualitative traits of the grape clusters. This type of protected cultivation has proved to be particularly suitable for early-maturing cultivars and seedless grapes; these latter, in particular, may improve the berry quality, and any case; avoid the competition of seeded varieties with bigger and savoury berries. The harvest date may be delayed by protecting the ripening clusters from meteorological adversities. This technique profits from the good preservability on vine which characterizes the grapevine cluster and improves this trait by protecting the berry sanitary status during ripening. In fact, starting from veraison, berry skin turns thinner and juice turns sweeter, thus berry becomes more sensitive to fungal attacks, especially after the wetting caused by the late summer rains. To preserve the crop, the vineyard roof in covered with plastic films. Under the best conditions, which are with healthy grapes and low rainy season, the berry integrity may be guaranteed for a long period, and the harvest may be delayed by 3 months, without reducing yield and berry carpometric characteristics. For best results, it is very important to maintain berry hydration and turgidity as long as possible, since the grape freshness is the goal to achieve and the attribute which distinguishes this product from that coming from the cold storage. The protected cultivation to delay harvest is particularly convenient for late seeded cultivars, especially for grapes having an appropriate skin thickness, abundant pruine and a big berry with a relatively low surface: volume ratio, which is all traits helping to limit berry transpiration and thus favors its final hydration status.

Wolfe and Brown (1992) stated that each of the grape cultivars [*Vitis* species, (L) Batch], 'Concord', 'Himrod', 'Challenger', 'Reliance', 'Glenora', 'Moored', planted June 1983, and 'Mars' planted Spring, 1987, was trained to the 4-cane Kniffin (KN) and the Geneva Double Curtain (GDC) systems. Yield per vine, pruning weight, number of nodes, cluster weight, number of berries per cluster, berry weight, and per cent soluble solids were recorded. Vines from 'Reliance' trained to the KN system produced fruit with significantly higher per cent soluble solids than did vines trained to the GDC. No significant differences in per cent soluble solids were observed between the two training systems for the other cultivars. 'Concord' produced more kg/vine of pruning weight when trained to the KN system than when trained to the GDC. Pruning weight did not differ significantly between the two training systems for the other cultivars. Cultivars more productive (yield/vine) on the GDC trellis were 'Concord', 'Himrod', 'Reliance' and 'Moored' whereas 'Challenger' was more productive when vines were trained to the KN system. No differences between the two training systems were observed for 'Glenora' or 'Mars'.

Ferree *et al.* (2002) reported that the own-rooted 'Seyval Blanc' vines were evaluated over 5 years in the following training systems: bilateral cordon, bilateral cordon sylvos (sylvos), upright cordon-spur pruned and upright cordon-cane pruned. Vines in all systems were spaced 1.8 x 2.7 m and were shoot thinned to 40 shoots and cluster thinned to one cluster per shoot. Cumulative yields over 5 years of the upright cordon systems were higher than vines in the sylvos system. Average weight of cane prunings was higher in the upright cordon vines than in all other systems. Vines trained to the sylvos system had more canopy gaps, less *Botrytis* bunch rot, and higher juice soluble solids than other systems in some years. The chemical constituents of the must and wine or evaluation by tasting showed little influence of training system.

Reynolds and Justine (2009) opined that the training a grapevine involves a manipulation of vine form. The type of training may lead to differences in total leaf area and the percentage of leaf area well-exposed to light. Consequently, the ability for a grapevine to photosynthesize efficiently depends upon its training system and the accompanying light microclimate of its leaves. In addition to altering the light microclimate of the canopy, training may impact numerous other variables such as fruit bud differentiation, cluster exposure, vine water status, and leaf transpiration. Modification of vine training systems to achieve balance between vine vigor and yield has led to divided canopy systems that might simultaneously increase yield and improve fruit composition through optimization of canopy light microclimate. Consequently, many training systems have been identified as being capable of improving wine quality through a combination of enhanced canopy and fruit microclimate.

Prakash and Chadha (1998) studies on the distribution pattern of root density and activity using radioactive 32P were carried out in four grape cultivars namely,

Thompson Seedless, Gulabi, Black Champa and Bangalore Blue as influenced by three systems of training *viz.*, bower, kniffin and head systems. Of the four cultivars studies, root density was found to be the highest in cultivar Bangalore Blue followed by Thompson Seedless, Gulabi, Black Champa. All the cultivars had greater root density nearer the trunk at 50 cm radial distance and at a shallow depth of 0-15 cm in the soil. In all the cultivars the highest root density was observed in the vines trained on bower system. Bower trained vines showed a greater root spread as compared to those on kniffin and head systems. Root activity was the highest in cultivar Gulabi as indicated by greater absorption of 32P followed by Thompson Seedless and Black Champa. The lowest activity was noticed in the cultivar Bangalore Blue. In all the cultivars Thompson Seedless and Bangalore Blue root activity was more nearer the trunk (40 cm radial distance) at a depth of 15 cm under all the training systems. On the other hand, in the cultivars Gulabi and Black Champa, the activity was more widespread even up to 120 cm radial distance and up to, 30 to 45 cm depths. Even though, the total root activity under different training systems did not differ significantly, the spatial distribution of root activity varied. In general, both root density and activity were found to decrease with increasing depths and radial distances in all the cultivars except in Black Champa. Mathematical functions worked out for fitting root activity as a function of depths and distance indicated that only in Bangalore Blue and Thompson Seedless the spatial distribution was system followed by kniffin and head systems.

Palanichamy *et al.* (2004) reported that the Pusa Navrang - a teinturier grape hybrid -was pruned at three different levels *viz.*, 4, 6 and 8 buds per cane retaining uniformly 12 canes per vine on the 'head' system. Among the three pruning treatments, the highest yield (8.16 kg/vine) and the maximum number of bunches (36.2/vine) were obtained with pruning at 6 - bud level. The maximum bunch weight (234 g) along with its size (20.5 x 15.3 cm) was recorded at 4 - bud pruning. Similarly, the berry weight (1.5 g), TSS (17.8 per cent) and juice percentage (74.5 per cent) were the highest in the same treatment. Looking at the superior quality of grape berry at 4 - bud level, specially juice percentage and TSS as this variety is recommended for juice and coloured wines, it is advocated that comparatively higher yield obtained in 6 - bud level can be forgone.

Dwoght and Gerald (1992) stated that each of the grape cultivars [*Vitis* species, (L) Batch], 'Concord', 'Himrod', 'Challenger', 'Reliance', 'Glenora', 'Moored', planted June 1983, and 'Mars' planted Spring, 1987, was trained to the 4-cane Kniffin (KN) and the Geneva Double Curtain (GDC) systems. Yield per vine, pruning weight, number of nodes, cluster weight, number of berries per cluster, berry weight, and per cent soluble solids were recorded. Vines from 'Reliance' trained to the KN system produced fruit with significantly higher per cent soluble solids than did vines trained to the GDC. No significant differences in per cent soluble solids were observed between the two training systems for the other cultivars. 'Concord' produced more kg/vine of pruning weight when trained to the KN system than when trained to the GDC. Pruning weight did not differ significantly between the two training systems for the other cultivars. Cultivars more productive (yield/vine) on the GDC trellis were 'Concord', 'Himrod', 'Reliance' and 'Moored' whereas 'Challenger' was more productive when vines were trained to the KN system. No differences between the two training systems were observed for 'Glenora' or 'Mars'.

Zoecklein *et al.* (2008) observed the viognier grapes grown in northern Virginia and resultant wines were evaluated as a function of training system. Treatments included vertical shoot-positioned (VSP), Smart-Dyson (SD), and Geneva double curtain (GDC), with vines of all treatments spaced 2.4 m apart in 3.0 m wide rows. In addition to increased cluster numbers and crop yield, GDC training generally increased fruit zone sunlight interception and fruit exposure, while it decreased cane pruning weights per meter of cordon, compared with SD and VSP. Crop adjustments were made between bloom and veraison in six seasons, to result in average yields of 10.5 kg/vine (GDC), 9.9 kg/vine (SD), and 6.0 kg/vine (VSP), with the lower SD canopy bearing 30 to 40 per cent less crop than the corresponding upper SD canopy. Crop loads (yield/cane pruning weight) were generally between 4 and 12; GDC crop load approached 20 in three seasons, while SD approached 14 in one. Leaf area per crop ratio was determined one season and exceeded 1.8 m^2/kg of fruit for all systems. Fruit was harvested at similar Brix values, with differences in berry weight, pH, titratable acidity, and malic and tartaric acids among treatments generally not significant. Volatile compounds were analyzed using headspace solid-phase microextraction GC-MS. Fruit showed consistent differences in linalool, α-terpineol, β-damascenone, and *n*-hexanol concentrations among training systems. SD had the highest concentration of most free volatiles quantified in both juice and wines, while GDC wines frequently had the highest concentration of phenol-free glycosides. Triangle difference sensory testing demonstrated differences between GDC and SD in wine aroma and flavor and between VSP and SD in flavor. GDC wines generally had higher fruity and floral aromas compared with the other systems.

Mariagiovanna *et al.* (2012) opined that the enhancement of grape aroma and the investigation of the relationships between grape and wine composition are important given the growing interest in wines with distinctive varietal character. Each training system involves specific grape growing conditions, which affect the concentration of volatile metabolites of grape. The aim of this work was to evaluate the influence of three training systems, including little tree, bilateral Guyot, and four rays, on the volatile composition of Primitivo grape and on the volatile composition and sensory properties of the corresponding wine in Apulia region of Italy. Free and bound volatile fractions were evaluated by gas chromatography coupled to mass spectrometry. Sensory analysis was performed by a panel of trained judges. Results showed that the little tree and bilateral Guyot training systems improved the accumulation of aroma precursors and volatile compounds in grape and wine and the sensory characteristics of the wine.

Bemizzoni *et al.* (2009) stated that the effects of training systems established as vertically shoot-positioned spur-pruned low cordon (SPC), single high-wire cordon (HW), single Guyot (SG), and vertically split double Guyot (DG) were tested over five years (2003–2007) on *Vitis vinifera* L. cv. Barbera vines planted at 2.5 m interrow and at 0.9 m, 1.2 m, and 1.5 m within-row spacings. In the spur-pruned (SPC and HW) systems, total shoots per count node were about three-fold the values recorded in SG and DG, while vine capacity as total leaf area throughout the trial was highest in SPC

(6.48 m^2) and lowest in SG (2.93 m^2), although the latter bore the most vigorous shoots. Vine capacity measured as leaf area decreased linearly with increasing within-row spacing. Yield per meter of row decreased by ~20 per cent at 1.5 m as compared to 0.9 m vine spacing. No significant differences were found in the yield per vine across training systems, as higher shoot number in SPC and HW was offset by higher shoot fertility and cluster weight in the cane-pruned systems. Must composition at harvest was similar among SPC, HW, and SG, while DG produced grapes of overall inferior quality. Vine spacing had no effects on grape composition. Results indicate that in Barbera similar crop potential and quality expression can be achieved in either cane- or spur-pruned training systems when properly managed. Vine spacing at 0.9 m within-row is advisable as it ensures 20 per cent higher yield per hectare at very similar grape quality across training systems.

6

Girdling and Thinning

For taking grapes to good production along with growth regulators, girdling and thinning are also necessary.

- ☆ For girdling and thinning there is no alternative. Girdling should be done in time and in a proper way. The injury caused by girdling should heal in time.
- ☆ Thinning should be carried out every week.
- ☆ Between bunch thinning, bunch tipping, berry thinning and flower thinning, there is distinct difference, therefore, proper care of bunch and leaf is very important. For bunch tipping, how many leaves to be maintained after fruit set and whether leaf has to be kept then encouraging the bunching, take care of all these points will make difficult work easy.

Girdling

To increase the sugar development in grape bunch and to increase the bunch weight, girdling is done. Girdling is done after GA is used. Girdling helps in increasing the yield by 1.5 to 2 times, that is why girdling is a must.

Girdling Means What?

On any part of a grapevine if a circular cut is given it is called girdling. This is done on canes, on T's (main arm) and on main stem (trunk). Cane girdling have lots of benefit but by running girdling on vine many bunch will have positive effect and from trunk girdling the entire bunch will get the benefit.

1. Cane girdling has to be done on every bunch and it requires a lot of time, lot of labour and a lot of money. On an average girdling of 40,000 canes is required.
2. Arm girdling is done on the canopy. On every vine 2 to 4 T are there and on an average 3000 arms are there in an acre. Arms (T) are at a two meter height, the girdling will be very cumbersome and also using of girdling plier is not easy; ultimately the labours are not happy with this kind of girdling.
3. Trunk girdling is done very easily with an average closer spacing if about 1800 vines are available so by engaging more labour in 4-5 days this work can be completed in time. In a wider spacing again it will be very easy. But, in all bunches equal balance is not available.

When Girdling has to be Done

The objective of girdling is to improve yield quality. Immediately after October pruning, once fruit is set, girdling is done. If this process is done in a hurry or early the purpose is not served. After girdling, the food manufactured in leaves definitely goes to branches and will be available to bunches. Before girdling tipping sprouts are a must. The food which is not moving towards bunches will move towards the terminal growth. The vigorous growth affects the bunch growth. When the grapes are at flowering stage the vine nutrition availability will be limiting; naturally the demand for food/nutrition increases. In seedless grape varieties, at fruit set stage the GA and other growth regulators, if already applied and girdling is done, will have very good effect.

If girdling is done early the fruit set percentage increases and the bunch will be too compact. Within a bunch for formation of the berries every berry development space availability is difficult hence may result in berry getting flattened and finally the berry length reduces and the diameter increases. If this process is reversed then the weight of the bunch and one per cent sugar also reduces and the quality of berries at distal and proximal deteriorates.

At present situation if early girdling has to be done then do it on trunk and if late then do it on canopy area. In our place the height of the trunk is 1.5 to 2 meters and the girdling at 25 cm above ground has to be avoided. When soil is raised and water comes in contact with the girdling zone, then the fungus moves in plants through injury. If watering is done by drip then it is excellent and very important.

In a normal situation the centre of the trunk girdling is advised. Girdling at a height is cumbersome. At the time of girdling near the bunch if the distance is more, then girdling results get delayed. After October pruning if the girdling is done at 6th bud and the same if you do at 10th bud then the results are encouraging. In 4-arm kniffin system the lower arm shows good results than the upper arms. While doing girdling another point one has to keep in mind is that the total bunches on vine, total number of branches and sterile branches should have optimum leaves. If more bunches are there, then the result is late. If the total number of bunches, the sprouting and leaves are more than the result will be late but it will be very good.

Care While Doing Girdling

Irrigation of Land

While girdling operation is on there should be sufficient water in the soil. If the soil is dry then barks come out easily. Too much wetting of land is not advisable since the healing process is not uniform and leaves get yellowish. If formation of other leaves actively slow down then the girdling effect is not visible. For good results the soil should be wet.

Bunch Care

Once the fruit sets in and bunch tipping is completed then girdling is advised. Around 70-80 per cent of bunch setting is necessary to go for girdling; girdling should also be completed in the same day.

Application of Fungicide on Girdling Places

If girdling is done closer to ground and the soil is too wet then application of fungicide becomes necessary because the injury caused by girdling provides place to enter the fungus. Dead arm, Botrydiplodia, mealy bug, stems borer may create problems. Usually attention is given towards care of pests and diseases but not towards the trunk. Therefore, before the disease/insect are noticed, spraying of chemicals viz Blitox, Captaf, Ziride, Dithane-M-45, Bavistin, Nuvan, Carbaryl. Malathion etc., and mixture pasting is necessary and this work should be completed on the same day of girdling.

Proper Time Injury Healing of Girdling

Girdling in time and in a proper way should be done and callus formation within 15 days takes place and in another one week the injury healing takes places. Deep girdling not only takes more time for healing but inside it will be hollow. Girdling nearer to ground heals late than the girdling which is nearer to canopy. Under such situation it means during favourable situation utmost cares of bunch is necessary and before girdling apply hormonal paste. For callus formation apply Indole acetic acid or Indole Butyric acid (IAA/IBA) of 1000 ppm. On the surface of girdling injury, cow's compost + urine help in healing of injury. Through compost and urine we cannot control insects/diseases, the callus formation starts from top of the surrounding but the fungus starts from below the injury nearer to ground. Therefore fungicide/hormone should be applied to these particular places. In cow compost CoC, Nuvan, mixing is economical. For one kg compost mix 25g fungicide + 20g/milli insecticide and paste it. On the emergence of new leaves the sap flow starts through the trunk. For initiation of new growth we have to go for application of fertilizer. New leaves which are dark green in colour and full in size are capable of manufacturing food and for their maintenance, nutrient management is necessary. At least 15 days time is required for maturity and within 40-45 days, photosynthesis capacity exists which means 55 to 60 days after October pruning; the photosynthesis activity will be maximum then it gets slowed down. For taking this benefit girdling has to be done within 60 days of pruning. On vine, active photosynthesis in leaves continues for 42 days whose benefit we have to harvest, The new sprout, when it is soft, the carbohydrate (food) moves towards the terminal end then it moves down to trunk then again through branches moves towards bunch. This natural process sometimes gets delayed but to start early in time girdling is necessary.

During girdling, growth will not stop and the sap will not move towards the bunch but it moves towards the terminal portion and girdling effect will not be observed. Before fruit set it means during bunch emergence the growth checks and girdling process will have to be completed then fruit set increases within the bunch which will be too compact. Fruit set is easy and comes to harvest early also. At the time when the bunch gets ready early then bunch weight will not be much and also sugar per cent less. In a bunch if the setting is not uniform and girdling is over the berries get the benefit of girdling. The benefit on an average is noticed before girdling if bunch thinning and tipping is attended to.

When the grape bunch starts developing after girdling the bunch quality can be improved. Within a vine, if more number of bunches are there or if more fruit setting in a bunch is observed, then bunch development has to be restricted which further delays sugar development and in order to overcome these two problems girdling helps because bunch and leaves number may not increase. The girdling effect is visible within a week. After girdling increases in fruit weight and quality cane is observed then necessary fertilizers and hormonal use can be avoided.

Thinning

In a grape all the berries within a bunch, all the bunches within the vine and all the vines within a garden should be uniform and for the purpose of export, thinning is very important.

Thinning Means what?

The common meaning of thinning is some excess berries from a bunch are removed and counted number of berries in a bunch we have to keep. For every thinning operation we have to see how many bunch per vine are there and what should be the leaf area for every bunch; all these points should be kept in mind and then only we should do the thinning.

Types of Thinning

Following are the types of thinning:

1. Leaf thinning
2. Bunch thinning

1. Leaf Thinning

(*a*) Tipping of sprouts

(*b*) Removing of leaves from branches

(*c*) Removing of branches.

2. Bunch Thinning

(*a*) Removal of few full bunches

(*b*) Pinching of bunch
(*c*) Berry thinning

While doing thinning, leaf or bunch thinning one has to attend.

When Thinning is Done?

For getting quality bunch once the bunch emerges on vine and fruit setting is complete then bunch are counted, immediately thinning work starts for getting good result. After fruit formation the thinning is not beneficial, rather it is harmful.

When leaf and bunch thinning starts, don't apply any nutrition or hormones manually but to stop the entry of disease/insects both fungicide/insecticides can be used.

1. Removal of Excess Growth

After October pruning there are two types of growth one grows along with bunch secondly another grows without bunch. This growth should be removed immediately at the beginning stage itself; otherwise the supply of nutrient of bunch will be less which results in weak bunch.

While October pruning is in operation, if you keep more canes and more number of buds (10) per cane or pruning is done by keeping more canes on vine or early pruning is done, then more vegetative growth is observed, the growth of which have to be removed as early as possible. Usually in a cane, at terminal 3-4 buds position, more growth is observed. On vine the fruit (bunch) load is at 5-7 buds, excess growth at tips is removed after October pruning to stop new growth and at the bottom of 3-4 buds apply Bordeaux paste.

2. Tipping of Bunches

The bunches at the tip are sour and small in size. Only the long bunches are not advisable for packing in boxes. For this purpose by thinning you have to give a good shape to the bunches. For tipping of bunch uses scissors only and do thinning with 10 per cent tipping which is a must and maximum up to 30 per cent bunch tipping is advisable.

In grape bunch after fruit set, the berry nutrition management and its growth start very fast. New leaves expansion completes and food manufacturing starts. On vine every bunch and berries of bunch, balanced nutrition is the responsibility of leaves on the vine. The food manufactured by leaves goes to leaves, and some will go for the emergence of the new leaves and some portion of the food goes to branches, roots, bunches etc. If the food availability in plant is less than the actual vegetative growth is observed. Then the branches will be longer and to stop this growth we have to pinch it only by means that can save the food. Along with this girdling it also helps in storing the food by leaves and will be utilized by the bunches. Girdling checks the flow of sap to roots and bigger portion goes to the bunches. There it helps in improving the bunch size and berry size. Checking of branch growth should be done earlier.

This process on vine improves food manufacturing capacity in the leaves and in other words girdling is a check post to stop the downward flow of nutrients that helps the bunch to develop. Excess removal of growth makes the vine empty. Then sometimes stopping of growth is not possible under such situation; excess vegetative growth should not be stopped immediately; their growth has to be stopped at 6-7 leaves which help in food manufacturing.

Sometimes the growth with or without bunch will be stunted or immediately they become stunted and after pinching we won't get the required leaves, the leaves become smaller in size. Under such situation before pinching, girdling should be done first. On vine some branches are vigorous and some are stunted. Hence in the vigorously growing branches pinching is done to stop the growth then immediately girdling should be attended to. For the growth of the terminal end, girdling should be done with utmost care and for that purpose, for single girdling one acre is advocated and within 4-6 days callus formation and further flow of sap should take place.

Second important point is after thinning, looking to the number of branches the number of berries per bunch, you have to increase the number of leaves. For increasing the yield and more number of bunches for good quality, less number of bunches are required. If you are not keeping equal proportion of bunches and leaves then the number of bunches will be more in vine.

Proper number of bunches per hectare or acre are to be taken into consideration and how many bunches per vine are there, how many canes on a vine and on each cane how many bunches and number of leaves are there all these we have to keep in mind.

For example on one cane ½ kg bunch should have 150 berries and that cane should have 15 leaves; that means for one leaf 10 berries are required. The leaves behind the bunch should be 6 and in the bunch 9 leaves, so totally 15 leaves are necessary.

In Thompson Seedless more than 500 berries will be there per bunch. For retention after flower thinning, 30 per cent thinning of berries or tipping of bunch, 30 per cent thinning or so from the total number of berries, approximately 1/3 berries we have to keep for optimum berry production. After a week the tipping is done, keep required number of leaves so as to have good aeration.

For bunch nutrition the food comes from leaves but whatever the food manufactured it won't go to leaves but to fulfill the vine nutrition requirement it goes to different parts. On vine sometimes branches without bunch growth will be there and by keeping 6-7 leaves tipping is done for utilization of manufactured food by vines.

7

Irrigation

Water is also important input in grape. Water in root zone of the vine is held around the surface of the soil particle as a film. When vine roots cannot absorb adequate water to meet the evapotranspiration from vine surface, vine suffer from waterstress. Irrigation is a vital cultural practice in grape growing regions. In newly planted vine yards the young trees are irrigated by allowing water in to a small circular basin of 50 cm radius. At present more than 90 per cent gardens are in drip irrigation in Northern Karnataka.

In scheduling of irrigation for vine yards, there are two approaches time schedule and demand schedule. In time schedule irrigate the garden once in 5-7 days in summer and 10-12 days in winter. Demand schedule is based on water needs of vines which vary with stage of the growth crop load, water holding capacity of the soil and aridity in the atmosphere and method of irrigation followed. While determination of time and quantity of water in each irrigation are important in surface irrigation.

Among the methods of irrigation, ring irrigation, furrow irrigation, sprinkler irrigation, drip irrigation.

Quality of Irrigation Water

It is the dissolved constituents; quality of irrigation water has not only bearing on the performance of grapevine but also on physico-chemical properties of soil on a long run, rendering it saline or alkaline. Dissolved salts are present in dissociated form as ion in water. The cations present are calcium (Ca^{++}), magnesium (Mg^{++}), sodium (Na^{+}) and potassium (K^{+}) while anion are bicarbonate (HCO_3^-), carbonate (CO_3^-), chloride (Cl^-) and sulphates (SO_4^-), some time anion like borate (BO_3^-) and nitrate (NO_3^-) are also present. Irrigation water with a very low salt content may pose problem to overcome salinity hazard salt tolerant root stock like Dogridge, Salt Creek (Ramsey) 1613, or St. George could be employed. Bi-carbonate in (HCO_3^-) are common in irrigation water, they are in combination with sodium, potassium, calcium, magnesium, chlorides and boron are found irrigation water, the common chlorides are soluble and not fixed in soil, higher chloride inhibit growth, grape vines are highly sensitive to chlorides toxicity. The symptom appears and burning intensifies with increasing atmospheric temperature during Feb-March. Irrigation water containing less than 4 mg/litre of chloride is safe for irrigation, if chloride are more than 15 me/litre that water should not be used, while the water in the range of 4-15 me/litre can be used cautiously, in sandy soil water with chlorides are less problematic than in heavy clay soils.

Spray anti-transpirants like 4 per cent Kaolin reduce moisture stress. Yield upto 20 tonnes/ha are harvested by giving 25,000 litre of water per day through drip during peak stage as against the 75,000 litres/day/ha.

Klein (2006) stated that the improving nutrient use efficiency can be achieved by applying nutrients when uptake efficiency is at its highest level and by avoiding nutrient leaching by rain and by excessive irrigation. Efficient irrigation is a major factor in nutrient use efficiency. A considerable gap exists today between our scientific knowledge on irrigation and its implementation in everyday practice. A new irrigation controller (Common Sensor), developed and tested in practice, was found to implement an Autonomous Threshold Tension Irrigation (ATTI), whereby the plant itself is activating the irrigation, eliminating the need from the farmer to make any (correct or erroneous) decisions regarding when and how much to irrigate. The autonomous irrigation was found to improve irrigation efficiency and water use efficiency. The new battery operated irrigation controller continuously monitors and records the soil water tension, opens a hydraulic irrigation valve at a pre-set soil water tension (0 to -50 kPa) and controls the depth of irrigation by closing the valve at an appropriate time to control depth of irrigation. The continuous record of soil water tension and the irrigation interval recorded can be plotted for immediate inspection and downloaded for further analysis. The success of the controller, which incorporates a combination of hardware, software and method of operation, has been proved in practice and substantiated by scientific measurements (including in grapes) during the last 4 years. Irrigation experiments were carried out in potted soil-less culture of fig grown in a screen house and in a field irrigation experiments with olive. Simultaneous measurements of soil water tension and photosynthesis in grape and fig showed an absolute correlation ($r^2 = 0.9384 - 1.000$), indicating a direct control of photosynthesis and stomatal conductance by soil water tension. Drainage in potted plants was reduced considerably by irrigation at low soil water tension. Neutron scattering measurements in the olive field experiments showed an integrated water use: initially low volume irrigation by the Common Sensor, until complete depletion of winter soil water and subsequently relying for water consumption solely on the irrigation controller. Neutron scattering measurements showed a complete elimination of drainage in drip irrigation. Elimination of drainage in a heavy yielding 'On' year approximately doubled irrigation efficiency in olive (reduced input by 50 per cent, compared to the recommended rate) when soil water tension was maintained at 0 to -10 kPa. Maintaining soil water tension at 0 to -20 kPa further reduced water use by 75 per cent, indicating a substantial increase in water use efficiency. Olive fruit size was reduced 12 per cent by irrigation at a threshold of -20 kPa, compared to -10 kPa but oil yield increased 6 per cent. Using the Common Sensor irrigation controller, irrigation scheduling and volume of water applied are autonomously regulated by the plant, eliminating errors in irrigation. The autonomous frequency of irrigation by the Common Sensor controller is a function of plant water consumption, the set threshold of soil water tension and the soil water available before the set threshold is reached. The effect of available soil water on the frequency of irrigation is influenced by the irrigation system layout (number of laterals and drippers). Our data and the experience we gained using the new controller indicates that it is almost impossible to predict irrigation needs, even by scientists carrying out all the required measurements, let alone by the farmer using an empirical approach. We found it possible and practical to trigger an autonomous irrigation by the plant. It was found that the autonomous irrigation, using an appropriate technology and method, has a major control on plant performance (growth and potential yield) through maintenance of an accurate and steady threshold of soil water tension.

Sharma *et al.* (2006) opined that due to lack of scientifically standardised irrigational schedule for grafted vines the growers have been irrigating the vineyards arbitrarily. Where irrigation water is easily available, the growers tend to over irrigate. On the other hand, the growers in Sangli and Solapur regions of Maharashtra state have to transport the water in tankers from distant places due to lack of natural water resources. This involves both cost and energy and each and every drop of water needs to be used judiciously. The aim of the present study was to get maximum response to the applied water based upon pan evaporation rate for Thompson Seedless vines grafted on Dog Ridge rootstock. For standardizing the irrigation schedule for grafted vines grown on heavy soils, eight different irrigation schedules (treatments) were tested. The vines were irrigated at variable rates of replenishment of pan evaporation at different growth stages for consecutive three years (2000 to 2003) to develop the grower friendly irrigation schedule. The irrigation rates for different growth stages and the stage at which the moisture stress has little or no effect were identified. The new schedule standardised by irrigating the vines at 60 per cent rate of replenishment of pan evaporation at shoot growth and berry setting to harvest reduced the irrigation water requirement by 52 to 53 per cent compared to check treatment (irrigating the vines at 80 per cent and uniform rate at all the growth stages). The highest water use efficiency was recorded by irrigating the vines as per irrigation schedule but the yield and berry and bunch weight were significantly less. Irrigating the vines at reduced rates (20 per cent rate of replenishment) of irrigation at bud differentiation, berries set and shatter stage saved considerable amount of water without affecting the yield and quality of the grapes. Irrigating the vines at 50 per cent or less rate compared to check treatment during shoot growth, berry growth and development stages although resulted in higher water use efficiency, reduced the yield, bunch number and biomass significantly.

Boraiah and Srinivas (2001) reported that the field experiment was conducted at the Indian Institute of Horticultural Research, Hessaraghatta, Bangalore during 1999-2000 to study the "Response of grapes to microirrigation under different evaporation replenishment rates". The fruit yield of grape was higher with drip irrigation (30.20 t ha^{-1}) compared to microjet irrigation (29.94 t ha^{-1}). The yield attributing parameters like number of branches per plant (90.75), average bunch weight (473.33 g), 100 berry weight (380.99 g), berry diameter (22.23 cm^2) and berry length (31.30 cm^2) were higher with drip irrigation. Increasing the evaporation replenishment increased the fruit yield and other yield parameters like, the fruit yield was higher with 100 per cent evaporation replenishment (31.98 t ha^{-1}) which was on par with 75 per cent evaporation replenishment (31.53 t ha^{-1}), but it was lower with 50 per cent evaporation replenishment (26.69 t ha^{-1}). Evaporation replenishment rates did not influence the growth parameters like leaf area of index leaf, fresh weight of index leaf, dry weight of index leaf, dry matter content of index leaf and pruning weight although they were higher with 100 per cent evaporation replenishment. The total soluble solids was marginally higher with drip irrigation (15.43 per cent) compared to microjet irrigation (15.37 per cent), but the titrable acidity did not vary with the method of irrigation (0.34 per cent). The sugar: acid was higher with drip irrigation (45.11) compared to microjet irrigation (44.86). Irrigation scheduled with an evaporation replenishment rate of 75 per cent either with microjet or drip irrigation resulted in higher yields. The water use with 75 per cent evaporation replenishment was 945.75 mm.

Stevens and Douglas (1994) describe the three dimensional variation in L_v within a quarter of the planting area of Colombard grapevines on Ramsey rootstock grown under drip and full-cover microjet irrigation. Under drip irrigation roots were concentrated under the vine row, whereas under microjet irrigation roots were evenly spread across the planting area. The maximum L_v were 1.2 and 0.6 cm/cm^3 and the estimated total root lengths per vine were 32 and 26 km for drip and microjet irrigated vines, respectively. Under drip, 56 per cent of the variation in L_v could be accounted for as a function of depth and radial distance into the row, and under microjet, 45 per cent of the variation in L_v could be accounted for as a function of depth. Twenty five per cent of the vine roots were in soil with air filled porosity at field capacity of 6 per cent or less. Based on the variation of root length per unit area (L_a) across a quarter of the planting area and between vines, we concluded that selection of a location at which the L_a would be representative of that in the entire irrigation unit is feasible in microjet irrigated vines but not in those irrigated with drip. We suggest that the absence of a location representative of L_a confounds the scheduling of drip irrigation based solely on measurements of soil moisture.

Stevens and Prior (1994) reported that the Potted Sultana vines were waterlogged for the first 0, 1, 3, 5 or 7 days of a two week cycle designated W0, W1, W3, W5, and W7, respectively. The vines were harvested at the end of the fourth cycle. Shoot elongation rates over the first two cycles for W0 to W7 were 56, 51, 34, 38 and 35 mm/day, respectively. Vine leaf and stem dry weights at harvest were 37, 32, 22, 21, and 19 g/vine for W0 to W7, respectively. Photosynthesis and stomatal conductance were reduced both during and following waterlogging but leaf water potential was unaffected. Waterlogging reduced leaf blade N, P and Ca, increased Na and Mg, and did not affect K, Cl, Mn, Zn and NO_3-N.

Steven and Harvey (1996) observed over two consecutive seasons, daily rates of change in the water content of the top 120 cm of soil were calculated for sample sites located at 50 cm from the butts of 12 Colombard vines on Ramsey rootstock which were part of a controlled deficit irrigation experiment. Records were selected where the rate represented a change due solely to depletion of soil water by grapevine evapotranspiration (DS_{ET}). Soil water availability was quantified as a logarithmic transformation of the absolute value of the root-weighted measure of soil matric potential. DS_{ET} was negatively correlated with soil water availability and positively correlated with reference crop evapotranspiration (ETo). At very high values of soil water availability, *i.e.* soil matric potential between -8 and -16 kPa, and after the canopy was fully developed, DS_{ET} had a 1:1 relationship with ETo. When root-weighted soil matric potential was greater than -16 kPa, the mean values of the ratio of soil water depletion rate to reference crop evapotranspiration (DS_{ET}/ETo) varied from 0.1 to 1.2 for 30-day periods between bud burst and leaf fall. Considering the full set of measurements, DS_{ET}/ETo was highly negatively correlated with soil water availability for all 30-day periods except those less than 60 days from bud burst. In the period 150 to 179 days after bud burst, 70 per cent of the variance in DS_{ET}/ETo was accounted for by regression on soil water availability for values of root-weighted soil matric potential between -8 and -100 kPa. During this period a fall in root-weighted soil matric potential from -10 to -42 kPa was associated with a decline in the value of DS_{ET}/ETo from 1.3 to 0.6. At both high and low values of ETo, the ratio of DS_{ET}/ETo fell linearly with declining soil water availability. The changes in values of midafternoon stem water potential and stomatal conductance associated with this decline in DS_{ET}/ETo are discussed.

Stevens and Cole (1987) reported that a range of 6 soil matric potential regimes was established in a microjet irrigated vineyard (Colombard scion on Ramsey rootstock) by varying the soil water deficits at which irrigations were applied. The average amount of water applied per season to the most and least frequently irrigated treatments was 1342 and 740 mm, respectively, and the midsummer irrigation interval was 4 and 18 days, respectively.

The plant water potential experienced by vines in any of the treatments was found to be a function of the log of the absolute value of the root weighted soil matric potential (log ê RWSMP ê). In response to decreasing soil water potential, shoot water potential decreased and stomatal resistance increased, indicating a direct response of the plant to soil water potential.

Indices of the plant water availability for the season were calculated by summing the daily values of the log ê RWSMP ê for each treatment during the season. The value of these indices (stress indices) increased as plant water availability decreased.

In both seasons (1984/85 and 1985/86) the vine yield and berry weight significantly decreased as the value of the stress index increased. The only must component which had a significant interaction with the stress index over both seasons was the malate concentration, which decreased as the value of the stress index increased. None of the other must components measured, ie Brix, pH, titratable acidity and the concentration of tartrate and potassium, had a significant interaction with the stress index in both seasons.

Our results demonstrate that vine yield is sensitive and that, except for malate concentration, must composition is insensitive over the range of soil water availabilities in this experiment. The irrigation treatments applied were representative of the range of irrigation practices encountered in the Riverland vineyards. Hence in some vineyards low soil water availability during part of the irrigation cycle may be causing a loss of yield.

Stevens *et al.* (1999) reported that the mature field-grown grapevines, Colombard on Ramsey rootstock, grown in a semi-arid climate, were irrigated with saline water during any 1 of 4 growth stages within the season: pre-flowering, during berry development, during berry ripening and post-harvest. At other times, plots were irrigated with river water (EC 0.5 dS/m) as was the control throughout the season. Saline water (EC 3.5 dS/m) with a high SAR was produced by addition of sodium chloride brine. Soil cation exchange capacity was 14 C mol/kg and, at the end of the trial, the soil exchangeable sodium percentage in the control was 6 per cent, in the treatment salinised pre-flowering 13 per cent, during berry development 20 per cent, during berry ripening 20 per cent and post-harvest 19 per cent. Treatments were applied for 6 consecutive seasons. Vines were highly productive with the average yield in the control equal to 62 t/ha of grapes. Saline irrigation caused significant, but small, declines in yield in three seasons, in pruning weights in 2 seasons and in berry weights in 4 seasons. Effects on growth, once established, often persisted unchanged through one or more subsequent seasons of saline irrigation. The growth stage shown to be most sensitive to saline irrigation was berry development; saline irrigation during berry development reduced the yield by 7 per cent and during berry ripening by 3 per cent and pre-flowering it reduced the berry weight by 1 per cent, during berry development by 6 per cent and during berry ripening by 4 per cent. The amounts of irrigation applied in each of the 4 growth stages were not equal and hence treatments did not receive equal additions of salt. Normalising data to remove this effect showed that the rate of yield decline per unit dS/m increase in the seasonal average salinity during berry development, 7 per cent dS/m, was 3-fold greater than the 2 per cent dS/m during berry ripening. We conclude that this scion/rootstock combination, grown under these conditions, can maintain high productivity despite 'slugs' of saline irrigation.

Steven and Harvey (1994) opined that in semi-arid South Australia, vigneron concern with soil salinity has led to over-irrigation which causes temporary waterlogging following irrigation. The relative effects of salinity and temporary cyclic waterlogging on growth were estimated from two experiments on potted Sultana vines. Growth of vines, with free draining root zones, irrigated with solutions containing NaCl at 10, 20, 40, and 60 mM was 89, 76, 58, and 51 per cent of vines irrigated with a solution containing 2 mM NaCl. When vines irrigated with a 2 mM NaCl solution were subjected to cyclic waterlogging for 1, 3, 5, and 7 days in a 14 day cycle, growth was 43, 29, 30, and 26 per cent of that in vines with free draining root zones.

Steven *et al.* (1996) reported that the effect of foliar salt uptake on potted grapevine growth and ionic composition was investigated in a split plot trial. The main plot was a two by two factorial consisting of separately irrigating the roots and foliage with non-saline or saline (NaCl 25 mM) solutions. The sub-plot was a four by two factorial consisting of four grape cultivars (*Vitis vinifera* cvs.) on their own roots or Ramsey (*Vitis champini* cv.) rootstock. Saline foliar irrigation over 27 weeks reduced total vine growth by 14 per cent while saline root irrigation had no effect. Leaf Na and Cl concentrations were elevated by both saline foliar and saline root irrigation. The increases in concentrations with saline foliar irrigation were four times those with saline root irrigation. Leaf K concentration was reduced by saline foliar irrigation and increased by saline root irrigation. With saline irrigation of both roots and foliage the Cl and Na levels were highest in the leaves of 'Shiraz', but with saline irrigation of only the roots 'Sultana' had the higher levels of leaf Cl and 'Shiraz' the highest leaf Na. Saline foliar irrigation had no effect on the concentrations of Na, Cl and K in the roots. In 'Sultana', saline foliar irrigation did not affect the leaf concentrations of N, NO_3-N, P, Mg, Zn and Cu. It increased the leaf concentration of Fe, and decreased that of Mn. Rootstock modified the effect of salinity on Fe concentrations. The B concentration was decreased by saline irrigation of either the foliage or the roots, but not by saline irrigation of both. In roots, saline foliar irrigation increased B in own-rooted vines, but not in those on 'Ramsey' rootstock.

Stevens and Harvey (1990) observed that the transient soil salinization of irrigated Colombard/Ramsey grapevines was produced by substituting, for a number of different two month periods in the growing season, saline water (EC 3.5 dS/m) for irrigation water drawn from the River Murray (EC 0.5 dS/m). Irrigation with saline water produced a transient increase in the root weighted saturated soil solution (RWECe). This increase caused a decline in leaf water potential and an increase in leaf petiole sodium and chloride contents. There was no statistically significant effect on vine vegetative growth (pruning weights) or yield. These results are contrary to those predicted by the Murray-Darling Commission model which is designed to quantify vine yield losses due to salinity.

8

Nutrition Management

Grape yield and nutrition management have close relationship and is very important; also it is the quantity which will have effect on quality and optimum fruit load which brings uniform sugar, acid ratio that ultimately decides the quality. In our place grape is cultivated but with two distinct pruning practices, *i.e.*, April, to promote growth and October pruning for production of bunches to take every year high and quality production of grapes and nutrition management is very important.

Climatic Relationships/Nutrition Management

1. To get a good grape bunch, pruning is done between rainy season and winter. But in Karnataka we find three distinct phase of climate *i.e.*, summer, rainy and winter seasons. The establishment of flower primordia in grape takes place before winter but the development of bunch starts in winter; it's ripening starts from the end of winter season. So it requires two different pruning methods to get the crop. If you think grape bunch development takes place in winter and its nutrition management, your work should be done in that period. In winter the development is slow compared to other seasons; during winters relative humidity and temperature are less but the light is medium. For the development of bunch three stages are required:
2. After pruning, sprouting, flowering, and bunch development for the establishment of flower; it requires 6 weeks.
3. After flower establishment, the berries and bunch development starts; it requires 6 weeks time.
4. Third stage of the development stops and the ripening period starts and to

complete this stage it requires 6 weeks. It means from October pruning onwards from flowering to harvesting 4½ months time is required. For a good grape production as mentioned above in three or four stages the nutrition management in a very good manner is very important.

(*a*) First stage – sprouting, sprout development, flower production. It takes 4 weeks.

(*b*) Second stage – flowers bunch emergence, flowering, fruit set. It takes 4 weeks.

(*c*) Third stage – berry development, bunch development. It takes 4 weeks.

(*d*) Fourth stage – Fruit ripening. It takes 4 weeks.

Overall it requires 4 months. For every stage of development that takes place right from leaf, flower and bunch onwards and other parts, one has to understand about nutrition activities.

Post-October Pruning First Stage

Before October pruning remove all the leaves and prior to that the establishment of flower primordia, nutrients applied for the development of framework of the canes etc., is important. After October pruning emergence of new sprouts, new leaves, nutrient requirement, management of deficiency symptoms, depends on the availability of nutrients in the soil. The climate, NPK and trace element availability of water-all these factors are congenial, then sap flow starts. After vine canes are pruned, at cut portion the water starts coming out. If this situation is not observed then availability of water in root zone is less and air is more. One should understand that properly. That kind of situation *i.e.* increases in water and decrease in air at root zone does exist.

Third part is use of fertilizer and hormones. New growth that we call as white root hair growth are necessary, for that purpose phosphoric acid and uric acid and hormone -like Indole acetic acid are required and in due course of time we see amino acid is produced in the soil. Very often farmyard manure and compost is required and use of green manure is equally important. The living organism plays an important role to work on green manure, so it should get well-mixed in the soil. Vermiculture and vermicompost difference one should keep in mind. Azotobacter, rhizobium, PSB, bio-fertilizer are available in the market. The availability of C: N ratio in soil is important. Therefore, before pruning and production of white root hairs, application of fertilizer and compost is equally important. After pruning early application of fertilizer will help in the improvement of soil nutrient.

After October Pruning 2nd Step

At this stage new growth development and production of full leaves are observed and the leaves are green in colour with good leaf area and age of the leaves will be around 30 days; that means they are mature enough to do the job of photosynthesis; at the same time the sunshine hour helps the dormant bunch to develop, and in soil you may find depletion of nutrients. Suddenly the vines may show hungry symptoms as more nitrogen is required.

This situation is observed at the beginning of the 2nd step of October pruning *i.e.*, 25-30 days of pruning; nitrogen requirement one has to fulfill completely, if nitrogen nutrient is applied early it will help the vine. Sometimes if nitrogen is applied in the beginning more nitrogen is available then it will have bad effect. How much nitrogen, in which form and in how many split doses through soil and foliar, one has to decide whether only nitrogen is to be applied or other fertilizer is to be applied.

Petiole analysis is a landmark in understanding the exact requirement when vigorous sprouting starts; 6th leaf or its petiole analysis is done for nitrogen, which maybe in the range of 1.5 to 2 per cent.

Based on Petiole Analysis Nitrogen Requirement

Sl.No.	*Petiole (N)*	*Per Hectare N Required*	*Per Acre N Required*
1	2.5 and above	Nil	Nil
2	2.4 to 2.5	10 kg	4 kg
3	2.3 to 2.4	20 kg	8 kg
4	2.2 to 2.3	30 kg	12 kg
5	2.1 to 2.2	40 kg	16 kg
6	2.0 to 2.1	50 kg	20 kg
7	1.9 to 2.0	60 kg	24 kg
8	1.8 to 1.7	70 kg	28 kg
9	1.7 to 1.8	80 kg	32 kg
10	1.6 to 1.7	90 kg	36 kg
11	1.5 to 1.6	100 kg	40 kg
12	1.4 to 1.5	110 kg	44 kg
13	1.3 to 1.4	120 kg	48 kg
14	1.2 to 1.3	130 kg	52 kg
15	1.1 to 1.2	140 kg	60 kg
16	1.0 to 1.1	150 kg	60 kg
17	1.0 to 1.0	160 kg	64 kg

Means the nitrogen quantity you supposed to apply out of which 50 per cent nitrogen at second stage you have to apply menas after 4 weeks 50 per cent. Nitrogen you have to apply. This 50 per cent quantity of N to be applied in two weeks for two times and every week 4 times if two times a week, 8 times. Alternate day 16 times or daily 28 to 32 times. If drip irrigation facilities available then apply through drip also.

Requirement of Magnesium alongwith Nitrogen

For the production of Chlorophyll in leaves nitrogen is required but magnesium is equally important. Ammonium sulphate 100 kg + 15 kg magnesium sulphate together can be used or go for Petiole analysis to assess the magnesium sulphate requirement.

After October Pruning 3rd Step

At this stage of fruit development the demand for nitrogen and phosphorus increases. At the beginning nitrogen and 0.4 per cent phosphorus is required in the petiole. For that purpose the nitrogen quantity (remaining 50 per cent) alongwith phosphorus should be applied. If the lime stone exceeds > 8-10 per cent then apply 7:10:0 (500 kg) and if lime stone is medium in soil then apply 18:46 (DAP) fertilizer (100 kg) per hectare for 55 to 60 days; after October pruning this quantity you have to apply.

After October Pruning 4th Step

At this stage the growth of leaves and fruits stops but ripening starts. The berry colour starts changing. To help this activity application of potash becomes necessary. About 75-90 days after October pruning, apply the required quantity of nitrogen, phosphorus, and potash or only potash.

Per Hectare

1 tonne compost (3-4 cart loads)

+ 300 kg 19:19:19 and 125 kg sulphate of potash Per Acre

400 kg compost (1 cart load)

+ 300 kg 19:19:19 + 50 kg SOP

At this stage nitrogen should be applied in big doses. Application of nitrogen fertilizer should be stopped when the bunch starts ripening and the new growth starts. If you apply fertilizer early without potash (*i.e.*, NP) it will have good effect on fruit development and early set bunch shows more water berry symptoms. So alongwith DAP or individual application of N and P add sulphate of potash. If you apply MOP it will affect the quality of bunch and if chloride is more in irrigation water, in such a situation apply sulphate of potash alongwith compost.

At every stage of application of fertilizer one has to keep in mind that at this stage which fertilizer is required, and how to fulfill the requirement.

Ist Stage – (One month) – nitrogen and magnesium

IInd Stage – (Two months) – nitrogen and phosphorus/Urea Phosphate.

IIIrd Stage – (Three months) – nitrogen, phosphorus, potash/urea phosphate + potash – lastly potash.

Micronutrient Requirement

In grape nutrient management, NPK is very important but how to fulfill the requirement of NPK, in what form it should be applied are very important. Alongwith NPK, the secondary important nutrients *viz.*, calcium, magnesium and sulphur are also equally important. In Karnataka the grape growers, among the secondary elements, use more of magnesium. If magnesium is applied in more quantity, more yield is expected; this has been impressed in the farmers mind and this is true also. It doesn't mean by reducing the required fertilizer or just skipping the application of fertilizer, one can expect higher yield. Most important is to understand the importance of secondary elements and how they help in increasing the yield. By micronutrient application (1) how to increase yield and (2) quality improvement; this understanding is important to farmers.

For grape development and yield all secondary elements are important. Every individual element effect is different *viz.*, zinc, iron manganese, copper, molybdenum and boron are also important. After October pruning, 4 steps are already discussed like the use of micronutrient are important during 2 and 3 steps, for leaves growth and fruit development etc., When leaves show symptoms of developing zinc and manganese deficiency these two are required in large quantity. At fruit stage ferrous and boron are required in large quantity. Copper is required at development stage and molybdenum is required at establishment of flower primordia.

Once ferrous and zinc deficiency are seen there is change in leaf colour. Boron and manganese deficiency have correlation with leaf shape and leaf colour to some extent. But, in flower/bunch, micronutrient deficiency is not visible. It may not affect the yield but hamper the quality. Therefore, it is a must to use micronutrient in grape cultivation.

After pruning the new growth when it emerges comes with force and sometimes slow growth is observed. Manganese and zinc deficiency results in slow growth. When the growth is not uniform there will not be uniform flower primordia establishment. The injury caused by pruning has to be healed. Otherwise some branches may die and the food has to be diverted to good and healthy canes which ultimately result in improper nutrition of bunch of berries. Boron deficiency causes improper fruit set. Fruit set will be early but slow and stops immediately, which affects the berries on vine, which set early, is slow growth. Early set berries become bigger and the late set berries become smaller in size and at the tip the shoot berries will be more. Leaf boron and molybdenum deficiency results in early filling of sugar which causes berry growth to stop and at ripening phase sink starts, and in some berries totally sugar filling wont takes place, which results in water berries and further reflect in mummification.

Lime Nodules Affects Soil and Micronutrients

In Karnataka lime nodules are more in the soil and with soils having 8-10 per cent lime content, quality grape production becomes very difficult. Ferrous content of lime nodules will not be available to leaves and leaves show deficiency symptoms in younger leaves *viz.*, the leaves will be whitish or yellowish in colour. This can't be helped, which further causes physiological disorders with the symptoms being observed on leaves only and on bunch or berries and later on, any precaution taken is not going to help.

Micronutrient Requirement

If you get more yield then you won't get export quality. Per acre 10 tonnes or per hectare 25 tonnes yield is not important. If you get the produce of 25 tonnes what if it comprises 20-25 per cent for export, the use of micronutrient becomes essential. For getting export quality, micronutrient, nitrogen, phosphate, potash, are used in a balanced manner. The requirement of micronutrient is felt at new growth whenever deficiency symptoms are exhibited; at the same time pinching is advocated; so one should not think that the problem is solved. At the time when major nutrient like nitrogen and potash for vigorous growth magnesium requirements becomes more under such situation micronutrients are also required in large quantity. If you apply then it will affect and reflect on quality.

Testing and Apply

By petiole analysis the exact requirement of nutrients can be assessed and after diagnosis the period can be calculated. Simple soil analysis will not be helpful. If one has to go for leaf analysis which requires time and by which time when the critical vine requires nutrients it is late. Even then if we apply micronutrients and our orchard improves leaves may show improvement or they may resemble like improved ones but it will not reflect on bunch. If problem is noticed then micronutrients are required. Therefore advance application is always better.

Foliar Application on Leaves

Alongwith soil bio-fertilizers, especially compost and green manures are applied and through that micronutrients requirement will be fulfilled to some extent but that will not be sufficient. One has to compensate through soil micronutrients. At both pruning stages application of micronutrients helps in improving the soil fertility reserve. Then application of micronutrient to leaves is not necessary. When the new sprouts come up, zinc. manganese, are required. At fruit development and ripening boron, molybdenum are required. Copper application is required as a catalyst but in lime-rich soils ferrous availability is improved by applying chelated form or manganese and zinc which are to be applied together. While doing foliar spray on leaves, first spray at sprouting, second spray when the bunch is in flowering stage, third spray at full fruit set and after spraying when the bunch is in flowering stage, spray at full fruit set and after girdling and if necessary, spray after third spray and the requirement of micronutrient is almost over. While you are going for spraying but by mistake if you apply, more there is possibility of chloride injury to leaves and when the bunch is in flowering stage micronutrient spray will be helpful in future.

Application of zinc, (30-40 ppm.) manganese (20-25 ppm.) ferrous (25-35 ppm,) is beneficial but excess boron is harmful; always apply in low concentration but you may not get the result and in early application of boron don't apply, it will not be beneficial.

1. Foliar application of different micronutrient by mixing can be applied to leaves or ready-made mixture may be used.

Schedule of Fertilizer after October Pruning for Higher Yield of Grape

Sl.No.	*Time/Vine Care*	*By Soil Application/Hectare*	*By Foliar Spray*
1.	3-4 weeks after pruning/at new growth	Nitrogen 50 kg, Magnesium 5 kg,	Zinc, magnesium (50 to 100 ppm)
2.	4-8 weeks after pruning, at bunch growth flowering, berry setting	Nitrogen 50 kg, phosphate 50 kg	Boron, zinc (75 to 150 ppm)
3.	8-10 weeks after pruning berry development	Nitrogen 50 kg, phosphate 50 kg, potash 50 kg	Ferrous copper (100 to 150 ppm)
4.	10-12 weeks fruit ripening starts	Potash 100 kg	Boron, molybdenum (150 to 220 ppm)

2. Chelated micronutrient should be used with care.
3. Along with micronutrient hormones cannot be used.
4. While applying fertilizer through soil and foliar means, there should be some gap.
5. At flower primordia, by petiole analysis and as per diagnosis change the programme.
6. Once the bunch ripening starts stop using foliar sprays of micronutrient.
7. Every year this programme should not be changed on large scale.

For every farmer to get a good quality grape, a proper beginning is necessary. Grape orchard should have drip irrigation facility and one should have detailed information about irrigation management.

Water Management by Drip

Grape will be continuously growing throughout the year and that development has to take place in a proper manner and to get higher yield in grape two times pruning is done. First pruning in April and second pruning in October and later on the water requirement of vine from pruning is done to get full vegetative growth (canopy cover). Establishment of flower primordia to (60-120 days) harvest and again from October pruning till harvest are critical as different problems crop up. Therefore water management scheduling is very important.

Water Management through Drip in Grape

For grape cultivation one has to consider the following points while going in for drip irrigation:

1. Rate of Evaporation

Daily how much evaporation loss has occurred is expressed as USWB class. A pan evaporimeter is used to measure it and it is expressed in mm/day and daily at 8:30 it is recorded. When drip is used for irrigation and by chance there is a rainfall is not necessary. If rainfall is less, then daily evaporation loss, from rainfall has to be deducted from daily evaporation loss and keeping in mind the balance in evaporation loss irrigation is taken up.

Requirement of Effective Rainfall by Drip Irrigation

Day	*Evaporation mm/day*	*Rainfall mm/day*	*Effective Rainfall/ (mm)*	*While irrigation Consider Evaporation Loss (mm)*
3.7.06	7.4	2.9	2.9	4.5
4.7.06	8.3	–	–	8.3
5.7.06	8.10	10.2	8.0	0.0

If the rainfall is more than expected, looking to soil condition, don't irrigate for 48 to 72 hours. Later on keeping in mind evapotranspiration loss and rainfall, go for irrigation.

2. Evaporation Factor

There is little controversy about pan-evaporimeter because it is made up of iron sheet and it is influenced by the surrounding atmosphere/existing situation. If the level of water in pan-evaporimeter is less, it is considered that the loss of water will be more. In our situation the factor will be 0.7 to 0.8.

3. Yield Factor

As the canopy spread advances, the requirement of water also advances. One square foot area should give 400g fruit from one bunch. One bunch growth requires 2600 to 2700 sq. cm of leaf area, means 6 x 4 foot distance in an acre have 1800 plants. And on every vine 25 bunches, with 400g weight, per acre 18 tonnes and per hectare 40-45 tonnes yield is expected. Hence the leaf area and water requirement have co-relation. As per FAO book on page number 24 to 33 detailed information is given for a new grape garden, the yield factor ranges between 0.25 to 0.95. But when the canopy is fully covered, on the pendals this factor from pruning to bunch emergency arises and for harvest 0.6 to 0.90 should be considered.

4. Wetting Area

Through drip irrigation system, 30 per cent soil should be wet. On an average grape garden should have 60 cm soil layers and it should hold 40 per cent available soil moisture for getting good crop.

Water Requirement of a Grapevine

Evaporation loss and role of evaporation factor will play an important factor in assessing the water requirement of a vineyard and how much water is required can be calculated by yield factor.

At micro-level/drip irrigation method how much water is required for a vine can be calculated by = A x B x C x D x E (litre/day)

A: Daily total evaporation loss (9 mm)

B: Evaporation factor 0.7

C: Grapevine growth factor

D: Grapevine yield/Sq. meter.

E: Soil wetting area (per cent)

(*Note*: For Alternate day watering, two days evaporation loss has to be taken into account).

How Many Times Irrigation is Required

Once you calculate the water requirement, later on for irrigation of grapevine, by drip irrigation how much we have to keep watering is very important.

Meanwhile the field area measurement and how much drippers are required:

For Example

A = Daily evaporation loss - 10 mm

B = Evaporation loss factor -0.70

C = Grapevine vine factor - 0.90

D = Vine spacing wetting are - 1.83 M x 1.22 M = 40 per cent = 0.40 one day

Water requirement/vine = One day evaporation loss x open pan evaporation. (mm) meter (litre/day)

Development stage x grape area (M^2) x wetting area.

Grape vine water requirement = 10 x 0.7 x 0.9 x 2.23 x 0.40 = 5.65 litre/day.

Different Buds of Evaporation, Grapevine Yield, Factors and Water Requirement for Grapevine

Evaporation Loss	*Grape Vine Spacing (Square Foot)*											
	6' x 4'				*8' x 5'*				*10' x 5'*			
	Yield Factor				*Yield Factor*				*Yield Factor*			
	0.6	0.7	0.8	0.9	0.6	0.7	0.8	0.9	0.6	0.7	0.8	0.9
3	1.2	1.4	1.5	1.7	1.9	2.2	2.5	2.9	2.4	2.8	3.2	3.5
4	1.5	1.8	2.0	2.3	2.5	3.0	3.4	3.8	3.2	3.7	4.2	4.7
5	1.9	2.2	2.5	2.9	3.2	3.7	4.2	4.7	3.9	4.6	5.2	5.9
6	2.3	2.7	3.0	3.4	3.8	4.4	5.0	5.6	6.6	5.5	6.3	7.0
7	2.7	3.1	3.5	4.0	4.5	5.0	5.6	6.6	5.5	6.4	7.3	8.2
8	3.0	3.5	4.0	4.5	5.0	5.9	6.7	7.5	6.3	7.3	8.4	9.4
9	3.4	4.0	4.5	5.6	5.7	6.6	7.5	8.5	7.1	8.2	9.4	10.6
10	3.8	4.4	5.0	5.7	6.3	7.3	8.4	9.4	7.9	10.5	11.7	–
11	4.2	4.8	5.5	6.2	6.9	8.1	9.2	10.3	8.6	10.0	11.5	12.9
12	4.5	5.3	6.0	6.8	7.5	8.8	10.0	11.3	9.4	11.0	12.5	14.9
13	4.9	5.7	6.5	7.4	8.2	9.5	10.9	12.2	10.2	11.9	13.6	15.2
14	5.3	6.2	7.0	7.9	8.8	10.2	11.7	13.9	11.0	12.8	14.6	16.4

Drip for how much time to keep?

One dripper/hour = 3.75 litre

Therefore, Drip running period hours = 1.50 (One and half hour)

What Type of Dripper and How Many Drippers Required

While fixing drippers the following points one has to take into consideration.

1. Dripper should be sturdy/durable
2. Dripper hole should be bigger because it won't get clogged.

How many dripper and which type is required depends on the type of soil, how much land to be wetted, light soil requires low output dripper of 4 litre/hr. Two drippers on both the sides of vine and in a short time irrigation is made possible. But for medium and heavy soils, for one grapevine it requires one dripper of 4 litre/hour and it should be fixed at 20-30cm apart. Between two vines one or two drippers at equal distance should be fixed. By dripper how much water is given the following table:

Relationship of Soil, Dripper and Wetting Area

Sl.No.	*Type of Soil*	*By Dripper and Horizontal Area (cm)*	*Depth of Wet Soil (cm)*
1.	Heavy	50-60	100-120
2.	Medium	40-50	80-90
3.	Light	20-30	40-50

The wetting area depends on the soil type, of dripper, method of quantity of irrigation of water applied, the distance between two dripper. The vertical 20-40 cm depth of wet soil is sufficient for grapevine. Excess watering helps in removing the salt from the soil, which moves the salt away from the soil. Overlap watering should be 20-30 per cent as per the research findings.

For Overlap Watering By Dripper Wetting Area?

(Distance between two dripper) = Meter x 0.7 to 0.80

Micro–irrigation by all methods can be given every alternate day and main reasons are if more than required time the vines are irrigated, the loose soil (water and air) is filled up which spoils the vines, water and air balance which results in non-availability of oxygen to vines. Therefore alternate day irrigation helps in reducing the disease also.

Pruning and Water Management

In grape, two time pruning is done, and at different stages of pruning, different water quantity is required and while giving irrigation by drip little changes are required.

April Pruning

Once grape harvesting is complete grape garden should not be tampered for 8 days before April pruning and 10-15 days afterwards April pruning dripper should be stopped. The other work of the orchard can be attended to. Take up inter-cultivation; apply compost and bio-fertilizers. In the beginning, keep dripper for a longer time on soil to a saturation point and later drip irrigation x 0.6 is run on. After April pruning and after complete work is over, the grapevine requires less; water because number of leaves are less if watering is done more than the required amount then excess growth is observed. After pruning 30 to 60 days, the leaves growth hastens. During this period the flower primordia is established and at that time more watering is required and during this period more temperature, persists the evaporation loss through soil, the transpiration loss through leaves will be more, hence the quantity of water given should be increased. In afternoon, to suffice the water requirement, this period yields factor 0.75 that should be considered. Vine growth is very fast and canopy is fully covered; in 60-80 days the flower primordia also gets developed. The following yield factor for irrigation should be followed:

Weeks after Pruning	*Days after Pruning*	*Yield factor*
1-2	0-15	0.60
3-4	16-30	0.75
5-6	31-45	0.90
7 and above	45 and above	0.90

October Pruning

15th October is the best time for October pruning when the soil is too wet because then excess growth takes place. Therefore, for 8 days keep the drip closed after October pruning and later after 15 days the drip has to be kept open, 0.70 yield factor at 0.60 with light irrigation be given. Then after 20 days, 0.70 yield factor and then later for 25 days, by taking 0.80 yield factor, drip irrigation be given and when the water filling starts in berries from 75-80 days for that purpose by taking drip yield factor (0.9) the watering has to be continued. If watering is excess or less it leads to water berry and too much watering leads to bigger water berry and berries cracks, and rot. Once the harvesting is over stop drip irrigation. Follow the following drip irrigation chart as mentioned below:

Sl.No.	*Weeks after October Pruning*	*Day after October Pruning*	*Yield Factor*
1	1	0-8	No watering
2	2 to 3	9-13	0.70
3	4 to 6	24-42	0.80
4	7 to 9	43-68	0.80
5	10 and above	61 and above	0.90

It looks that water requirement vary from month to month *i.e.*, it should be between 3500 to 11,000 litre per day. Water is required more in May (11,000 litre/day) and is very low in November (3500 litre/day). Through drip the grape vine requires 40 to 50 cm of water. Through regular watering, if you consider the water requirement, only 50 per cent is needed by drip method. If you follow the above method good buds can be expected and yield naturally will be good.

Drip and Fertilization

In Karnataka at present the irrigation data shows that the watering through flood method is only 12.6 per cent. In the near future the water application through drip is having lot of scope because in sugarcane through flooding the yield has come down by 50 per cent. (Fig-22)

Grape Vine (6' x 4') Drip/Ac/Day Water Requirement

Month	*Daily Evaporation Loss (mm)*	*Required Water (litre/day/Ac)*	*Month Daily Evaporation Loss (mm)*	*Daily Evaporation Loss mm*	*Required Water Litre/ AC/day*
April	11.05	7300	October	5.80	4000
May	11.82	10900	November	4.30	3500
June	8.93	8400	December	3.72	3400
July	5.85	6900	January	3.85	4000
August	4.85	5000	February	4.48	5700
Sept	4.82	5000	March	8.05	8300

(Dr. Subhash Shinde, Shetakari March 16)

Schedule of Drip Irrigation in Grape Cultivation

Sl.No.	*Days*	*Vine Management*	*Water Requirement*
	April		
1.	Hard pruning (before 8 days) 7-14 days	–	Stop watering
2.	15th April hard pruning	–	Stop watering
3.	Days 16-30 April (15 days)	Sprouting	Take 0.6 yield factor and irrigation by drip
	May		
4.	Days 1-15	Canopy spread 50 to 60	–
5.	Days 16-31 (16 days)	Leaf growth 75-100 per cent	0.75 per cent is yield factor taken and drip irrigation given.
	June to Sept		
6.	Day 1 June to 30 Sept	Est. of flower primordia and maturity of cane	0.9 yield factor taken and drip irrigation given

Contd...

Table–*Contd...*

Sl.No.	Days	Vine Management	Water Requirement
	October		
7.	Day 1 to 15 October pruning and previous 15 days.	–	Stop watering
8.	15 October pruning	Keep buds and prune	–do-
9.	Day 16 to 23 October (8days)	–	–do-
10.	Day 24 October to 7 Nov (15 days)	Opening of buds and bunch sprouting	0.6 yield factor
	December		
11.	Day 8 Nov to 27 Nov (25 days)	Fruit set completes and berry develops.	Take 0.8 yield factor and irrigate
12.	Day 23 Dec to harvest	Filling of berry water	Take 0.90 yield factor and irrigate

Fertigation Means what?

Method of fertilizer application through drip is called fertigation and following are the advantages of fertigation:

1. Fertilizers are applied to the nearest point of the roots and in equal proportion.
2. Fertilizer alongwith water is applied to the feeder roots.
3. As per the requirement the quantity and grade can be selected.
4. Fertigation reduces cost of labour, electricity and time. The insecticide and fungicides also can be mixed with fertigation to reduce labour cost. Soil structure is not hampered.
5. By fertigation 25 per cent fertilizer can be saved. In that condition pH is 4.5, except in case of multi-K- which is having a pH of 9.5. They are acidic fertilizers and good for sodic, alkaline and calcareous soils. By fertigation more utilization of fertilizer is possible and they won't go beyond root hairs.

Type of Fertilizers Applied Through Drip

1. **Liquid Fertilizers:** They fully get dissolved in water with low concentration and required in large quantity. Hence more transportation cost.
2. **Solid Soluble Fertilizers:** They fully and easily dissolve in water and required in less quantity. Therefore it results in less transport cost and also it occupies less area. Soluble liquid and solid forms of fertilizers are pure and pH will be 4.5 to 5.5 which helps in reducing soil pH. They should be

dissolved in water in proportion of 1:10 or 1:15. One has to take care while going in for fertilizer application. Checking the drippers and fix-up of filters. They should have 0.08 to 0.15 mm holes. If the fertilizer granules are bigger than 0.08 to 0.15 filter helps in retaining them. Hence the dripper holes won't clog. They are costly example. Urea, SOP, potassium nitrate, mono-ammonium phosphate are suitable to use alongwith water (1:10 to 1:15) and they should be passed through a cloth.

Fertigation System

Fertilizer alongwith water through drip are applied under pressure.

1. Through ventury, fertilizer alongwith water gets dissolved and applied.
2. Here the fertilizers are dissolved in a tank and through pump it is pumped in the main valve of the drip.
3. Delivery/flow bypass Tank: Here where the water is coming and going out through two small pipes connected to the main line and between a choke valves is connected. One entry pipe is connected to the tank and continuously fertilizer gets dissolved in water.
4. Fertilizer management: Based on crop, soil type, stage wise the fertilizers are applied.

Care of Drip System While Fertigation

1. Keep filter clean.
2. Fertigation pump and tank examination.
3. Every month clean the dripper.
4. Observe level of salt in dripper; whether they are clogged, so clean them with acid or chlorine. Check up the pressure also.

I- Fertigation During April Pruning

Month	*Stage*	*Every Week/Dosage*		*Per Month Dosage*
March	Rest period	13:13:13	6½ kg	25 kg
April	From pruning to twelfth leaf	10:36:10 + Urea	6¼ kg 1.5 kg	25 kg 6 kg
May	Establishment of flower primordial	13:13:13	6¼ kg	25 kg
June	Maturity of cane	5:15:30+ 0:0:51	12½ kg 6¼ kg	50 kg 25 kg

Total Nitrogen, Phosphorus and Potash/Ac
Nitrogen 17 kg, phosphorus 23 kg, potash 37 kg

II - Fertigation during October Pruning

Month	*Stage*	*Every Week/Dosage*		*Per Month Dosage*
October	Pruning to flowering	13:13:13 + Urea	6¼ kg 4.5 kg	25 kg18 kg
November	Flowering to girdling	13:13:13 + Urea	6¼ kg 3 kg	25 kg 12 kg
December	Girdling to berry development	5:15:30 + 0:0:51	6¼ kg 6¼ kg	50 kg 25 kg
Total NPK/AC	Nitrogen 23 kg	P_2O_5 14 kg	Potash 34 kg	–
Yearly NPK/AC	Nitrogen 40 kg	P_2O_5 37 kg	Potash 71 kg	–

Notes:

1) As per soil testing change the schedule
2) Foliar spray should be in early hours or evening and use sticker
3) Alongwith foliar spray, fungicide can be mixed.
4) Because of any reason, if growth of vine or bunch stops then spray 14:14:14, 4 to 5g/litre of water 2 to 3 times every week.

Sl.No.	*Spraying Time*	*Grade*		*Dosage*
		II- April Pruning		
1.	3-6 weeks	6-7 leaf stage	1 Microsole	50 g/AC
2	4-5 weeks	7-8 leaf stage	1 Magnesium nitrate	500 g/Ac
3.	5-7 weeks	8-10 leaf stage	2 Calcium nitrate	500 g/Ac
4.	7-8 weeks	Before pinching	3 Microsole	100 g/Ac
5.	8-10 weeks	After pinching	1 Magnesium nitrate	500/Ac
6.	10-12 weeks	Cane maturity	2 Calcium nitrate	500 g/Ac
		II- October Pruning		
1.	3-4 weeks	Immature bunch of parrot colour	Microsole	50g/Ac
2.	4-5 weeks	Greenish bunch	Magnesium nitrate	500g/Ac
3.	5-7 weeks	Bunch development	Calcium nitrate	500 g/Ac
4.	7-8 weeks	Before girdling	Microsole-B	100g/Ac
5.	8-10 weeks	Berry of Jowar size	Microsole-B	100g/Ac
6.	10-12 weeks	Berry of pea size	Microsole-B	100g/Ac

Nutrient needs at different stages of growth

Sl.No.	*Stage and Days after Pruning*	*Relative Need*		
		N	*P*	*K*
A.	**After Back Pruning**			
i.	Shoot growth period (1-30 days)	High	Medium	Nil
ii.	Fruit bud differentiation (31-60 days)	Low	High	Low
iii.	Fruit bud development (61-120 days)	Nil	Medium	High
iv.	Cane maturity (91-120 days)	Nil	Nil	Nil
B.	**After forward Pruning**			
i.	Shoot growth period (1-40 days)	High	Low	Low
ii.	Flowering and berry set (41-70 days)	Medium	Medium	Medium
iii.	Berry growth (71-110 days)	Low	Low	High
iv.	Berry ripening (111-140 days)	Nil	Nil	Medium
C.	**After Harvest**			
i.	Rest Period (1-15 days)	Low	Low	Low

Fertigation Schedule

Sl No.	*Stage*	*Formulation*	*Dose (Kg/ha/day)*	*Nutrients (kg/ha/day)*		
				N	*P_2O_5*	*K_2O*
A.	**After Back Pruning**					
	1-30 days	20-10-10	1.5	9.0	4.5	4.5
		Urea	1.0	13.8	0.0	0.0
		12-16-0	2.5	9.0	45.8	0.0
	61-90 days	0-0-51	2.5	0.0	0.0	46.0
	91-120 days	13-0-46	2.0	7.8	0.0	13.8
				39.6	50.3	64.3
B.	**After Forward Pruning**					
	1-40 days	19-19-19	2.0	15.2	15.2	15.2
		Urea	1.5	27.6	0.0	0.0
	41-70 days	12-61-0	1.0	3.6	18.3	0.0
		0-52-34	1.0	0.0	15.6	10.2
	71-110 days	13-0-46	2.0	10.4	0.0	36.8
		0-0-51	2.0	0.0	0.0	40.8
	111-130 days	0-0-51	3.0	0.0	0.0	30.6
				56.8	49.1	133.6
C.	**After Harvest**					
	1-15 days	19-19-19		2.9	2.9	2.9
	Total			**99.3**	**102.3**	**200.8**

Required Doses of Fertilizers in Grapes

		N	P_2O_5	K_2O
1.	Required nutrients (kg/ha)	300	500	1000
2.	Through organics (40 per cent)	120	200	400
3.	Through inorganic nutrients in soil (30 per cent)	090	150	300
4.	Through fertigation (20 per cent equivalent to 30 per cent)	90	100	200

Manurial Schedule

A. Organics (200-200-400)

Source	*April-June*				*October-December*			
	Dose	*N*	P_2O_5	K_2O	*Dose*	*N*	P_2O_5	K_2O
FYM	30 t	75	30	180	30 t	75	30	180
Bone meal	1 t	10	120	–	–	–	–	–
Oil cake	–	–	–	–	1.5 t	30	15	45
Total		85	150	180		105	45	225

B. Inorganics (150-150-300)

Source	*April-June (60-75-100)*				*October-December (90-75-200)*			
	Dose	*N*	P_2O_5	K_2O	*Dose*	*N*	P_2O_5	K_2O
Super Phosphate	250	0	40	0	250	0	40	0
DAP	75	13.5	34.5	0	–	–	–	–
Urea	100	46	0	0	120	55	0	0
28-28-0	–	–	–	–	125	35	35	0
SOP	200	0	0	102	400	0	0	204
Total		59.5	74.5	102		90	75	204

A field experiment entitled "Effect of Nitrogen of Growth, Yield and Fruit Quality of Grape (*Vitis vinifera* L.) cv. Thompson Seedless" (Singh *et al.*, 2003) was conducted during two consecutive years *i.e.* 1996 and 1997 at Horticulture Farm, Rajasthan College of Agriculture, Udaipur. Four levels of nitrogen (0.125, 250 and 500 g/vine) were used for treatment. The results indicated that an application of nitrogen @ 500 g/vine proved to be effective in improving shoot length, shoot diameter, number of leaves/shoot, leaf area, number of bunches, size of bunch, yield/vine and yield/ha. The chemical fruit quality parameters such as juice content, acidity were recorded maximum in this treatment but sugar content was minimum at this level and total soluble solids content of fruit was not influenced by the application of nitrogen.

Thakur *et al.* (2006) reported that to improve the quality of Perlette grapes and to enhance its maturity the 'technology', *i.e.*, Flower bud thinning (70 per cent) with brush (1st week of April) + Two GA_3 (40 ppm) dips (1st after fruit set at 4 mm berry size and 2nd six days latter) + Girdling (removing 4 mm ring of bark in between the two GA_3 dips) has been recommended to the grape growers of North India. An experiment was carried out to ascertain the effect of foliar sprays of potassium salts (Potassium Sulphate and Monopotassium Phosphate) on the quality of Perlette grapes. All the treatments, *i.e.*, technology alone or in combination with foliar sprays were on a par as far as fruit yield is concerned, but significantly lesser than control. The bunch length and bunch breadth were not affected significantly by any of the treatments. The highest bunch weight (344 g) was recorded with Technology + monopotassium phosphate (1.5 per cent), it was followed by Technology + monopotassium phosphate (1.0 per cent). The maximum TSS (18.46 per cent) and minimum acidity were recorded with Technology + Potassium Sulphate (1.5 per cent) and it was on a par with Technology + Potassium Sulphate (1.0 per cent) and Technology + monopotassium phosphate (1.5 per cent). Fifty-four days were required from full bloom to harvest in case of grapes with Technology + potassium sulphate (1.5 per cent) and Technology + monopotassium phosphate (1.0 per cent and 1.5 per cent) as compared to 64 days in control.

The influence of foliar applied Polyfeed and Multi-K (KNO_3) on growth, yield and quality of Bangalore Blue grapes were carried out during summer and winter seasons of 1998 at University of Agricultural Sciences, Bangalore. The salient results revealed that, out of ten treatment combinations of Polyfeed and Multi-K, the treatment with 0.2 per cent of 19:19:19 NPK applied thrice at 45, 60 and 75 days after pruning recorded highest growth rate, *i.e.*, production of maximum number of leaves, cane length, number of productive canes and leaf area during both the seasons and production of maximum number of bunches per vine and yield; however, foliar application of 0.1 per cent of 19:19:19 NPK + 0.1 per cent 14:10:34 NPK + 0.1 per cent KNO_3 at different intervals recorded the maximum inter nodal length, cane girth and highest juice content during both the seasons. Vines receiving maximum K in spray form recorded highest total soluble solids (TSS) and total sugars (19.22 °Brix and 19.03 per cent).

Shikhamany and Sharma (2006) stated that a nutritional surveys conducted by the National Research Centre for grapes, Pune, sodium content of petioles of vines on Dogridge rootstock and in Vertisols was more than on own root and in Alfisols. This was in contrast to potassium contents. The average (all the samples pooled together across the soil and stocks) K content ranged from 0.36 per cent to 3.68 per cent with a mean value of 2.15 per cent ±0.76 S.D and the Na content ranged from 0.03 per cent to 1.05 per cent with a mean value of 0.28±0.27 S.D. Vines grafted on Dogridge rootstock showed preference for the absorption of Na, while own rooted Thompson Seedless for K. Both petiole potassium and sodium were positively correlated with yield. Optimum level of petiole K was less (51.45 me/100 g) in Vertisols on rootstock, as compared to Alfisols on own root (82.81 me/100 g). Contrarily, optimum level of Na was more (46.88 me/100 g) on rootstock in Vertisols than on own root in Alfisols (7.4 me/100 g). On the other hand, total content of (K + Na) was almost same on rootstock in Vertisols (95.3 me/100 g) and on own root in Alfisols (96.2 me/100 g) suggesting the substitution of K by Na. Potassium and sodium were positively correlated in Alfisols, while negatively in Vertisols. Antagonism between K and Na revealed that efficiency of applied K would be less when petiole levels of Na are high in Vertisols. Synergism between K and Na was also observed in Alfisols. High sodium content was not found toxic in vineyards under study, as long as it's content of petioles did not exceed 46.88 me/100 g and the Na/K ratio by 0.7. Higher ratios of sodium/potassium above 0.70 were associated with 'Black Leaf', marginal chlorosis and necrosis of leaves and yield reduction. Antagonism between K and Na revealed that efficiency of applied K would be less when petiole levels of Na are high in Vertisols. Higher rates of potash application are recommended in such situation, particularly in Vertisols on Dogridge rootstock to restrict the uptake of sodium and safeguard against these deleterious effects.

Petiole nutrient standards available for Thompson Seedless on own root need not be applicable for the same cultivar grafted on rootstock due to variation in the ionic preference of the roots and nutrient interactions (Sharma and Shikhamany, 2006). With the aim of determining the critical levels and/or optimum ranges of $NO_{3\text{-}}$N, N, P, K, Ca and Mg for Thompson Seedless on Dogridge rootstock, surveys were conducted in 3–5 year old vineyards of Thompson Seedless in Maharashtra. Regression analysis for the relationship of petiole nutrient contents separately at bud differentiation stage and full bloom with yield revealed that the critical nutrient ranges for N, P, K, Ca and Mg at bud differentiation stage were respectively 1.20–1.53, 0.39–0.47, 0.59–0.60, 0.73–1.03 and 0.88–1.28 per cent on dry weight basis. Critical values at full bloom for the same nutrients were, respectively, 1.44–1.80, 0.28–0.36, 1.61–2.95, 0.51–0.81 and 0.58–0.87 per cent.

Field experiment was conducted (Patil *et al.*, 2006) with a view to elucidate the effect of integrated nutrient management on the productivity and quality of Thompson Seedless grapes using Randomized Block Design with seven treatments replicated thrice at the Regional Agricultural Research Station, Bijapur under telephone system for two years from 1998–2000. The treatments consisted of T_1- Recommended dose of fertilizers, *i.e.*, 300:500:1000 kg NPK/ha + Recommended dose of farmyard manure, *i.e.*, 35 t/ha (RDF+RDFYM); T_2- Recommended dose of fertilizers alone (RDF); T_3 Recommended dose of farmyard manure alone (RDFYM); T_4- Recommended dose of fertilizers wherein 50 per cent nitrogen supplied through farmyard manure (RDFF); T_5- Recommended dose of fertilizers wherein 50 per cent nitrogen supplied through vermicompost (RDFV); T_6- Recommended dose of fertilizers wherein 50 per cent nitrogen supplied through green manure (Glyricidia) (RDFG) and T_7- Recommended dose of fertilizers wherein 50 per cent nitrogen supplied through poultry manure (RDFP). Number of bunches per vine was significantly higher in RDFV (45.58) as compared to other treatments, while 40.87, 40.44, 40.07 and 38.37 bunches per vine were found in RDF + RDFYM, RDFG, RDFF and RDFP, respectively, which were on par with each other. Bunch weight was significantly higher in RDF + RDFYM (255.84 g) as compared to other treatments. This was followed by RDFV (235.00 g), RDF (235.00 g), RDFG (228.34 g) and RDFF (225.84 g), which were on par with each other. Berry diameter was more in RDF + RDFYM (13.77 mm), followed by RDFG (13.50 mm), RDFV (13.40 mm), RDFF (13.27 mm), RDFP (13.24 mm) and RDF (13.24 mm) which were on par with each other. Highest berry weight (122.34 g) was recorded in RDFV followed by RDF + RDFYM (121.84 g) and RDFG (117.50 g) and these were on par with each other, but, significantly differed with other treatments. Bunch yield per vine in Thompson Seedless grapes was significantly increased by various INM treatments. Bunch yield was highest in RDFV (10.26 kg) followed by RDF + RDFYM (9.56 kg) and RDFG (8.76 kg), which did not differ significantly with each other, while these were superior over other treatments. RDFV recorded significantly higher yield per hectare (35.61t/ha) as compared to other treatments. The next best treatment was RDF + RDFYM (33.17 t/ha). The pooled data on total soluble solids to acid ratio indicated the superiority of RDFV (40.58) followed by RDFG (40.21). Total soluble solids to acid ratio in other treatments did not differ significantly with each other.

Sharma *et al.* (2006) reported that the fertigation is being followed in Maharashtra but the fertigation schedules followed by the growers are arbitrary due to lack of availability of scientifically worked out fertigation schedules. None of the aspects related to fertigation in grapes have been investigated under Indian conditions particularly for grafted vines. The study was aimed at: i) to compare the efficiency of fertilizers applied through irrigation water (fertigation) with conventional method (direct soil application), ii) to work out the NPK doses required for vines grafted on Dog Ridge rootstock using fertigation and iii) to standardizing the fertigation schedule by applying different doses of NPK at different vine growth stages. The experiment was conducted for the four cropping seasons under double pruning system in drip irrigated vines during the years 2000–2004. The vines received saline irrigation (EC = 1.70–1.80 dS/m). Three different fertigation schedules (S-I, S-II and S-III) and three rates (80, 60 and 40 per cent of the recommended dose for direct soil application method) of NPK were tested along with the conventional method of fertilizer application (direct soil application method). Application of the NPK fertilizers through irrigation at the rate of 40 per cent produced the yield equivalent to 100 per cent NPK directly applied to the soil (recommended dose). Application of higher rates of fertilizers through fertigation did not improve the yield significantly demonstrating that only 40 per cent of the recommended NPK dose was sufficient when applied through irrigation water. In general the fertigation treatments produced higher yields than the conventional method of application. Highest nutrient use efficiency in terms of kg yield/kg of applied nutrient was obtained with fertigation. The average NPK use efficiency was highest under fertigation schedule- I (40 per cent NPK dose). This fertigation treatment on an average resulted in 173.5 per cent increase in fertilizer use efficiency in case of N and K and 173.6 per cent increase in case of P over conventional method of fertilization.

Bhandari and Randhawa (1978) conducted nutritional survey of the apple orchards of Shimla hills. Three cultivars of delicious group of apples were taken for comparison using two recommended leaf sampling procedures. The investigations have revealed the prevalence of Zn, Mo and B deficiencies with marginal levels of Cu. The three cultivars and the study were at par with each other for all micronutrients except B and Mo, where Red Delicious and Stalking Delicious differed significantly with Golden Delicious. However, locations significantly affected leaf contents of all the micronutrients except, Cu.

Bhandari and Sharma (1978) twenty three representative apple orchard of Simla districts were surveyed for their nutritional status. Red delicious, Stalking Delicious and Golden Delicious cultivars were taken for the survey using spur, leaf and midterm shoot leaf sampling procedures, 50, 38, 26, 25, 6 and 1 per cent of the leaf samples fall in deficient to low range for P, Ca, S, K, N and Mg, respectively. Whereas, rest of the samples had adequate to high levels. Red and starking delicious cultivars differed significantly with golden delicious for N and S contents. The locations significantly affected all the nutrient elements, except Ca.

Arora and Sekhon (1981) studied the relationships of DTPA extractable Zn, Cu, Fe and Mn with pH, $CaCO_3$, organic matter, clay and silt have been examined in the samples collected from 23 soils series of Punjab. The correlation studies have indicated that pH, silt and organic matter influence the extraction of these micronutrients with DTPA. The dominant influence of pH on extractable Zn and Mn suggests that DTPA suggests that DTPA-extracts largely the inorganic mineral forms of these nutrients.

Muneshwar Singh and Sekhon (1991) estimated available status of Fe, Mn, Cu and Zn in twenty soil series occurring extensively in intensively cultivated areas of the country with DTPA extract of surface (0-15 cm) soils. Among four micronutrient cations, deficiency of Zn is expected in six soil series and adequate in 14 soil series. Iron, Cu and Mn were found to be adequate of each of the 20 series. Only few (4-5 per cent) samples in Masitawali and Mazodar were low in estimates of available Fe.

Arora *et al.* (1992) collected and analyzed eighty samples of leaves from the middle of non-fruiting shoots from 69 orchards located in five districts of the state at peak harvest stage for Ca, Mg, S, Zn, Cu, Fe and Mn. The results have shown that the concentrations of S and Zn much below their critical levels. While, 58 and 20 per cent of samples respectively were in the deficient range.

Yashoda and Avasthe (1995) revealed that the total micronutrients and DTPA extractable Cu and Mn and hot water extractable B had a significant negative correlation with altitude whereas DTPA-Zn and Fe were positively correlated.

Khan *et al.* (1997) thirty soil samples representing different parent materials of Bangladesh were studied for the distribution of total and DTPA-extractable Zn, Cu, Mn and Fe in the surface horizons. They revealed that the DTPA extractable Zn, Cu, Mn and Fe showed high correlations with their total amounts indicating a genetic relationship of these soils with Zn, Cu, Mn and Fe.

Patiram *et al.* (2000) soil samples of 0-20, 20-40 and 40-60 cm depth and leaf samples from non-bearing twigs were collected from the 32 mandarin orchards of Sikkim to study the micronutrient cation status. They revealed that the orchard soils were sufficient in available Cu, Zn, Fe and Mn and leaf analysis indicated one-third orchards to be low in Zn and Mn, sufficient in Cu and high to excess (80 per cent) in Fe. Available micronutrient cation content decreased with increasing soil depth and did not correlate to their concentration in the mandarin leaf. Soil pH had the negative correlation with available Fe and Mn and positive with Zn and Cu.

Gathala *et al.* (2004) revealed that the status of N, P, S, Fe and Zn in pomegranate orchards of Jaipur district is low. These nutrient elements need to be taken care and application of these would improve the plant nutrition level and hence, the growth and fruit yield. Very low percentage of orchards with low N status indicates that the orchards are in general well supplied with N. It seems that leaf analysis is useful tool for determining status for fertilizer use.

Jackson (1986) found that plants with low berry numbers made more leaf area than those with high numbers. Low berry number, high leaf area, growth rate and high berry sugar and pH levels and low acid levels.

Janaki *et al.* (2002) were conducted an experiment with grape cv. Muscat at Mathapatti (Coimbatore). They revealed that combined application of 2.0 kg B plus 0.2 per cent foliar spray registered a maximum yield of 24.6 t per ha in winter and 26.15 t per ha in summer season. This application also helped to increase the brix and total sugar of berries and reducing acidity.

Monga *et al.* (1990) revealed that the N, P, K levels of 500:750:500 g per vine per year produced significantly more number of bunches (167), fruit yield (53.21 kg/ vine) than that of other treatments.

Srivastava and Soni (1989) reported that application of 600 g each of N and P along with basal dose of 35 kg FYM per vine under Udaipur agro-climatic condition for obtaining early and higher yields of Perlette grapes.

Ahlawat and Yamadagni (1988) observed that the shoot length, number of leaves per shoot and bunch weight Perlette grapes was decreased due to application of K which might be due to toxic effect of chloride ion containing muriate of potash. Application of 500 g N per vine resulted in better sized bunches and yield. Whereas number of bunches and yield remained unaffected by K application.

Dabas and Jindal (1985) found that all the concentrations (0.1, 0.2 and 0.3 per cent) of boric acid and magnesium sulphate increased grape fruit buds over control. None of the treatments except 0.3 per cent $MgSO_4$ would significantly reduce the bud mortality. Almost all the treatments increased TSS and reducing sugars while decreased acidity except in higher concentrations.

Sindhu *et al.* (1999) observed that the maximum bunch weight, yield and total soluble solids of fruit were found in vines which received 1.5 and 0.5 per cent urea and zinc sulphate spray, respectively. Acidity decreased with increase in concentrations of urea and zinc sulphate spray.

Cheema *et al.* (2003) observed that berry thinning was more effective than cluster thinning in increasing berry size, total soluble solids and reduction in juice acidity and extent of short berries in the clusters. Exposing the clusters to direct sunlight through canopy management had some effective improving the fruit quality.

Rana and Sharma (1999) revealed that the growth, yield, weight of cluster, volume of cluster, weight of berry, volume of berry, juice percentage, specific gravity of juice and total soluble solids were increased with 0.05 per cent of $FeSO_4$ spray. However, pH of juice acidity, total sugars, reducing sugars, non-reducing sugars and sugar/ acid ratio were not affected by iron sprays.

Bhatia *et al.* (2001) conducted an experiment at Hissar (Haryana) by which the studies carried out on guava in which K_2SO_4 (0.5, 1.0 and 1.5 per cent), $ZnSO_4$ (0.5,

0.75 and 1.0 per cent), H_3BO_3 (0.3, 0.5 and 1.0 per cent) and water (control) were prayed on the trees during winter season crops at the stage when the fruit was of walnut size and second after 15 days. All the nutrients increased fruit weight and the yield which was maximum (73.0 kg/tree) with H_3BO3 1.0 per cent. The total soluble solids and sugars were more with H_3BO_3 followed by K_2SO_4. Ascorbic acid was found maximum with K_2SO_4 (182 mg/100 g).

Rajpal Singh *et al.* (2001) investigated at an experimental orchard of Department of Horticulture, Haryana Agricultural University (Hissar). An aqueous solution of different chemicals *viz.*, urea (1, 2 and 3 per cent), $ZnSO_4$ (0.25, 0.5 and 1.5 per cent), NAA (10, 20 and 40 ppm) and ascorbic acid (10, 25 and 50 ppm) were applied as foliar spray of ber plant. They revealed that the foliar spray of urea and $ZnSO_4$ significantly increased the N and Zn contents and decreased the P and K content of ber leaves. P and K contents increased significantly with K_2SO_4 spray.

Pauline *et al.* (2004) revealed that the application of micronutrients B (0.1 per cent), Fe (0.1 per cent) and Zn (0.2 per cent) as foliar spray on papaya cv. Ranchi did not show any significant variations in the two years experiment on the reducing sugar and b-carotene content of the fruits. The total and reducing sugar were found maximum with the application of Fe in both the years; however, in general, non-reducing sugar was recorded highest with the application of B + Zn. The acidity content of fruit did not show any set trend with the micronutrient application.

Surender Kumar *et al.* (2004) observed that the bunch of grapes cv. Delight was applied with aqueous solution of cycocel, GA and kinetin at berry set stage. Among different growth regulators, cycocel 4000 ppm and 5000 ppm extended the shelf life of bunches and reduced physiological loss in weight. However, other quality parameters such as total soluble solids and acidity were not significantly affected by any of the growth regulator treatments.

Singh and Gupta (2000) conducted an experiment at two different locations of guava orchards; the soil was clay loam and sandy loam with the pH of 7.75 to 8.00. And were collected soil samples (0 – 150 cm) to evaluate the micro and secondary nutrients. It is concluded that 2 to 3 spray of 0.25 per cent $ZnSO_4$ with 2.5 per cent urea and 0.5 per cent iron sulphate are necessary for immediate relief, soil application of $ZnSO_4$, $FeSO_4$, gypsum and also manure is essential to check decline in fruit bearing capacity of guava trees.

Malik *et al.* (2000) found the maximum number of Kinnow fruits and yield per tree which received 1.5 and 0.4 per cent urea and zinc sulphate spray, respectively. The maximum of juice and total solids were found in trees sprayed with 1.0 and 0.8 per cent urea and $ZnSO_4$, respectively. Ascorbic acid content however decreased with foliar spray of urea and $ZnSO_4$.

Hooda *et al.* (1990) reported that the growth interns of number of leaves and vine height decreased at soil saturation salinity (ECe) of 6 dS m^{-1}. The number of shoots per vine however remained unaffected. At high soil salinity (10 dS/m) the percentage reduction in number of leaves and number of shoots per vine in various cultivars (Perlette, Delight and Beauty seedless) ranged from 79 to 89 and 85 to 92 per cent, respectively. These cultivars remained unaffected upto soil salinity values of 4 dS m^{-1}.

Walker *et al.* (1997) reported that the leaf area and priming wood was reduced as salinity increased, but these parameters were greater in Sultana and Ramseyet of grape varieties in all salinity levels (0.43, 1.7 and 3.4 dS/m) because of higher inherent capacity for biomass production.

Boora *et al.* (2002) carried out a field experiment to study the influence of NPK as growth, yield and quality of sapota cv. Cricket Ball indicated significant increase in trunle girth, leaf area with application of 500 g N per tree gave maximum yield (12.33 kg/tree) with highest fruit weight (63.98 g). A combined application of 800 g N, 400 g P_2O_5 and 200 g K_2O per tree per year resulted in the highest fruit yield (13.82 kg/tree), quality parameters *i.e.*, TSS, TSS/acid ratio and total sugar were improved by higher levels of N. The response of P and K was found non-significant in all the parameters.

Chokha *et al.* (2002) concluded an experiment at ICAR, New Delhi to study the effect of foliar application of macro and micronutrient and their combination on yield characters and fruit quality of grape cv. Perlette. They revealed that the nutrient treatment improved average fruit weight per bunch significantly over the control though Mg (0.02 per cent), Fe (0.2 per cent) and B (0.4 per cent) were experimentally better than others and also observed that no unripe/shoot, berries and total sugar, acidity, juice and tannin content was also better for micro and macro nutrient sprayed vines. Thereby, these micro and macro nutrients improving the fruit yield and quality of cv. Perlette.

Gill *et al.* (1987) reported that the highest (0.73 per cent) acidity was recorded in berries from bunches treated with 30 ppm GA at 50 per cent flowering in Thompson Seedless, while the lowest acidity was recorded in berries from bunches treated with 50 ppm at berry set stage. Al-Dujaili *et al.* (1987) showed that 75 ppm GA applied at full bloom in Thompson Seedless resulted increase in total sugars.

Ram and Bose (2000) conducted an experiment at Kalimpong, the soils of the Mandarin orchard was sandy loam with a pH of 5.6 to 5.8 and contained total available N (0.18 per cent), P_2O_5 (10 kg/ha) and K_2O (79 kg/ha). They revealed that effect of Mg and micronutrients was significant on plant height, stem girth and spread of canopy and the effect of these nutrients were more pronounced when applied in combinations. The Mg + Cu + Zn showed the maximum increase in height, stem girth and spread of the plant. The treatment with Mg and micronutrients except Cu resulted in higher Mandarin fruit yield than the control.

Sinha *et al.* (1999) carried out an experiment under black soils in the horticultural garden (Rahuri, Birsa Agricultural University). The effect of plant growth regulators (2, 4-D, GA3) and micronutrients (zinc and boron) was studied on the fruit set drop, retention and cracking of litchi. Higher levels of micronutrients (0.8 per cent zinc and 0.4 per cent boron) along with 10 ppm 2, 4-D and GA3 showed marked effect on fruit retention. This treatment also helped in reducing fruit drop and cracking of litchi.

Kunwar and Sinha (1987) conducted a survey of three grapes (*Vitis vinifera* L.) cv. Himrod, Pshuyler white and Perlette grown at eight different sites in Srinagar valley of UP hills to assess their micronutrient status and fruit quality. The results showed that Pschuler white had highest N (1.14 per cent), K (2.96 per cent), Ca (0.80 per cent) and total sugar (12.12 per cent) contents where P (0.51 per cent), Mg (0.39 per cent) and TSS (16.120 brix) was noted in Perlette but acid attributes of fruit (Acidity 1.89 per cent) in Himrod. Significant negative correlations were observed between P and Ca (r = -0.504) but significant positive correlations between N and TSS (r = +0.463).

Awasthi *et al.* (1985) conducted a survey of the mandarin orchards of Nagpur area in Himachal Pradesh to assess the macro and micronutrient status of the trees and the soil for foliar diagnosis. The observations revealed that the declining condition of the mandarin orchards was associated with the low N, K and Zn status of the foliage as well as low K and medium to high P status of the soil.

Ahlawat *et al.* (1990) collected soil and leaf samples from 20 ber orchards of 6 to 20 years of age with average growth and productivity. The soil samples of 75 per cent orchards were deficient in organic carbon whereas 90 per cent with respect to phosphorus. However, potassium concentration which ranged from 2.12 to 3.55, 0.12 to 0.20 and 0.94 to 1.62 per cent, respectively.

Sohan Singh *et al.* (1985) studied the removal of nitrogen, phosphorus and potassium by four year old lower trained grapevines of cv. Perlette at village Jassian (Ludhiana). The removal of these three elements was observed to be 28.50, 3.34 and 10.00 kg per acre, respectively for a crop load of 11 tonnes. The ratio of recommended doses of fertilizer and the removal by the vine is 12:1, 100:1 and 25:1 for nitrogen, phosphorus and potassium, respectively.

Morris *et al.* (1987) studied in commercial vineyards near Althus, Arkanas. All vines were own rooted and grown in a fine sandy loam. They revealed that the application of K fertilizer in addition to that already recommended for the vineyard did not affect the yield except for the Gewurztraminer cultivar. Excessive K fertilization and cluster thinning tended to increase petiole and fruit K and fruit pH in must cultivars. The per cent soluble solids showed little response to the treatments except in the tendency for cluster thinning to slightly improve soluble solids in Carbernet Sauvignon and the two French American hybrid cultivars. The leaf analysis data based on the critical nutrient contents helps in reducing the complexity in making recommendations. Petiole NO_3-N, total N, P, K, Ca and Mg contents of 0.26, 1.54, 0.32, 2.24, 0.96 and 0.32 per cent on dry weight basis were found to be critical with yield determination of 6.6, 10.1, 5.7, 19.5, 4.9 and 11.4 per cent, respectively (Shikhamany *et al.*, 1985).

Neilsen *et al.* (1987) reported that petiole N and P concentrations were increased by their respective fertilizer additions. In general, soluble solids and titrable activity in Foch grape fruits were not consistently affected by N, P, K fertilization, although juice pH was increased in 2 to 3 years at the highest rate of N and K fertilization and in first year at the highest rate of P fertilization in the British Columbia region.

Bhargava and Chadha (1988) reported that nutritional surveys are more useful to obtaining the effects of applied doses on yield and quality and in developing a leaf nutrient guide than through experiments conducted under controlled conditions. For an ideal nutrient guide, the following five ranges for each element can be considered for classification of nutritional status of the vines namely deficient, low, optimum, high and excessive interms of yield and fruit quality responses.

Shikhamany *et al.* (1989) observed that higher N application enhanced Ca, Mg and Mn status of the vines by reduced the concentration of NO_3-N, total N, P and K in the petioles. Higher level of applied potash was associated with increased contents of NO_3-N, K and Ca, but with reduced contents of Mg and Fe. A complementary effect of the levels of N and K_2O was observed on the petiole Fe content in Thompson Seedless grapes under Bangalore conditions.

Brar *et al.* (1979) revealed that the absorption of N was highest in Anab-e-Shahi and lowest in Thompson Seedless. While Himrod was most efficient in absorbing P. Thompson Seedless absorbed highest Zn as compared to other cultivars. There was no significant difference in the absorption of K, Fe, Mn and Cu among the different cultivars.

Ahlawat *et al.* (1982) investigated the absorption pattern of various cultivars and found that absorption of N was highest in seedless while Round and Madeleine angevine and lowest in Perlette. Himrod was most efficient in P absorption. Maximum calcium and magnesium content was found in Perlette and Banqui Abyad. The effect of fertigation with 128 kg per ha of nitrogen of various fertilizer sources containing 0, 50 or 100 per cent ammonium ions in apple cv. Anna, it was observed that fertilizer source containing nitrate only (0 per cent NH_4^+) increased yield by 15 per cent and growth by 10 to 14 per cent as compared with fertilizer containing 50 to 100 per cent NH_4 (Klein and Spieler, 1987).

Bravdo and Hepner (1987) reported that NO_3 nitrogen of soil content was significantly affected by fertilizer treatments and its concentration increased with depth and distance from the emitter. The K content of the soil was significantly increased by K fertilization, but contrary to the NO_3-N distribution, K was mainly concentrated near the emitters. The level of K was very low especially in the non-fertilized plots in grape vines.

Nutrient applications were found in general to increase the growth of vines which was observed under black soil with pH of 8.21 (Lavania and Singh, 1977). High N status was associated with increased growth and it was further increased when either P or K was added to N.

Khandagale (1977) reported 300 g N to be optimum for three year old vines spaced at 3.0 m × 1.5 m under the condition of Rahuri (Maharashtra) with shallow black soil. The computed yield was maximum with 0.65 kg N, 1.22 kg P_2O_5 and 0.91 kg K_2O per vine. For Himrod cultivar of grape the optimum combination was 0.4 kg N, 1.2 kg P_2O_5 and 0.8 kg K_2O for a mature vine spaced at 3.0 × 3.0 m^2 (Sanghavi and Nijjar, 1978).

Under conditions of Bangalore for Anab-e-Shahi vines of less than five years age a combination of 500 kg N, 125 kg P_2O_5 and 750 kg K_2O per ha per year was found to be optimum. However, the optimum combination for the vines of more than five years of age was 500 kg N, 500 kg P_2O_5 and 1000 kg K_2O per ha per year (Anon., 2004).

Thompson Seedless under Ludhiana (Punjab) conditions (Nijjar and Singh, 1979). The optimum levels of N, P and K proposed by Gaikwad *et al.* (1981) for Thompson seedless under Rahuri (Maharashtra) conditions were 1.91 to 2.04 per cent N, 0.32 to 0.40 per cent P and 2.30 to 2.62 per cent K. The critical levels of nitrate nitrogen, total nitrogen, potassium and magnesium in the petioles of Thompson seedless were 0.263, 1.54, 2.24 and 0.325 per cent, respectively (Shikhamany *et al.*, 1982).

Maharkar and Patil (1984) observed the vineyard best performance to have leaf petiole contents of N, P, K, Ca, Mg, B, Zn and Fe to be 1.44, 0.4, 1.48, 2.06 and 0.65 per cent and 42, 59 and 189 ppm, respectively under conditions of Maharashtra. Petiole samples were collected at flowering stage will determine nutrition for quality rather than yield (Bhargava and Sumner, 1987).

Rana and Chandel (2003) conducted a trial to study the effect of biofertilizer (*Azotobacter* and *Azospirillum*) and nitrogen levels alone and in combinations on growth, yield and fruit quality of strawberry cv. Chandhler. They observed that maximum plant height, number of leaves per plant, leaf area, fruit size and fruit weight, TSS and total sugar content with 100 kg N per ha. The maximum plant height, yield, fruit size and weight were recorded when the plant inoculated with *Azotobacter* along with 60 kg N per ha. However, the maximum TSS content was observed with *Azotobacter* inoculation along with 80 kg N per ha.

Yesh Pal Singh *et al.* (2003) conducted an experiment in high density orchard of mango cv. Dashehari under Pantnagar soil and climatic condition. They revealed the effect of zinc (30 kg/ha $ZnSO_4$), boron (20 kg/ha borax) and copper (10 kg/ha $CuSO_4$) markedly increased the yield and quality characters but combined application of Zn + B and Zn + B + Cu was found most effective in increasing the yield and physico-chemical characters of fruits.

Naresh Babu *et al.* (2004) observed that the effect of different doses of nitrogen (60, 120, 180, 240 and 300 g/plant/crop) applied in single dose (2 months after planting), two split doses (2 and 4 MAP), 3 split doses (2, 4 and 6 MAP), 4 split doses (2, 4, 6 and 8 MAP) and 5 split doses (2, 4, 6, 8 and 10 MAP) were studied on sandy loam soil. Application of 240 g N per banana plant out yielded all other nitrogen doses. Significantly, higher pseudostem height (214.63 cm) and girth (59.95 cm), number of leaves per plant (10.95), number of fingers per bunch (76.32) and chemical characteristics interms of total sugar (15.36 per cent) and sugar acid ratio (34.13) were also observed with this level of nitrogen. Significant improvements in growth, yield and quality parameters except ascorbic acid were recorded on case of four split applications as compared to single application.

Saraswathy *et al.* (2004) indicated that the sapota trees give with $ZnSO_4$ 50 g and borax 25 g per tree (soil) and two foliar sprays of zinc sulphate 0.5 per cent and borax 0.3 per cent during 50 per cent flowering and pea stage of fruit development in addition to the recommended dose of organic manures and N, P and K registered the highest value for growth, yield and quality parameters. The study also revealed that application of borax or zinc sulphate singly or in combination gave remarkable increase in the yield over the control. The same trend was also noticed for growth and quality parameters.

Singh *et al.* (2005) observed that the foliar sprays of plant bio-regulators and nutrients to the full bearing apple trees of cv. Starking Delicious at 25 to 30 per cent bloom resulted significant differences in the physical and chemical characteristics of fruits. The treatment of GA3 + kinetin (0.5 + 0.5 ppm) proved most effective for increasing fruit diameter, total soluble solids and colour (70.30 mm, 10.570 brix and 0.21 OD, respectively) whereas for other quality parameters the mixure of plant bioregulators and nutrients in various combinations was found superior.

Rattanpal *et al.* (2005) investigated the potassium nitrate (2.5 per cent and 5 per cent) and potassium sulphate (2.2 per cent and 4.4 per cent) alone and in combination with 2, 4-D (20 ppm) were sprayed 30, 45, 60 and 75 days after full bloom on six year old Kinnow tress on their effect on physical parameters of fruit was observed. The texture of the experimental soil was loamy sand having pH of 8.2 to 8.4 with low organic carbon and ECe. The K availability in top 15 cm soil was 153 kg per ha. They were concluded that KNO_3 5.0 per cent + 2, 4-D 20 ppm sprayed at 60 days after full bloom proved best with maximum increase in fruit size.

Dudi *et al.* (2005) conducted a trial at farmers field on 12 year old Kinnow orchard to study the influence of four levels each of nitrogen (N) (0, 375, 750 and 125 g/plant) and FYM (0, 50, 100 and 150 kg/plant) on fruit quality and yield. Individual application of 750 g N and 100 to 150 kg of FYM per plant effectively increased the total soluble solids, sugars, ascorbic acid, juice content and fruit yield, however, the highest level of N resulted reduced quality of fruits. Rind thickness also increased with higher levels of applied N, which is undesirable. Application of nitrogen and FYM in combination improved the quality as well as the yield of fruits.

Sharma *et al.* (2005) revealed that the application of phosphorus upto 400 g per plant and 200 g zinc sulphate per plant increased the stem girth, plant height, plant spread, shoot length and leaf area of Ber cv. Umran. Maximum plant height (5.64 m) and leaf area (30.2 cm^2) were recorded from the plant receiving 400 g P_2O_5, which were significantly higher than other treatments of phosphorus except 600 g P_2O_5 per plant. The plant supplied with 600 g P_2O_5 produced maximum shoot length (131.3 cm) and increase in stem girth (6.06 per cent). For different growth parameters, the effect of 200 g and 300 g zinc sulphate was at par but better than control as well as 100 g zinc sulphate.

Bhobia *et al.* (2005) conducted an experiment at Hisar during winter season on seven year old trees of guava hybrid cv. Hisar Surkha. The soil of experimental field was sandy loam having pH 8.0, EC 0.24 dS m^{-1}, OC 0.39 per cent and available N, P_2O_5 and K_2O 180, 58 and 370 kg per ha. They recorded plant growth interms of height and spread was maximum with the application of 800 g nitrogen through inorganic source, whereas, fruit yield (85 mg/plant) was found maximum with the application of 40 per cent nitrogen through inorganic (urea) + 60 per cent N through organic (FYM) source. The leaf nitrogen was highest when 100 per cent nitrogen was supplied through inorganic, whereas P and K contents were maximum when whole nitrogen was added organically. Leaf composition of 1.76 per cent N, 0.20 per cent P and 0.87 per cent K was associated with highest yield.

Tanjeet Singh Chahal and Bal (2005) conducted an investigation at Amritsar. The soil of experimental Cape gooseberry orchard was sandy loam, the chemical status found was pH 8.4, EC <0.8, total N 0.28 per cent, available P 16 kg per ha, available K 175 kg per ha and OC 0.40 to 0.75 per cent. They revealed the N and P individually at 15 g per plant and potassium at 10 g per plant to be the most efficacious treatments. Whereas, the combination of N, P and K at level 10, 10 and 5 g per plant, respectively was the most judicious treatment for improving the plant growth characters and maximizing the fruit yield.

Sharma and Sindhu (2005) studied the effect of potassium salts superimposed on thinning and GA3 and girdling in combination on quality and maturity of grapes cv. Perlette, K_2SO_4 at fruit set and various stages super imposed on recommended technology resulted in 2.4 per cent and 4.2 per cent more TSS over recommended technology and control, respectively. Recommended technology advanced the maturity period by six days over control, KNO_3 delayed the maturity by 2 to 3 days over K_2SO_4.

Himanshu Kumar Singh *et al.* (2005) conducted on experiment comprising three fertigation treatments from drip and conventional application of N fertilizer and irrigation as control at Rajendra Agricultural University (Bihar) to study the effect of fertigation on papaya cv. Pusa Delicious. The soil texture was sandy loam; climate is humid and subtropical climate. They observed that the quality of fruits in terms of length, circumference and weight under fertigation treatment was significantly superior to conventional application of fertilizers and irrigation. Maximum yield of papaya was obtained with treatment of drip irrigation with 947 litre water per plant per year + 100 per cent recommended dose of N fertilizer this also gave 43.0 per cent higher yield in comparison to control.

Ravi Kher *et al.* (2005) revealed that the GA3 90 ppm was found to be the most effective in increasing the weight, specific gravity and decrease the total acid content of guava fruits. Maximum firmness was obtained with CCC 600 ppm, however, it was at par with GA3 90 ppm. Maximum value for TSS, total sugars, non-reducing sugar and TSS acid ratio with NAA 60 ppm, whereas maximum reducing sugar and ascorbic content were recorded with NAA 50 ppm and GA3 60 ppm, respectively.

Sindhu *et al.* (2002) conducted a study on eight years old Perlette grape vines. The soil of the field was sandy loam in texture, having pH 7.9 EC 2.0 mmhos/cm, OC 0.65 per cent and N 180 kg, P 18.3 kg and K 273 kg/ha. They showed that the bunch weight, bunch size and acidity of fruit increased whereas total soluble solids decreased with increasing doses of N application from 92 to 908 g N per vine. The increase in the application of P increased the acidity and decreased the TSS, however bunch weight, size, berry weight and berry size increased upto 908 g K_2O per vine increased bunch weight, bunch size, berry weight, berry size and TSS while it decreased acidity. Sugars generally accumulated and acidity decreased with the advancement towards ripening.

Beattie and Forshey (1954) estimated the correlation between N and fruit yield as 0.329. They conducted that a petiole N level of 0.77 per cent in mid July (one month after bloom) was required for concord grape yield of 3.5 t per acre.

Seeliger and French (1971) reported that when P was broadcasted as superphosphate gave excessively high available P levels in top 6" soil but yield responses were delayed for 8 years until the slow moving P reached the 6" to 12" soil.

Wasnik and Bhargava (1981) revealed that continuous use of acid forming nitrogenous fertilizers were found to be reducing the soil pH from 7.2 to 3.9 during the experimentation period of five years.

Srivastava and Soni (1989) reported that application of 600 g each of N and P along with basal dose of 30 kg FYM per vine under Udaipur agro-climatic condition for obtaining early and higher yields of Perlette grapes.

Singram and Prabhu (2001) revealed that the foliar application of $ZnSO_4$ 0.5 per cent + borax 0.2 per cent combination applied twice during vegetative and full bloom stages excelled others increasing the shoot length and number of internodes, number of leaves per shoot and quality characters with highest juice per cent, total soluble solids (TSS), total sugars, sugar/acid ratio and lowest acidity.

Sanjay Kumar and Pathak (1992) noticed that spraying of 1.0 per cent urea proved most effective in improving the size and weight of bunches (325 g/bunch). The highest yield (8.45 kg/vine) was recorded with application of urea followed by 0.4 per cent foliar spray of $ZnSO_4$ (8.3 kg/vine). Further, the best quality berries were also obtained with 0.2 per cent $ZnSO_4$ application.

Prabhu and Singaram (2001) revealed that the combined foliar application of $ZnSO_4$ @ 0.5 per cent and borax @ 0.2 per cent increased weight of cluster (245 g/cluster), volume of cluster (173 ml), weight of berry (3.01 g) and quality characters with higher juice per cent, TSS, total sugars and lower acidity in grape fruit.

Beniwal *et al.* (1992) revealed that foliar spray of urea improved berry size, bunch weight and yield of Perlette grapes, 1.5 per cent urea was found to be effective for increasing bunch weight, bunch volume, berry characteristics and yield. Spray of 0.5 per cent urea and 1.0 per cent potassium sulphate proved effective in increasing total soluble solids and reducing sugars. Acidity increased with urea spray whereas potassium sulphate decreased acidity.

Suleman Mohammed *et al.* (1993) showed that the application of K improved bunch weight, colour of the berries and yield, medium level of K (400 g/vine) and lowest number of canes (50/vine) improved the quality characters.

Sindhu *et al.* (2002) showed that the bunch weight and size and acidity of fruit increased whereas TSS decreased with increasing the dose of N application from 92 to 908 g N per vine. The increasing application of P increased the acidity and decreased the TSS and K application increased bunch weight, bunch size, berry weight, berry size and TSS while it decreased acidity.

The *Azotobacter* is an aerobic, heterotrophic, asymboiotic free living nitrogen fixing bacteria, isolated and described by Beijerinck (1901). In addition to fixing nitrogen asymbiotically, it is also known to produce plant growth hormones and fungistatic substances. This organism grows well in nitrogen free medium utilizing molecular nitrogen for its cell protein synthesis. The dead cells on subsequent mineralization contribute towards the nitrogen availability of plants.

Different N free media are used for the isolation, cultivation and maintenance of *Azotobacter* with different carbon sources. Brown and Burlingham (1968) suggested the use of starch medium as a selective medium for isolating *Azotobacter*. Other media which could be used for the isolation of this organism are Waksman No. 77, Ashby's mannitol phosphate agar medium, Jensen's medium and Burk's nitrogen free medium.

The ability of the members of the family *Azotobacteracae* to grow in a medium with no combined nitrogen helps in their selective isolation. An organic carbon source and phosphate are usually the minimum nutrients required for the development of *Azotobacter* under natural conditions (Becking, 1981). Using this as a base, a number of simple enrichment procedures were designed for the successful isolation. The nutrient solution method originally is a good and well established method to enrich and isolate *Azotobacter chrococcum* from soil and water samples. Soil paste method (Becking, 1981) and sieved soil plate method using silica gel or good quality agar are the few other methods involved in the isolation of these organisms.

Reinhold *et al.* (1987) isolated and described *A. halopreferens* from roots of kallar grasses grown on salt affected soils of Pakistan. This species showed some adaptation to salinity and showed remarkable growth and nitrogen fixation at 40°C with 0.25 per cent salt in the growth medium. A *Pectinolytic azospirilla* was isolated from rice roots by Khammas *et al.* (1989) in Iraq and named as *A. irakense*.

Agarwal and Tilak (1988) studied the establishment of *Azospirillum brasilense* in the roots of *Eleusine coracana* and reported the establishment of bacteria endobiotically in roots, mainly in the cortical intracellular space and vascular tissue. *Azospirillum* from the roots of wheat cultivars was isolated identified and as *Azospirillum brasilense* and *A. lipoferum* and further reported that *A. brasilense* as the most potent nitrogen fixing species. Jaskowska (1995) investigated the occurrence of bacteria in the rhizosphere of rye, wheat, barley and corn and found 51 per cent of the samples contained *Azospirillum*. Mahesh Kumar (1997) isolated *Azospirillum* from the roots of *Bombusa vulgaris* and *Dindrocalamus strictus* grown on different soils.

Investigations were carried out on the isolation, identification, genetic diversity of *Azospirillum* and their inoculation effect on the growth and flowering of *Gallardia* by Gadagi (2000) who reported that *Azospirillum* as one of the common inhabitant of ornamental plants.

Vasanthakumar (2003) isolated ten *Azospirillum lipoferum* and nine *A. brasilense* strains from the endorhizosphere of solanaceous crops and studied their beneficial traits like N – fixation, P-solubilization and production of plant growth promoting substances.

The standard kjeldahl analysis gives an estimate of nitrogen fixation. The fixed nitrogen is estimated by the amount of nitrogen obtained in a nitrogen free liquid medium by growing these N2 fixing PGPMs are also known to produce growth promoting substances (Brown and Burlingham, 1968) and fungistatic substances which render these organisms to screen for efficiency.

The inoculation of bacteria, bacterization on to the plant seeds and roots, is reported to enhance plant growth (Sakthivel and Gnanamanickam, 1986).

Tamibar and Okon (1993) reported tryptophan conversion to indole-3-acetic acid via indole-3-acetamide in *A. brasilense* sp. 7 and the associate soil bacteria of the genus *Azospirillum* are known to promote plant growth, probably via secretion of phytohormones including indole-3-acetic acid. Auxin production in two phosphate solubilizing *Pseudomonas* sp. cultures isolated from the rhizosphere of wheat and rye cultivated in liquid medium.Highest auxin amounts produced being 0.29 and 1.89 mg IAA equivalent per litre culture medium (Leinhos, 1994).

Prathibha *et al.* (1995) observed 2.4 to 18.23 mg N per g of carbon source utilized by *Azospirillum* strains from cotton genotypes cultivated in northern parts of Karnataka. Similarly,

Maheshkumar (1997) isolated six *Azospirillum* strains from the roots of bamboo plants and found them to fix 15.6 to 22.4 mg per g carbon source utilized.

In the experiment conducted by Tamilvendan and Purushothaman (1996), *Azospirillum lipoferum* and *Azospirillum brasilense* showed nitrogen fixation in the range of 7.54 to 24.53 mg of nitrogen per gram of malic acid after seven days at 28°C under static conditions. Of the 88 *Azospirillum* isolates, 55 per cent were identified as *A. lipoferum* and 41.57 per cent as *A. brasilense.* These represented isolates from the rhizosphere/endorhizosphere of different ornamental plants. The nitrogen fixing capacity of these isolates ranged from 1.4 to 20.54 mg N per g of malic acid.

Among the free living nitrogen fixing bacteria *Azospirillum* is considered to be more efficient with nitrogenase properties comparatively better than the other nitrogen fixers. Nitrogen fixation is the first major mechanism for the enhancement of plant growth by *Azospirillum* (Prasad and Govindarajan, 2001).

Sivaprasad *et al.* (2003) observed 11 to 21 mg of N fixed per gram malate utilized by 50 *Azospirillum* strains associated with rice plants in acid sulphate soils in Kerala. Among several diazotrophic bacteria isolated from diverse crop plants, *Azospirillum* isolates showed highest amount of N fixation ranging from 2.55 to 22.4 mg per gram of 'C' source added. The amount of N fixed by *Beijerinckia* isolates found to vary between 2.55 to 4.74 mg per gram of 'C' source used.

Rao and Sinha (1963) isolated some bacterial and fungal cultures from the rhizopshere of wheat and tested in culture media. Soil inoculation with pure cultures of organisms increased the acid extractable P_2O_5. Under sterile and non-sterile conditions some of the organisms were identified as *Bacillus megaterium, Esherichia fruendi* and *Bacillus circulans.*

Phosphate solubilizing microorganisms belonging to the genera *Bacillus, Pseudomonas, Citrobacter, Enterobacter, Rohnella* and *Serratia* have been isolated in the recent years from the soils and rhizosphere of crop plants by various workers (Halder *et al.*, 1990, Krishnaraj and Gowda, 1990, Illmer and Schinner, 1992, Leinhos and Vacek, 1993, Kim *et al.*, 1998 and Maheshkumar, 1997).

In the recent years, Gaind and Gaur (1991) isolated *Bacillus megaterium, B. brevis, B. circulans* and *B. subtilis* from the rhizosphere of oat and arhar. Surange and Kumar (1993) isolated phosphate solubilizing bacteria from legume root nodules. Solubilization of rockphosphate in non-sterile soil by *Rhizobium meloloti* strain TAL 1372 was found to be effective.

Illmer and Schinner (1992) isolated *Pseudomonas* sp. and *Penicillium* sp. from forest soils and found them to solubilize high amounts of insoluble inorganic phosphates. Narsimen *et al.* (1994) reported that *Aspergillus aculaetus* was best at solubilizing tricalcium phosphate (94 per cent), dicalcium phosphate (54.5 per cent) and aluminium phosphate (31.8 per cent). However, *Aspergillus niger* was best in solubilizing the iron phosphate (36.8 per cent) and rock phosphate (56.6 per cent).

Singh and Kapoor (1994) obtained 27 isolates of bacteria capable of producing clear zone of solubilization on tricalcium phosphate. Isolates were obtained from rhizosphere and non-rhizosphere soils under various crops in Hissar, India. Solubilization of phosphate increased with increase in incubation period upto 15 days.

Nahas *et al.* (1996) found that the fungal isolates were more active than bacteria in solubilizing insoluble phosphate. Secondly, the fungi presented more acid phosphatase activity than the bacteria and conversely the bacteria produced more alkaline phosphatase activity than the fungi.

Fluorescent pseudomonads are well recognized as plant growth promoting rhizobacteria, P-solubilizers and as biocontrol agents against plant pathogens. They are rapid and aggressive root colonizers of cereals, pulses, oilseeds and vegetables (Johri *et al.*, 1997).

Gupta *et al.* (2002) described *Pseudomonas* species as a potent phosphate solubilizer while they were developing heavy metal resistant mutants of phosphate solubilizing *Pseudomonas* NBRI 4014.

Neelam and Meenu (2003) reported high tricalcium phosphate solubilizing ability of *Pseudomonas* sp. (TP2) isolated from rhizosphere of *Trigonella*.

The genus *Pseudomonas* encompasses arguably the most diverse and ecologically significant group of bacteria on the planet. They were included under gram negative, rod shaped, chemoheterotrophic bacteria with polar flagella and are characterized by the yellow green iron chelating low molecular weight siderophores called pyoverdines or pseudobacterin that floresce under UV light (Suryakala *et al.*, 2004). They are common inhabitants of rhizosphere and are the most studied group within the genus *Pseudomonads*. They can be visually distinguished from the other Pseudomonas species of soil has the ability to produce water soluble yellow green pigments. They comprise of *Pseudomonads aeroginosa*, the type species of the genus, *P. aureofaciens*, *P. chlororaphis*, *P. fluorescens*, *P. putida* and the plant pathogenic species *P. cichorii* and *P. syringae* (Dwivedi and Johri, 2003).

Several soil microorganisms including phosphate solubilizers and nitrogen fixers are known to produce plant growth promoting substances (PGPS). Barea *et al.* (1976) isolated 80 phosphate solubilizing bacteria from rhizosphere of crop plants out of which 43 produced IAA, 29 produced gibberellins, 45 produced cytokinin like substances and 20 of them were able to synthesize all the three types of harmones. Production of IAA and GA to a considerable extent by P solubilizing *Bacillus polymyxa*, *Bacillus pulvifaciens* and *Pseudomonas striata* was reported by Sattar and Gaur (1987). P-solubilizing bacteria isolated from the rhizosphere of wheat and rye plants produced auxin type of PGPS, when they were grown in liquid culture medium supplemented with tryptophan (Leinhos and Vacek, 1994). The amount of auxins in the culture filterates varied from 0.01 to 3.98 mg IAA equivalent per litre of culture medium.

The species of fluorescent Pseudomonads are again grouped into different biovars and subgroups based on similarity in biochemical tests (Barett *et al.*, 1986). Thus, rapid identification of potentially and economically viable bioagents is possible through various methods of biochemical characterization.

Sattar and Gaur (1987) reported phosphate dissolving bacteria *Bacillus polymyxa* (H5), *B. polymyxa* (P1), *B. pulvifaciens* (R1), *B. pulvifaciens* (R5), *Pseudomonas striata* (27) and the fungus *Aspergillus awamori* (F18) producing auxins and gibberellins to an appreciable extent and the results suggested that beneficial effects of these organisms may also be due to production of growth promoting substances besides phosphate solubilization. GA are a class of phytohormones most commonly associated with modifying plant morphology by the extension of plant tissue, particularly stem tissue (Salisbury, 1994).

Veena (1999) recorded IAA and GA production by P-solubilizing *Enterobacter,Xanthomonas* and *Pseudomonas* isolated from rhizosphere of sorghum plants. Geeta (2001) studied twenty eight P-solubilizing bacterial strains for the production of PGPS. The production of IAA among the strains varied from 3.61-35.45 µg/25 ml of broth and that of GA varied from 1.33 - 13.27 µg/25 ml of broth.

Khalid *et al.* (2003) isolated thirty bacterial isolates from rhizosphere of wheat, grown at different sites and evaluated for their potential to produce auxins *in vitro*. Calorimetric analysis showed the production of auxins ranging from 1.1 to 12.1 mg per L. Further, the amendment of L-tryptophan to the culture medium stimulated more auxin synthesis which ranged from 1.8 to 24.8 mg/l.

Rashmi (2004) involving P-solubilizing *Serratia* isolates showed production of IAA in the range of 4.10 to 28.08 µg per 25 ml broth and that of GA in the range of 1.35 to 8.6 µg per 25 ml of broth.

Suneesh (2004) reported that all the 48 fluorescent *Pseudomonas* isolated from the moist deciduous forests produced IAA and GA in the range of 1.63 to 17.00 µg per 25 ml of broth and 0.72 to 5.27 µg per 25 ml of broth respectively.

Glick (1995) reported that *Pseudomonas* make up a dominant population in soil and rhizosphere and exert growth promoting influence on a variety of plant species on account of their strong competitive behaviour, colonization potential and sustainability. Seed bacterization with these organisms has emerged as a powerful technology to enhance plant growth and yield, besides providing protection against diseases.

Harris and Adkins (1999) reported that they have inoculated combination of fungal and bacterial isolates for biological control of damping off diseases caused by *Rhizoctonia solani*, which was shown 90 per cent control.

Sonawane and Konde (2000) studied the effect of diazatrophic and VAM fungi on grape vines and reported that there was significant increase in leaf area and early sprouting of the cuttings due to *Azotobacter chrococcum* inoculation.

Mello (2001) reported that compatible strains of *Pseudomonas fluorescens* suppresses *Fusarium* wilt of radish and strain combination was more efficient than single strain inoculation for controlling the *Colletotrichum orbiculare*, P. *syringae* pv. *Lachrymans* and *Erwinia tracheiphila* in cucumber.

Jisha (2002) reported the efficiency of combined inoculation of *Pseudomonas fluorescence* and *Trichoderma* in mixtures for plant growth promotion and disease suppression in different spice crops.

Tripathi and Johri (2002) studied the biocontrol potential of Fluorescent pseudomonads that are isolated and recovered from rhizoplane and rhizosphere of pea and wheat using *in vitro* inhibition of *Colletotrichum dematium, Rhizoctonia solani* and *Sclerotium rolfsii* and *in vivo* biocontorl action in maize against sheath blight caused by *Rhizoctonia solani*. Some isolates were having multiple disease control potential while some others showed biocontrol potential against a specific pathogen. This study indicated that the *Fluorescent Pseudomonads* are diverse with respect to their biocontrol potential.

Rajan (2003) reported that foot rot of black pepper (P*iper nigrum* L.) caused by P*hytophthora capsici* is a serious disease of the crop. The use of several antagonists like *T.harzianum, T. virens* and VAM fungi were studied and evaluated the effective biological disease suppression.

de Boer and Vander Sluis (2004) proved the inoculation the strains of *Pseudomonas* sp. in combination on the radish results in enhancement of suppression of *Fusarium* wilt.

Norisawa (2004) reported that the effect of fungus *i.e., Heberoconium chaebospira* which had significantly increased the suppression of clubroot and Verticillium yellowish wilt in chinese cabbage in the field.

Suryakala *et al.* (2004) isolated *Pseudomonas fluorescens* from rhizosphere of rice, wheat, pigeonpea, groundnut and chilli crops. The isolates were found to produce siderophores when grown in amino acid medium under iron deficiency. These siderophores were antagonistic to fungal pathogens like *Fusarium oxysporum, Alternaria* sp. and *Colletotrichum capsicii*. In a study conducted on the suppression of Verticillium wilt, eight isolates of *Pseudomonas fluorescens* and *Pseudomonas putida* obtained from roots of olive plants were tested in nursery. All the tested isolates were effectively controlling the *Verticillium* wilt under controlled conditions. All tested bacteria produced the green fluorescent siderophore pseudobactin *in vitro* while some *P. fluorescens* isolates produced either salicylic acid in succinate medium or HCN *in vitro* assays. The antagonistic activity of *P. fluorescens* and *P. putida* isolates against defoliating (D) and non-defoliating (ND), *V. dahliae* pathotypes varied with culture media. On PDA isolates of *P. putida* were more inhibitory to the pathogen than those of *P. fluorescens. in vitro* evaluation of 28 isolates of *P. fluorescens* against principal pathogens of leaf rot of coconut showed PS1 culture possessed excellent antagonistic potential against *Colletrotrichum gloeosporiodes, Fusarium solani* and *Exserohilum rostratum* with per cent growth inhibition of 74.86, 77.01 and 76.09, respectively in King's B medium (Joseph *et al.*, 2003).

Earnapalli *et al.* (2005) isolated several rhizobacteria, screened and selected promising strains effective against *A. solani* and worked on the biocontrol potential of ten efficient antagonistic bacteria against early blight of tomato in a glasshouse. Results revealed that seven strains possessed high biocontrol potential. Out of these, *Pseudomonas* strain B-25 was found to be the most promising strain with a disease control of 57.78 per cent and good plant growth promotion.

A number of bacteria including *Azotobacter, Azospirillum, Pseudomonas, Acetobacter, Burkholderia* and *Bacillus* can be considered as plant growth promoting rhizobacteria (Brown, 1974, Kloepper *et al.*, 1988, Bashan and Levanony, 1990. Plant growth promoting rhizobacteria can effect plant growth either directly or indirectly. There are several ways in which different plant growth promoting rhizobacteria have been reported to directly facilitate the proliferation of their plant hosts. They can fix atmospheric nitrogen and supply it to plants, synthesize siderophores that sequester iron from the soil and provide it to plants, synthesize several different phytoharmones that can act to enhance various stages of plant growth, they have mechanisms for the solubilization of minerals such as P and make it readily available for plant uptake and they may synthesize some low molecular weight compounds or enzymes that can modulate plant growth and development (Brown, 1974 and Glick *et al.*, 1997a and 1994b).

Among the nitrogen fixing bacteria *Azotobacter* and *Azospirillum* are considered to be more efficient with nitrogenase properties comparatively better than the other nitrogen fixers. Field trials conducted in many places with different crops have revealed that biological nitrogen fixation by root associated *Azospirillum* contribute significant amounts of nitrogen to the plants thereby saving valuable nitrogenous fertilizers. It has been demonstrated that *Azospirillum* inoculated plants take up minerals (N, P and K) from solution at faster rates than controls and consequently inoculated plants in the field, accumulate more dry matter, N, P and K in their stems and leaves (Okon, 1985). The results also revealed that *Azospirillum* inoculation benefits plant growth and increases yield of crops by improving root development, mineral uptake and plant water relationship.

Significant yield increase was observed in tomato plants with the inoculation of *Azospirillum* spp. (Bashan *et al.*, 1995). Inoculation of *Azospirillum* sp. to wetland rice under acidic condition improved shoot growth, straw yield and N uptake (Govindan and Bagyaraj, 1995). Salomone and Dobereiner (1996) observed significant increase in grain yield of maize on inoculation with *Azospirillum* spp. Hemawathi (1997) observed improved plant growth and flower yield of chrysanthemum on inoculation with *Azospirillum* spp. over the uninoculated control.

Maheshkumar *et al.* (1997) reported improved seedling growth of bamboo due to inoculation of *Azospirillum* sp. Increased plant weight, grain and fodder yield and savings of 25 per cent N fertilizer in *rabi* sorghum due to inoculation of *Azospirillum* sp. was reported Alagawadi and Krishnaraj (1998) reported increased grain yield in sorghum due to inoculation of *Azospirillum* strains ACD-15 and ACD-20.

Dobereiner *et al.* (2001) observed a significant increase in plant dry weight due to inoculation of wheat plants with *A. brasilense* SP-245 and *A. irakense* KBC1 in Belgium.

Kim *et al.* (1998) studied the interaction of phosphate solubilizing bacteria and VAM on plant growth, soil microbial activities and the production of organic acids and revealed that a significantly higher soluble phosphorus concentration was observed in treatment E (*Enterobacter agglomerans*) and E + G (*Glomus etanicatum*) on day 55 compared with control. The highest N and P uptake was observed in extract treated seeds and were subjected to accelerated ageing. Growth and yield were influenced by seed treatment. Plants from seed treated with disodium phosphate exhibited the best growth, dry weight and yield for tomatoes and brinjal.

Gaikwad *et al.* (1999) reported that there was significant increase in all the growth parameters, dry matter, N, P uptake, fruit yield of brinjal due to inoculation with phosphate solubilizing microorganisms along with either single superphosphate or rock phosphate. Kulkarni *et al.* (2000) reported that when chickpea seed was treated with Mussoorie rock phosphate and phosphate solubilizing bacteria, increased the seed and grain yield attributes.

Sapatnekar *et al.* (2001a) conducted a field experiment on greengram (cv. S-8). Among all the treatment combinations phosphate solubilizers (*Trichoderma viride, Aspergillus niger, Chaetomium funicola* and *Aspergillus fumigatus*) along with 100 per cent SSP proved to be the best treatment regarding maximum nodulation (150), phosphorus uptake (27.3 mg) per plant, P availability (28.8 kg/ha) in soil and grain yield (18.4 q/ha).

Sapatnekar *et al.* (2001b) conducted a field experiment on wheat (HD-2189) to study the effects of phosphate solubilizers with different phosphate sources and their levels on phosphorus uptake and yield of wheat. Among the two phosphate solubilizing culture, the fungal, culture *viz., Aspergillus awamori* has proved superior to bacterial culture *Pseudomonas* sp. 1. Among the three phosphate sources used, single superphosphate was superior over rock phosphate and bone meal. The treatment which received 100 per cent SSP + *Aspergillus awamori* recorded maximum phosphorus uptake by plant (14.59 mg), phosphorus content in grain (7.68 mg) and yield (42.5 q/ha). An application of phosphate solubilizers along with phosphate sources was found beneficial over phosphorus sources alone.

A significant increase in yield of cotton was recorded by Prathiba *et al.* (1994) due to combined inoculation of *Azospirillum* and *P. striata* as compared to single inoculation. A field experiment conducted by Shivakrishnaswamy *et al.* (2001) to study the efficiency of biofertilizers as single and dual inoculation of nitrogen fixing (*Azospirillum* ACD 15 and ACD 20) and phosphate solubilizing (*Bacillus polymyxa* and *Burkholderia cepacia*) bacteria on the performance of sugarcane growth and yield showed increased plant height, tiller number, leaf and stem dry matter at different stages of crop growth resulting in an additional cane yield contribution over single inoculations and uninoculated control. Combined use of *Azospirillum* and Phosphorbacteria as seed treatment in pearl millet could save application of 50 per cent of nitrogen and phosphorus fertilizers. The dual inoculation of *Azospirillum* and phosphate solubilizing bacterium (PSB-7) recorded the maximum fruit yield in tomato plants (Vasanth Kumar, 2003). Combined inoculation of *Azospirillum lipoferum* and *Pseudomonas fluorescens* gave 20 per cent increase in rice yield over uninoculated control in the presence of recommended dose of N and P fertilizers.

Savalgi *et al.* (1994) recorded higher nodulation due to dual inoculation with *Azospirillum brasilense* strains and *Rhizobium* spp. strain NC-92 than *Rhizobium* alone in groundnut under red loam soil condition and among *Azospirillum* isolates ACR-8 (local isolate) showed better performance than SP-7.

Tarafdar and Marschner (1995) observed that dual inoculation with *Aspergillus fumigatus* and *Glomus mosseae* enhanced biomass production and nutrient uptake in wheat (*Triticum aestivum*) supplied with organic phosphorus as Na-phytate.

Galal (1997) reported that dual inoculation with *Bradyrhizobium japanicum* and *Azospirillum brasilense* improved the crop growth and biological nitrogen fixation of soybean over single inoculation.

Sunaina and Ajay (2005) have studied the efficacy of dual inoculation PGPRs strains in improving the growth of potato under field conditions. Seed potato were treated with 10 cfu per ml bacterial suspension for 20 min and plated. Out of four PGPR strains tested, *B. subtilis* B5 (PRB) was most superior in all the growth parameters and gave the highest tuber yield as compared to only 392 g with control.

Combined inoculation of *A. chrorococcum, B. polymyxa* and *P. striata* showed highest grain yield and nitrogen and phosphorus uptake than single inoculations in both sterilized and unsterilized soils (Kundu and Gaur, 1980).

El-Mukkadam *et al.* (1989) observed improved nodule branching due to combined inoculation of *Azospirillum lipoferum, Azotobacter chrococcum* and *Rhizobium* spp. as compared to dual inoculation treatment.

Belimov *et al.* (1995) found that the mixed cultures of *Azospirillum lipoferum* 137, *Arthrobacter mysorens* 7 and phosphate solubilizing *Agrobacterium radiobacter* 10 produced a significant positive effect on the grain yield of barley. They suggested that the combined inoculation of all the three organisms provided a more balanced nutrition for plants and the improvement in root uptake of N and P was a major mechanism of interaction between plants and bacteria.

In a pot culture trial conducted by Veena (1999) reported maximum plant growth, biomass yield and nutrient uptake in sorghum in the treatment receiving consortia of eight organisms comprising bacteria, fungi, actinomycete which was almost equivalent to application of 75 to 100 per cent recommended dose of chemical fertilizer s. Combined inoculation of four beneficial organisms was also found superior over single, dual or triple inoculation of beneficial organisms. Field experiment conducted by Devananda (2000) reported that maximum plant growth, yield and nutrient uptake in pigeonpea was obtained in the combined inoculation of *Rhizobium, Azospirillum* and *P. striata* (PGPRs).

Veena *et al.* (2002) have studied the influence of consortia of beneficial rhizosphere microorganism like nitrogen fixers, phosphoate solubilizers and major groups of general micrflora isolated from the rhizosphere of sorghum plants in comparison with single, dual, triple and multiple inoculations of most efficient organisms as well as with different levels of NPK fertilizers on growth and nutrient uptake of sorghum. The results had revealed that increasing the complexity of consortia with more number of beneficial organisms enhanced the growth, biomass and nutrient uptake of sorghum plants significantly which was most equivalent to application of 75 to 100 per cent recommended dose of chemical fertilizers.

Salakinkop *et al.* (2003) reported that combined inoculation of *Azospirillum brasilense,* VAM (*Glomus fasciculatum*) and P-solubilizing *Pseudomonas striata* and *Serratia* spp. increased the plant height, top root length, stem girth and weight of coffee seedlings at all stages of growth as compared to uninoculated.

9

Use of Hormones in Grape Cultivation

By Using Hormone

1. Grape yield can be increased
2. Grape quality can be improved
3. Grape keeping quality and storage life can be increased

Different Hormones

1. Ethrel
2. GA (Gibberlic acid)
3. CCC (Lihocin)
4. NAA (Planofix)
5. Brassinosteoride (Combine)
6. 2-Chloro-4 Pyridil-N- Phenyl Urea (CPPU)
7. 6 BA (Kinin)

While Using Hormone Understand the following

1. Understand grape bunch establishment
2. Quality hormone should be used

3. Understand the effect of hormones
4. Use correct dosage of hormone

For Grape Production Use of Hormones

Entire world uses hormones for getting higher yield in grape. In Thompson seedless unless we use hormones optimum grape production is not possible. After understanding it the use of hormones is begun. The yield is increased but quality is hampered and other problems may also be encountered. There will be chemical residue in fruits and after consumption it will affect the human body. Alongwith hormones the use of pesticide also affects the human health. In some places, after thorough examination in laboratory its results have revealed bad affliction on human health. Our grape consumers are increasing day-by-day and as also the European market is enlarging, there the consumers are worried about the effect of hormones on health.

On the one side the grape purchasers are rich and from the other side to fulfill the consumer's desire, the responsibility lies on growers. Under such a situation, we have to search a way to use the hormones to increase the yield to meet the demand.

Grape Production and Different Types of Hormones

At present almost all the grape growers use Gibberlic acid throughout the world. Chloromaquate, 4-CPA, 6BA, Ethrel, NAA, its usage helps in increasing the grape bunch size. The use of combine is also getting a momentum.

Crop Hormones Means What?

Crop hormone is a special type of organic chemical. It is effective and in a very short time the change can be observed with the food getting ready in a very short time mixed with crop hormones and growth regulators.

Types of Hormones

Auxins, Gibberlins, Cytokinins, growth, retardants, and Ethylene, likewise 5 types of hormones are used in grape cultivation.

Why to Use Hormones

1. To maintain vigour of vine
2. To increase or decrease the berry length.
3. To get bunch colour
4. For longer shelf life.
5. To enhance the leaf life
6. In fruits, sugar, juice, acids, seeds and their quantity there is quality modification.
7. To enhance the development of berry
8. To delay in ripening
9. For defoliation.

All these qualities are affected by hormones which help in increasing the yield of grape.

When to Use the Hormones

A. From April Pruning (hard pruning) to October Pruning: When the sprouts start growing, in order to initiate for proper food manufacturing, lot of money has to spent, then only the development of canes and the establishment of flower primordial takes place.

   Secondly, when the canes are growing, the flower primordia start moving towards the terminal end. Then the canes become barren. For growth of canes and branches either we have to increase or decrease the use of hormones like, CCC (Lihocin). They should be sprayed at 6-7 leaf stage, at 10-12 leaf stage and at 17-18 leaf stage. Likewise three times we have to use the hormones. Everytime taking into consideration the growth one has to decide the concentration of above said hormones. Everytime 500 ppm second time 700 ppm and third time 1000 ppm is needed to be applied sometimes due to hard pruning; if water availability is a problem then the growth will be slow but in rainy season the growth will be fast. Under such situation only one spray of 1000 ppm CCC is advisable.

B. Establishment of flower primordia and hormonal requirement: In our climate after April pruning, within one to 1½ months the flower primordia establishment starts but after one and half month it will be slow. After 4-5 buds in front of canes for 100 per cent flower primordia establishment we can use hormones.

For Perfect Establishment Use Uracil

Uracil is one type of hormone which is used for flower primordia establishment. In our climate after April pruning when the new growth starts, the flower primordia establishment takes place and that period starts from 30-35 day after April pruning and it continues for another 30-35 days and as long as the growth continues the flowers primordia also emerges.

For Grape Productivity

1. New sprout should be 30-35 days old for use of hormones.
2. This sprouting should have 6-10 buds.
3. Grape bud is a compound bud in which 2-3 buds comprises of leaf, tendril and bunch.

For our purpose 1 to 3 buds in new sprout of 30-35 days old is ideal with a compound bud. Once the primordia are established, it will last long and they develop and for that purpose uracil is required.

Use of Uracil

1. Spray uracil on new growth when it attains 7-9 leaves; means total leaves of uniform area

2. Second; uracil spray on 10-12 leaves on branches.
3. Third; spray when the growth stops; means 16-20 leaves on branches. (This work should be completed before pinching of branches)
4. If we are going for sub-cane, then uracil spray should be taken up at 6-7 leaf stage. When you pinch the cane then 6 or 7 leaves develop. Third time application is carried out when the second pinching is done. One gram of uracil is mixed in 3 litres of water, then add 17 litres of water, totally 20 litres of water means it will be 40 ppm. For one acre, first spray requires about 150 litre and it requires 7.5 g uracil./Ac, For second spray 10 g uracil/ Ac. means it is to be applied in 200 litres of water. For third spray, 250 litre of water is mixed with 12.5g of uracil. So totally for 3 sprays 600 litres of solution means 30 gram of uracil is necessary.

C. Establishment of flower Primordia and its period is important (cane maturity to development): There is a possibility of change in climate and so there is change in soil moisture, occurrence of pest and diseases which affect the establishment of flower primordia. If the sprouting is early or the growth is checked or the flower primordia melts, or the primordia moves towards terminal end then for better flower pirmordia, establishment of canes should be fully matured. For cane maturity C: N ratio should be optimum for leaf colour, size and their number which decides the size of flower primodia. Some time Anthracnose (Karapa) on quick growing branches affects the primordia than in axil, growth won't come under such situation. We have to see that the primordia are established by means of application 6 BA. This hormone contains Kinetin which can be dissolved in Ispropyl Alcohol with approx 5 ppm of two sprays are required. For one acre 2g and 6 BA is sufficient. Totally under favourable situation, lihocin uracil and 6 BA; two or three times it is sprayed for getting double yield.

D. *Control of post-emergence aerial roots on vine*: Even though watering is done, it is not available for plants, under such situations aerial roots appear on plant body, main trunk or stem. More aerial roots affect the hormonal balance on the canes, on buds hairy growth appears which increases the bud dormancy. In this condition after October pruning, the bud burst will be delayed; even application of Dormex becomes useless for that purpose. Before October pruning NAA of 20 ppm is recommended or per Ac 100 ml Plano-fix in 200-225 litre of water is sprayed. When there are no aerial roots before October pruning these sprays should be taken with NAA application @ 20ppm 3-4 week as a foliar spray.

From October Pruning to Harvesting

1. Before October pruning
2. While doing pruning – first step
3. After October pruning – second step
4. After October pruning – third step.

Before harvesting of grape: Take up 15 days before NAA-10 ppm or planofix spray which helps in reducing berry shattering/berry drop.

Other Hormonal Use

1. Ethrel

For defoliation Ethrel @ 1000 ppm to 2500 ppm is recommended. When pruning is to attend to without leaf removal then Ethrel is used after pruning for defoliation. Application of ethrel for defoliation results in accumulation of ethylene in vine but in coloured varieties like kishmish charani (sharad seedless), flame seedless, joyti seedless it results in colour development during ripening.

Second time application of Ethrel is needed when the grape ripening starts and it helps in case of coloured varieties for proper and uniform colour development. To apply ethrel @ 500 ppm, 200 litres quantity per acre is required. When the coloured varieties are to be sent for early marketing and then for sugar development, ethrel is useful. For varieties like kishmish charni, flame seedless, muscat, gulabi, ethrel is useful. In market Ethrel is available in different names– Ethrel, Ethopan, Thril, florel, Sefa, Cinon. The chemical name is 2- Flora, Ethyl Phosphonic acid. It dissolves early in water and it is available in a concentration of 39 per cent.

Ethrel 250 ml/100 litre of water – 1000 ppm

Ethrel 500 ml/100 litre of water – 2000 ppm

Ethrel 125 ml/100 litre of water – 500 ppm

Ethrel 625 ml/100 litre of water – 2500 ppm

2. Gibberlic Acid

Gibberlic acid is available to plants as Gibberlic which is helpful for cell elongation, bunch thinning, berry elongation and rachis elongation. After October pruning, bunch will be soft and at that time it is used. At the stages of small leaves and bunch of parrot colour, which turns to green colour, 10 ppm of GA has to be used on bunches which helps in elongation of bunch; at times 20 ppm GA has to be used for more than 20 ppm GA by which process the rachis, after elongation, turns upward or it leads to bending, or the bunch near the stalk breaks and falls down. The GA has to be used 25-30 days after October pruning, if used early then less results are expected. Gibberlic acid for second time has to be used before bunch flowering starts capfall. At this stage GA helps in elongation of berry stalk, swelling of berries and thinning helps in providing space for development of berries, thereby bunch will not be compact. All the berries should be uniform in size with excess shedding leads to loosen of bunch weight. For berry elongation, third time of GA spray is used. Within bunch to increase berry length and swelling, GA @ 25 to 40 ppm during 75 to 100 per cent flowering is recommended. If less cane of GA is applied, the berries become round in shape.

The diameter and length of berries should be in 1:2 proportion. When 60-75 per cent capfall-25 ppm, GA at 70-80 per cent capfall-30 ppm, 80-90 per cent capfall - 35 ppm, 90-100 per cent capfall 40-ppm, GA should be used.

How to Prepare GA Solution

One gram of GA is dissolved in 50 ml. acetone and add it to one litre of water it will then be one litre of stock solution.

1. 1 gm GA + 1 litre stock solution + 99 litres of water = 100 litre = 10 ppm
2. 1 g GA + 1 litre stock solution + 49 litres of water = 50 litre = 20 ppm
3. 1 g GA + 1 litre stock solution + 39 litres of water = 40 litre = 25 ppm
4. 1 gm GA + 1 litre of stock solution + 19 litres of water = 20 litre = 50 ppm
5. 1 gm GA + 2 litres of stock solution + 19 litre of water = 20 litre= 50 ppm
6. 1 gm GA + one litre stock solution + 9 litres of water = 10 litres = 100 ppm
7. 1 g GA + one litre stock solution-1000 ppm

3. Chloromaquate (Lithocin)

Chloromaquate, Cyocel, CCC and Lihocin are the synonyms available in the market. This is growth-retardant hormone and is widely used in grape cultivation. It is widely used, when rain occurs, new growth comes very fast and the growth in front of bunch will be very fast. In slow growth of the bunch, under such situation to check the growth, use of Lihocin is recommended. After October pruning when the bunch and bunch growth is soft Lihocin should be used. New growth at this stage is very fast with the availability of metabolite and continuously by growth with 5th and 6th bud bunch that comes out.

The very purpose of CCC is to divert the metabolites towards the bunch for proper growth by hindering the flow of sap towards the growing tip, which results in improvement of bunch growth.

How to Use CCC

After October pruning, when 5th, 6th buds are soft that means when its colour is yellowish or parrot colour CCC is sprayed because growth conditions are not same on all bunch of vines and when CCC is used at new growing points, growth gets slowed down and bunch develops. Food manufacturing power of leaves increases. When the metabolites move towards growing points it stops, the manufacturing power of new leaf increases this way and the metabolites moves towards the bunch in double quantity.

How Much to Use CCC

If used at proper time then 250 ppm is sufficient. Initially low dosage of 200 ppm, two days before 125 ppm, this process is delayed, then 375 to 450 ppm of third spray is required. However more than 500 ppm of foliar spray is harmful.

How to Prepare CCC

This is available at 50 per cent concentration, which is easily dissolved in water. Method of preparation of 100 litres of different concentration of CCC is as given below:

1. 100 ml Lihocon + 100 litres of water = 500 ppm

2. 50 ml Lihocon + 100 litres of water = 250 ppm
3. 40 ml Lihocon + 100 litres of water = 200 ppm
4. 75 ml Lihocon + 100 litres of water = 375 ppm
5. 60 ml Lihocon + 100 litres of water = 300 ppm
6. 90 ml Lihocon + 100 litres of water = 450 ppm

Precaution While Using CCC

1. While using CCC the soil should be wet.
2. While doing foliar spray, from leaves to growing point, from bunch to growing tip-the spraying is recommended.
3. On sprout when 6-7 leaves are available and later when 5-6 leaves are there, it is more effective.
4. If Anthracnose is expected then mix carbendizim alongwith fungicide and spray.
5. If flea beetle is observed then mix insecticide alongwith Lihocin and spray.

Note

1) If irrigation and Nitrogen are more, then control of disease and pest and will be difficult so after two days a second time spray.
2) If by mistake higher dosage is used, then in near future it will have bad effect.
3) Apply immediately 25 kg urea and irrigate fully.
4) Immediately 200 gm urea in 100 litres of water is to be mixed and take up foliar spray.
5) Take 10 ppm GA foliar spray.

4. 6BA

This comes under Kinetin or Kinin group and almost all farmers are using 6 BA. It is used for cell division, which further leads to berry development. Leaf is required for food manufacturing. It helps in increasing bunch weight and development of sugars. As soon as the girdling is over, 6 BA @ 5 to 7 ppm has to be sprayed. After GA spray if girdling is delayed then 6BA be sprayed. If the girdling effect is not visible immediately, then 10-12 ppm 6 BA spray is required.

Applications of 6 BA's are needed in two sprays. First 20 ppm and second one < 20 ppm. 6 BA has to be dissolved in isopropyl alcohol or in ammonium chloride. For berry elongation if GA application is delayed then the GA 10 ppm + 6 BA 10 ppm can also be taken as foliar spray.

1. For berry elongation, within 3-4 days, GA 40 ppm + 6BA 4 ppm is needed when bunch is nipped at bottom.
2. 8-10 ppm 6 BA spray on leaf and bunch.

5. NAA (Planofix)

Naphthalene acetic acid is utilized for specific purpose to prevent bunch drop or to stop small bunch melting with 20 ppm NAA 3 spray. At fruit set NAA application helps in getting good natural bloom. Before harvest to reduce the berry drop, you can prolong the harvest period for a week. 100 ml of Plano fix per acre is sufficient. If the crop load is optimum then other hormones are used in equal concentration then the Plano fix spray is not necessary. If you are expecting > 12.5 tones/Ac then NAA application for proper yield is useful.

6. Combine (Brassinosteoride)

It is a Brassinosteoride based hormone. It helps in cell elongation and cell division at quicker rate. It also helps to stand the adverse climatic situation.

1. Grape berry development.
2. It helps in grape pedicel thickness, so it helps in reduction of pre and post berry drop.
3. Berry and bunch weight increases.
4. Grape storage life enhances, keeping quality improvement during transit.
5. Uniform grape quality helps in export.

How to Use Combine

It should be used alongwith Gibberlic acid. It has to be used when berries have set. After fruit set the GA and combine should be in equal proportion, then again repeat and you will get good results. So 2-3 mixed spray of GA and combine are required.

Combine+GA ratio for Foliar Spray

1. At fruit set (before bajra-size berry) combine 100 milli + 3 g GA = 100 litre of water = 30 ppm
2. One week after dipping, second dipping has to be done. At this stage combine 100 milli + 4g GA = 100 litres of water (40 ppm) or after first dipping with 10 days gap combine 100 mill + 5 g GA = 100 litre of water (50 ppm)

Note

1) In a combine, GA totally dissolves, so dissolving GA in acetone is not necessary.

2) In readymade solution (sticking agent) for every 100 litres, apply 25 milli apsa -80 or any sticker.

6. CPPU

CPPU helps in cell elongation. It also helps in speeding the process of increase in berry diameter. CPPU along with GA or alone can be used. CPPU is used at a concentration of 1 to 2 ppm. If used at more than 2 ppm then the shape of berry gets spoilt and also there will be cracking, CPPU if used at early stage, will have effect on colour in the later stage. CPPU is used between fruit and a 10 mm berry size and used 3-4 times because during the period GA is used 2-3 times.

How Much to be Used/Acre

1. CCC

250 ppm/Ac 150 to 200 litres means 250 milli Lihocin in 200 litres of water or CCC 250 milli or chloromaquate ml in 200 litres.

CPPU + GA Dosage for Use

Name of Hormone	*Bunch Size/Size of Berry/Dosage*				
	4 mm	*6 mm*	*8 mm*	*10 mm*	*Total*
GA	30 ppm	40 ppm	50 ppm	60 ppm	180
CPPU	2 ppm	2 ppm	2 ppm	2 ppm	8
GA	10 ppm	50 ppm	50 ppm	50 ppm	160
CPPU	1.5 ppm	2 ppm	2.5 ppm	3 ppm	9
GA	35 ppm	40 ppm	50 ppm	50 ppm	170
CPPU	2.0 ppm	2 ppm	2 ppm	–	6

Or 5 ppm CPPU, 30 to 40 ppm GA together can be used. At present CPPU 0.1 per cent conc. solution is available, for that the following schedule is to be followed:

0.1 per cent CPPU Solution	*Already Prepared GA*	*CPPU (ppm) Milli*
100	100 litre	1
200	100 litre	2
300	100 litre	3
400	100 litre	4
500	100 litre	5

Use of Hormone

Sl.No.	*After October Pruning*	*Bunch Stage, Leaf Stage*	*Conc. of Hormone*
1.	Approximately 3 weeks	6-7 leaves bunch, parrot colour	CCC 200 to 250 ppm
2.	Approximately 4 weeks	Before bunch development	GA 15 to 20 ppm
3.	Approximate 5 weeks	Bunch development	GA 20 to 25 ppm
4.	Approximate 6 weeks	Bunch development	GA 25 to 30 ppm + 6 BA 5-10 ppm
5.	Approximate 7 weeks	Full bunch development	GA 25 to 30 ppm + combine
6.	Approximate 8 weeks	Full fruit set	GA 40 to 50 ppm + combine
7.	Approximate 12 weeks	Bunch ripening	NAA 10-20 ppm
8.	Approximate 14 weeks	2-3 weeks before harvest	15 to 25 ppm NAA

Sl.No.	Bunch Stage	Name of hormone	Conce of hormones
1	Early pre-bloom	Lihocin/CCC	200 to 250 ppm
2	Pre-bloom	Gibberlic acid	10-15 ppm
3	Pre-bloom	Gibberlic acid	15-20 ppm
4	50 per cent flowering	GA + 6 BA	20 to 25 ppm + 2 to 2.5 ppm
5	75 per cent flowering	GA + 6 BA	25 to 30 ppm + 2.0 to 2.5 ppm
6	Immediately after fruit set	Gibberlic acid + combine	30 to 40 ppm
7	Berry of jowar grain size	GA + combine	40 to 50 ppm

Main Purpose of Growth Regulators

At the present situation for export of grape good production is the main aim. On an average 12 tonnes per acre is easily expected, but 15 tonnes per acre is required. In grape, the quality means the length, TSS, colour and these three aspects are very important.

The berry shape means length and diameter should be same or they should be in equal proportion. A berry diameter of 18 mm and a length of 24 mm is possible by using growth regulator and alongwith that a sugar percentage of 20 is necessary. Once the bunch is ready in white varieties it will be milky/greenish but in coloured varieties the colour should be dark and uniform. Alongwith that the climate, the vine care, the bunch load, nutrition–all these points one has to take into consideration.

Approximately 3 weeks after October pruning one has to understand the bunch situation and while using hormone the sticker Apsa-80 25ml/100 litres will give good results of the hormone effects.

Ramteke *et al.* (2006) opined that the quality of grapes is determined by berry colour, size and pulp content. To achieve these quality parameters, the growers use various bio-regulators. Indiscriminate use of CPPU, one of the commonly used bio-regulator to increase berry size, leads to abnormal growth of the pedicel and sometimes causes berry scorching. The present experiment was, therefore, conducted to study the effect of CPPU on bunch and berry development of Thompson Seedless grafted on Dogridge rootstock. CPPU was used at 1 and 2 ppm on three levels of shoot length, *i.e.*, 3 to 6, 7 to 10 and >10 leaves above the bunch. The average area of leaf was 180–200 cm^2. The data on berry and bunch characters differed significantly except TSS and acidity of the berries. The application of 2 ppm CPPU on shoots with >10 leaves above the bunch showed the superiority over the rest of the treatments. The treatment of application of 1 ppm CPPU on >10 leaves above the bunch was also on par with the treatment of 2 ppm CPPU on >10 leaves above the bunch. Therefore, it is concluded that to achieve berry quality by using CPPU the number of leaves available per bunch (overall leaf area of the vine) should be at least 10 or more above the bunch. No abnormalities due to application of CPPU were found in any of the treatments.

The investigation was undertaken with a view to study the effect of cytokinins and brassinosteroid with GA_3 for yield and quality attributes and to standardize the application of these plant growth regulators at AICRP on grapes, MPKV, Rahuri during 2004–2005. The experiment was laid out in Randomized Block Design (RBD) with 3 replications and 10 treatments. After October pruning vines and bunches at specific stage of development were treated with respective plant growth regulators as per the treatments, *i.e.*, GA_3, brassinosteroid, CPPU, and benzyladenine. Application of GA_3 at 10 ppm at pre-bloom stage, 15 ppm at initiation of flowering, 25 ppm at 3 to 4 mm berry diameter stage in combination with I ppm brassinosteroid, benzyladenine at 10 ppm twice at 4 leaf and berry setting stage, respectively and last application of GA_3 at 25 ppm at 7 to 8 mm berry diameter stage was found better for maximum berry diameter (19.60 mm), berry length (21.58 mm), berry weight (3.36 g), number of berries per bunch (131.78), number of bunches per vine (15.41), bunch weight (447.85 g) and yield (6.79 kg/vine, 15.09 MT/ha) with attractive grayish yellow colour of berries. It was also given considerable amount of TSS (22.41 °B), lowest acidity (0.53 per cent), highest TSS: acidity ratio (44.31) and minimum days to harvest (132 days) were recorded in vines under control. It indicated that the application of GA_3 in combination with brassinosteroid and BA was found effective for cell elongation and cell division, which lead to increase berry size, yield and quality of Thompson Seedless grapes (Warusavitharana, *et al.*, 2006)

Ramteke, *et al.* (2006) conducted studies to standardize the dose of GA_3 and 6BA at 3–4 mm and 6–7 mm berry size stages after berry set, indicated that the application of GA_3 at 40 ppm + 10 ppm 6BA at 3–4 mm and GA_3 at 30 ppm + 10 ppm 6BA at 6–7 mm berry size stage was equally good in comparison with the higher concentration of GA_3 at 50ppm + 20 ppm 6BA at 3–4 mm and GA_3 at 30 ppm + 10 ppm 6BA at 6–7 mm berry size stage. Berry quality and shelf life were not significantly different among the treatments. Based on the results obtained it could be concluded that for increasing berry size and berry weight of Sharad Seedless grapes the application of GA_3 at 40 and 30 ppm along with 10 ppm 6BA at 3–4 mm and 6–7 mm berry size stage, respectively seems to be beneficial and economical.

A study was conducted by Justine *et al.* (2012) to determine the key environmental and viticultural variables affecting the concentration of 3-isobutyl-2-methoxypyrazine (IBMP) in Cabernet franc grapes. Berries were sampled from individual vines at 30 days after anthesis (DAA), 50 DAA, and harvest from 10 and 8 commercial New York State vineyards in 2008 and 2009, respectively. IBMP concentrations at 50 DAA were significantly higher in the warmer 2008 growing season (2008, 103 to 239 pg/g; 2009, 12 to 87 pg/g). However, in the cooler 2009 growing season a smaller per cent decrease in IBMP from 50 DAA to harvest was observed, so that IBMP at harvest was not significantly different between years (2008, 1 to 13 pg/g; 2009, 5 to 14 pg/g). IBMP accumulation up to 50 DAA and log-fold decrease of IBMP from 50 DAA to harvest was modeled as a function of >120 viticultural and environmental variables (122 in 2008 and 140 in 2009). Important variables identified for modeling IBMP at 50 DAA were those associated with vine vigor, which was positively correlated with IBMP accumulation. Cluster light exposure did not explain differences in IBMP accumulation across sites, but it was important for modeling smaller differences within some sites. IBMP decrease could not be satisfactorily modeled across multiple sites, but within sites the decrease was most consistently correlated with classic fruit maturity indices (total soluble solids [TSS], TSS*pH^2). The intensity of herbaceous aromas in wines produced from each site was not correlated with IBMP concentration, but multivariate models indicated that lower vine water status was the best predictor of increased herbaceousness.

Kerem and Bruce (2012) opined that the aqueous ethanol (5 per cent or 10 per cent v/v) was sprayed on Cabernet Sauvignon (*Vitis vinifera* L.) clusters at 11 and/or 13 weeks post-bloom to evaluate the impact on fruit maturity. Fruit ethylene concentration increased as a result of sprays. The week-13 spray fruit at 17 and 21 weeks post-bloom had higher color intensity, total anthocyanins, and total and phenol-free glycosides than controls, while differences in berry weights, Brix, TA, and pH were not observed. Wine total and co pigmented anthocyanins, polymeric pigments, total phenols, color intensity, and total and phenol-free glycosides were higher in week-13 ethanol spray treatments than control wines. Differences in volatiles were noted in musts and wines of treatments, compared to controls. Triangle difference discrimination testing indicated differences in wine aroma between control and 5 per cent week-13 ethanol spray and in flavor between 5 per cent week-13 and 10 per cent week-13 ethanol spray treatment.

Boll *et al.* (2011) reported that the GA-levels and turnover rates during early berry development were much higher and longer lasting in developing berries of Pinot gris compared to Sylvaner and were negatively related to their sensitivity to exogenous applied GA_3. Anthers and pollen are known to be major sources of endogenous GA_3 during pre-bloom, and as such higher pollen tube abundance could be responsible for higher endogenous GA levels during early berry development. Pinot gris and Sylvaner showed the expected differences in pollen tube abundance that corresponded to the observed endogenous GA levels. Furthermore, pollen tube analyses of a broad spectrum of additional cultivars with known GA-sensitivity support the hypothesis, displaying a broad range of pollen tube abundance that correlates negatively with the known sensitivity of the cultivars to GA_3 applications.

The observed variety-specific pollen tube abundance seems to lead to differential developmental patterns: postbloom, Pinot gris had no unfertilized ovaries while in Sylvaner up to 35 per cent of the ovaries were unfertilized. Furthermore, there were variety-specific differences in the degeneration of ovules: Pinot gris had on average 3-4 fertilized ovules per ovary while Sylvaner only had 1-2 fertilized ovules per ovary. However, during further berry development endosperm development was aborted in about half of the ovules in Pinot gris, while in Sylvaner all fertilized ovules developed into seeds. Also in shattered berries the percentage of unfertilized ovaries was much higher in Sylvaner than in Pinot gris. Altogether, variety-dependent effects were much stronger than treatment effects. The results imply that a simple microscopic pollen tube analysis of varieties with hitherto unknown gibberellin sensitivity might predict the sensitivity to exogenous GA_3-treatment including possible adverse side effects.

Manjunatha and Prakash (2003) observed that the good quality grapes invariably fetches premium price in the market. Apart from good variety, agro-climatic conditions, cultural and canopy management practices greatly influence the fruit quality. Other factors like crop regulation and use of plant regulator play very important role. An attempt was made in the present studies to reduce the pronounced influence of apical dominance in seedless grapes cv. Sonaka by adopting different levels of sub cane system coupled with growth regulator treatments after October pruning. The results revealed the importance of various growth regulators tested. Biozyme (0.25 ml/litre) + GA_3 (25 ppm) was found to be most effective in increasing the bunch size, berry weight than Quantum and BR. It increased the berry diameter up to 17 mm. Among the various sub canes tested, developing sub canes by pinching at five bud levels was found to be more effective than pinching at four and six bud level.

Lavee (1987) stated that the intensive vineyards tend to develop an excessive vegetative growth which often leads to massive shading. Such conditions might cause a reduction in productivity of the lower part of canes and whole vines. The severity of this problem increases particularly in the early years of vineyard development. Uniform bud break is critical for balanced growth and even maturation. The response of various vine forms and treatments to controlled bud break and vegetative growth were studied. Vertical vine forms were more productive than horizontal ones. However, shading in such vines was more severe due to expressed polarity. Dormancy breaking agents and particularly H_2CN_2 was found to control both time and uniformity of bud opening, and helped to overcome apical dominance. This effect was often carried through till maturation. Controlled pruning, sectorial chemically enhanced bud opening and physiological control of vegetative growth enabled overcoming the shading effect. In most cases 12 leaves above the bunch enabled normal fruit development. Chemical control of excess growth by means of ethephon applications caused a better light penetration to the vineyard and more uniform maturation. The effect of such treatments on time of maturation was only minor and insignificant. Chemicals reducing vigor by shortening the internodes had no effect on the long term fruiting ability. New approaches to vine training based on utilization of the apical dominance could be efficiently combined with chemical growth control to achieve an annual high production. Direct effects of growth regulators on fruit maturation, such as ethephon for coloration, gibberellins for size and auxins on bunch properties, are of significant importance for maintaining quality at high yields. Growth enhancement of vines could be achieved by gibberellic acid. This however did not lead to increased productivity in weakly growing vines, and reduced it in medium and vigorous ones. Preliminary results on growth control were demonstrated with paclobutrazol and on berry size with chloropyridylphenylurea. The relative efficiency of the different methods to maintain high quality in a heavily producing vineyard is discussed.

Wang *et al.* (2011) reported that the total thirteen quality characters of manicure finger grape which was treated by 21 different plant growth regulators in Pengshan county, Sichuan province were judged comprehensively according to the principle of fuzzy mathematics. The results indicated that the effect of growth regulators on the internal quality of manicure finger grape was larger than the external quality. In 13 quality indicators, the first four indicators which were vitamin C content, sugar acid ratio, soluble solid and total soluble sugar impacted greatly to the comprehensive characters. The best treatment on the comprehensive quality was the use of 5000 mg L^{-1} PBO on May 27 and 170 mg L^{-1} S-ABA on July 10 spraying on the trees and the best treatment on the external quality was the use of 5000 mg L^{-1} PBO +200 mg L^{-1} S-ABA spraying on the trees. The use of 10.0 mg L^{-1} Qibao spraying on the trees on May 27 was the best treatment on the internal quality.

Begum *et al.* (2007) study was undertaken to investigate the relative efficacy of four growth regulators, *viz.*, α-naphthalene acetic acid (NAA), indole-3-acetic acid (IAA), gibberellic acid (GA) and 2,4-Dichlorophenoxy acetic acid (2,4-D) at various concentrations (ppm) on the yield and some other quality characteristics of grape (Zakkao cultivar) at the experimental grape research vine of BCSIR Laboratories, Rajshahi during the period January, 2003 to July, 2004. Yield was found to be increased (33.11 per cent) by spraying with 05 ppm of 2, 4-D as compared to control and other growth regulator treatments. Fruits of treated plants showed increased total soluble solids (20.28 per cent) and vitamin (23.18 per cent) than that of untreated one. The titrable acidity (T.A.) of fruits (33.01 per cent) of the treated plants was reduced significantly as compared to control.

Oulkar *et al.* (2011) stated that a selective and rapid multiresidue analysis method is presented for simultaneous estimation of 12 plant growth regulators (PGRs), namely, auxins (indol-3-acetic acid, indol-3-butyric acid, and naphthyl acetic acid), cytokinins (kinetin, zeatin, and 6-benzyladenine), gibberellic acid (GA_3), abscisic acid, and synthetic compounds, namely, forchlorfenuron, paclobutrazole, isoprothiolane, and 2,4-dichlorophenoxy acetic acid (2,4-D) in bud sprouts and grape berries at the development stages of 2-3 and 6-8 mm diameters, which are the critical phases when exogenous application of PGRs may be necessary to achieve desired grape quality and yield. The sample preparation method involved extraction of plant material with acidified methanol (50 per cent) by homogenization for 2 min at 15000 rpm. The pH of the extract was enhanced up to 6 by adding ammonium acetate, followed by homogenization and centrifugation. The supernatant extract was cleaned by SPE on an Oasis HLB cartridge (200 mg, 6 cc). The final extract was measured directly by LC/MS/MS with electrospray ionization in positive mode, except for 2,4-D, GA_3, and abscisic acid extracts, which required analysis in negative mode. Quantification by multiple reaction monitoring (MRM) was supported with full-scan mass spectrometric confirmation using "information-dependent acquisition" triggered with MRM to "enhanced product ionization" mode of the hybrid quadrupole-ion trap mass analyzer. The LOQ of the test analytes varied between 1 and 10 ng/g with associated recoveries of 80-120 per cent and precision RSD <25 per cent ($n = 8$). Significant matrix-induced signal suppression was recorded when the responses for pre- and postextraction spikes of analytes were compared; this could be resolved by using matrix-matched calibration standards. The method could successfully be applied in analyzing incurred residue samples and would, therefore, be useful in precisely deciding the necessity and dose of exogenous applications of PGRs on the basis of measured endogenous levels.

Weaver *et al.* (1969) observed the effect of exogenous growth regulators on movement of assimilates into flowers and young fruits of 'Black Corinth' grapes was studied. Clusters were treated with growth regulator and after 0.5 hr to 5 days the leaves above the clusters were exposed to $^{14}CO_2$. Control shoots received $^{14}CO_2$ but no growth regulator. At harvest, counting and radioautographic techniques were used to ascertain amount and distribution of activity in clusters. Clusters were dipped in 4-CPA (4-chlorophenoxyacetic acid), GA_3 (gibberellic acid), or BA (benzyladenine). All berries were heavier than controls within 3 days. Total counts in the fruits were increased by 4-CPA, and the distribution of radioactivity among the sugar, organic acid, and amino acid fractions was usually altered by all treatments. In a time series experiment, within 6 hr after treatment of fruits with GA_3 there was almost an 8-fold increase in total counts relative to the control. After 12 hr there was about a 9-fold and 6-fold increase in counts in tartaric and malic acids, respectively, and in γ-aminobutyric acid, pipecolic acid, and valine increases of 56, 150, and 330 per cent. Radioactivity in fructose was increased 70 per cent in gibberellin-treated clusters over the controls. After 96 hr there were only about 1000 cmp per g fr wt in controls, but there were about 31,000 cpm counts in treated clusters. Treatment of clusters with gibberellin attracted less assimilates into the fruits when shoots had also been sprayed with gibberellin. Dipping portions of clusters in gibberellin increased the movement of ^{14}C assimilates into the treated portions. Hormonal control of mobilization is discussed.

This study was conducted during the period 2006-2008 in the Faliszowice Vineyard in the Sandomierz Upland (Magdalena, 2011). The experiment evaluated yield quality of 'Einset Seedless' grape after the application of gibberellic acid (GA_3) in the form of the growth regulator Arbostim 100 SL and 2-naphthoxyacetic acid (NOA) in the form of the growth regulator Betokson Super 050 SL. These growth regulators were applied during flowering by spraying the inflorescences or immersing them in whole at the following concentrations: 100 $mg \times L^{-1}$ GA_3 - the treatment was performed once at full flowering (70-80 per cent of open flowers); 50 $mg \times L^{-1}$ GA_3 - the treatment was performed twice: at the beginning of flowering (20 per cent of open flowers), and the second time at the end of flowering (80 per cent of faded flowers); 50 $mg \times L^{-1}$ GA_3 + 0.2 per cent NOA - the treatment was performed once at full flowering; and 0.2 per cent NOA - the treatment was performed once at full flowering. The study showed that the applied treatments had a beneficial effect on cluster and berry weight as well as on the number of berries per cluster in the grape variety 'Einset Seedless'. Gibberellic acid and the mixture of 50 $mg \times L^{-1}$ GA_3 and 0.2 per cent NOA were shown to have a significant effect on cluster and berry length; an exception were clusters and berries sprayed with 50 $mg \times L^{-1}$ GA_3 whose length did not differ significantly compared to the control ones. The applied treatments slightly affected berry width in the grape variety under study. The growth regulators applied were found to have an adverse effect on extract content in 'Einset Seedless' fruit.

10

Physiological Disorder in Grape

When grape is almost ready for harvest the physiological disorder starts exhibiting, *viz.*, pink berry, shot berry, mummy, water berry etc. We have to reduce the above problem and increase the quality.

Before the physiological symptoms start emerging, precautionary measures should be taken. While taking quality grape production, the symptoms appear slowly and by the time harvest completes its spreads on a large-scale, it results in reduction in yield but on a large-scale quality deteriorates and such quality affects bunches which are not fit for export and for local market ultimately the farmer won't get the price as per his expectation.

This kind of physiological disorders have no relationship with pests and diseases. The following are the commonly observed physiological disorders in grape production:

1. Small round berries – shot berries
2. Thickly accumulated berries – water berries
3. Drying of berries (mummification) – stalk necrosis
4. Reddish/pinkish berries – pink berries
5. Hard/stiff berry pedicel – stiffing stalk
6. Bunch scorching – scorching.

Among the above listed physiological problems one or more than one physiological disorders are observed in small or high proportion in a bunch.

1. The Cause for Physiological Disorders

While berry development and when its development stops the ripening starts. If there is a change in the climatic fluctuation then such physiological disorders are observed and develops. Especially when fluctuations in the temperature and relative humidity are observed.

2. Defective Soil and Water

If there is a problem in soil and water analysis reports like EC, pH etc., then the berry development will be affected. Due to which physiological disorders comes to the picture.

3. Improper Use of Fertilizer

Unless you understand the hunger of crops and which element, available nutrients and which fertilizer is suitable, you won't get proper yield. The improper foliar application of nutrients results in physiological disorders.

4. Improper Use of Growth Regulators

Early application of growth regulators, without its needs and wrong selection of growth regulators, physiological disorders are increased.

5. Improper Application of Insecticide and Fungicides

While controlling pests and diseases, wrong selection of fungicide, combinations of fungicides and wrong application method results in increase of physiological disorders.

6. Watering/Irrigation

Irregular/improper application of irrigation/watering during bunch development may increase the physiological problem.

Time and Application

While thinning, girdling, tipping, if there is any delay, physiological disorders are observed. Physiological disorders may occur due to many reasons and for every physiological disorder, control measures should be taken in time. If not taken, then the problem aggravates or the other physiological disorders symptoms are observed.

Physiological Disorder and its Control Measures: Climatic Factors:

1. Temperature

During bunch development if climate is not suitable such as low temperature, then night smoking is advised in closed confinement by erecting artificial curtains or wind breaks.

2. Relative Humidity

At berry development a RH of 60-70 per cent RH or more and at ripening period a RH of 50'-60 per cent or less is desirable. But at berry development (more than) 80 per cent relative humidity and RH < (less than) 45 per cent will effect berry development. If at ripening period RH increases then berry cracking is observed. To reduce relative humidity, stir the soil, reduce irrigation quantity and how much quantity of water you require for foliar spray depends on incidence of control of pest and diseases.

3. Quality of Soil and Water

In more number of places, the upper layer contains more of salt for example calcium, sodium, potassium and magnesium but during rainy season the salt gets leached away from the root zone. During berry development if there is shortage of water the salt gets accumulated in between the root zone, which spoils the nutrient uptake. If soil is good and water is bad, containing chloride, then also it will have lot of problems. In some soils the top layer becomes hard and such soils are problematic soils for grape cultivation because of improper aeration. In due course of time physiological disorders appear. In soils where more than 10 per cent lime nodules are noticed the production of quality grapes become very difficult. Therefore a period between April pruning to October pruning, apply green manure to reduce bad effect of lime nodules. In such grape garden after October pruning, while giving open well irrigation, first remove the existing open well water completely. Then recouped new fresh water which may be utilized for irrigation.

Such soils and open well water or bore well water after October pruning should be tested in the laboratory and as per recommendation, it should be used.

Soil nematodes and other harmful microbes should be kept under control.

4. Balanced Fertilizer Application

To avoid physiological disorders, apply fertilizers with care. As per soil testing report, ascertain the available nutrient in soil whether it is optimum or not for cultivation. Then select the fertilizer. In problematic soils and soil affected with lime nodules don't apply super phosphate, murate of potash and complex fertilizers *viz.*, 19:19:19, 15:15:15, 17:17:17 etc. In lime nodule; affected plot if you apply super phosphate, then it forms calcium phosphate that will get fixed in the soil especially phosphorus which is not available to develop bunches. With the application of murate of potash chloride deposits, it affects the bunch quality. If potash nutrient application is of more than required quantity then physiological disorders aggravate, and application of complex fertilizers may result in set-backs.

Therefore to get quality fruits and to avoid physiological disorders complex fertilizer application should be avoided.

Timely application of fertilizers with required quantity is important to observe crop condition. If growth is slow then petiole analysis becomes a must. Among the major nutrient (NPK), nitrogen in the beginning stage, phosphorus in the middle stage and later, potash is required and their balanced application from time to time is very important.

Water berries are noticed in a bunch under the situation when nitrogen is applied in large quantity or nitrogen is applied more than its requirement or applied at later stages when berries are of pea size. During flowering, the petiole should be examined thoroughly because if there is a deficiency of calcium stalk necrosis and stalk, stiffening may occur.

In most of the soils of Karnataka calcium is available in abundant quantity but it won't reach the bunches; as a result there will be more demand for potash, which results in reduction of calcium availability and uptake.

Deficiency of calcium before October pruning if found in vine then phloem becomes more. In the beginning, if availability of calcium is less and potash is more than Phloem along with Xylem function will be hampered which will not be visible to the naked eye. During nutrition, excess application of potash hampers cane growth. For grape quality, such as sugar percentage and colour development, potash is required. At the time of berry development, fertilizer without potash, are not desirable. Application of murate of potash in place of sulphate of potash increases the chloride level which will be a set-back to grape production and quality. If potash availability is more, then the demand for magnesium increases. When potash demand is fulfilled then deficiency symptoms of magnesium starts exhibiting, when demand of magnesium increases then calcium availability becomes problematic; then to fulfill the demand of soil for magnesium and calcium, apply calmag and take up foliar spray also. For soil application, apply calmag at October pruning and at fruit set take up foliar spray.

If requirement of leaf and calcium, magnesium is brought under control as calcium content in bunches and berry pedicel may not improve so easily since translocation of magnesium in leaf is easy.

When calcium problem is solved then physiological disorders can be rectified. Deficiency of zinc, boron increases the physiological disorder. During this period the climate should be mild with sunshine period of 12 hours and there should not be much difference between minimum and maximum temperatures. Night temperature, if it is 15°C, then day temperature should be around 30°C. If temperature is more, then take spraying of water. If night temperature is low then by smoking try to increase the night temperature. If sunshine hours are less then install tube light for that period.

Low temperature and fog will have bad effect on the berry development, only possible solution is to have wind-break plan with smoking of the garden. At berry ripening period, low relative humidity, medium temperature and full sunshine is helpful, which prevails after 15th January. But if ripening starts earlier than in the second fortnight of January it can cause problems. Under such situation spraying of some hormone will be of some help like CCC 250 ppm to 500 ppm and within completion of a week, Kinetin, Uracil or 6 BA 5 to 10 ppm foliar spray helps to overcome the shortage of sunlight.

Bunch Quality and Nutrient Management

One has to understand yield and fertigation, its co-relation to get good quality. If you apply less nutrients, yield is affected, the effect of nutrient on yield can be seen immediately but on quality it shows late.

Control of Nitrogen Application

Nitrogen is important to increase yield as well as quality. Excess application of nitrogen has effect on yield as well as on quality. Less application of nitrogen reduces yield but improves quality. Excess application of nitrogen reduces the potash uptake and ripening will be delayed. In a bunch, berries at the tip show improper sugar development with berries being bigger or smaller in size. The colour of berries at ripening shows light greenish colour rather than golden yellow. In a bunch one side of the berries are greenish and the other side are yellowish in colour. Late application of nitrogen causes pre-harvest and post-harvest berry drop problem.

After October pruning, application of optimum doses of nitrogen is very important at flowering, berry development and ripening stages.

At bunch development, if nitrogen availability is less then yield gets reduced. If nitrogen availability is more than the quality deteriorates. If nitrogen applied through leaves results will be quicker than soil application but when it is applied through soil it should be in split doses. For export quality grape, after October pruning, nitrogen has to be applied in ten stages that is before flowering for two times and at berry development four times. When it is in sink stage, three times and at the beginning of ripening one time; if necessary totally ten times. At the first stages irrigate through drip in two days interval. At second stage daily, third stage alternate day and fourth stage every fourth day. On the third and fourth stages nitrogen through foliar can be given.

Nitrogen Levels (per cent)

Bunch Stage	*Low Yield 20-25 ton/ha*	*Medium Yield 25-30 ton/ha*	*High yield 30-35 ton/ha*
Full bloom	1.20 to 1.75	1.70 to 2.50	2.10 to 2.50
Berry development (fast period)	1.50 to 1.75	2.10 to 2.50	1.40 to 2.98
Berry development (slow period)	1.30 to 1.50	2.0 to 2.25	2.20 to 2.60
Beginning of ripening, full ripening	1.20 to 1.40 1.00 to 1.10	1.80 to 2.0 1.40 to 1.60	1.80 to 2.20 1.40 to 1.80

Ammonium sulphate, urea can be used for foliar spray in ammonical and nitrogen forms.

Use more organic manure, zinc chelated, ferrous chelated as a foliar spray to overcome the problems of berries and pink berries. Give zinc for slow development of berry and at ripening stage boron demand is more. If soil temperature comes down then zinc and boron availability also get reduced. However during this period if you apply more of nitrogen and phosphorus, then availability of zinc boron also reduces.

Proper Use of Growth Hormones

Among the growth regulators, improper use of Gibberlic acid is also responsible for the physiological disorders. More use of GA at flowering stage results in thickening and hardening of berries and berry pedicel. So when GA is used in more concentration the physiological disorders may be noticed. After GA application demand for boron increases. Deficiency of boron is observed at terminal leaves and on berries. If you apply boron by keeping more leaves in front of the bunch and less berries in a bunch then more damage is noticed. When the berries start ripening use foliar spray of boron. If you apply more GA at this stage then take up spraying of lihocin or NAA. CCC will reduce the effect of GA and NAA to stop dissolving of berries, for which first spray of 200-250 ppm of CCC, later 25 to 40 ppm of NAA, is helpful.

Before the physiological disorder problem arises, careful or precautionary measures are necessary from fruit set to pea size berry development.

Quantity Attainment

Increasing grape productivity is one part while to improve the grape quality is another part, but to have both qualities at one go is very difficult. If the yield is low then quality definitely improves but is doubtful, whereas per hectare, what should be the yield one must be able to understand very easily. First, one has to understand what quality is. The grape which is going to be exported to a particular country and its quality parameters are required to be understood by the farmers.

The common understanding is to go for high yield with low quality but low yield with good quality is not proper. When you think of export then keep a target of 25-30 tonnes per hectare.

Grape Quality Parameters

1. Grape berry - long and oval.
2. Berry colour - greenish
3. Berry TSS - 20 to 22 per cent
4. Berry juice-70-90 per cent
5. Berry acidity-0.6 per cent
6. Berry sugar : acidity-36:1 to 40 :1
7. Berry peri carp - thin and crispy
8. Berry shape - triangular
9. Juice taste - pleasant taste
10. No. of berries in a bunch - 125 to 150
12. Berry pedicel- thick, long, greenish and strong
13. Berry thickness - 18 to 20 mm
14. Bunch weight - 400 to 600 gm
15. Bunch appearance - fresh and pleasing. The grape quality parameters are more but each item is of importance.

The post harvest quality parameters are:

1. After harvest, its shelf life is important.
2. Post-harvest berry drop should be less.
3. There should not be any spot or insecticide/fungicide

4. Chemical residues should not cross its limit.
5. Total grape bunches should suffice the variety in demand.
6. It should fit all the export parameters.

Suitable Climate for Grape Production

To get good grape per hectare tonnage, which grape variety we have to grow one has to understand and also in which season/month it should come for harvest to meet the export demand should be known. Season means climate *i.e.*, temperature, relative humidity and sunlight-these three factors are very important. Establishment of flower primordial, berry development and its ripening are three important development stages and each stage is totally different from each other and like-wise climatic requirements also vary. From flowering to fruit set it requires 15 days. During that period the temperature requirement should be minimum of 15 to 16°C and maximum of 30 to 32°C with relative humidity of 60 per cent with 11 hours of sunlight. Then later comes the berry development in which two stages are involved. First stage is of early development which is of three weeks then second stage of sink or rest period which is of few hours or of 2 to 3 days. Then again berry development starts which lasts for three weeks. So totally it is a period of 5-6 weeks.

Somkuwar *et al.* (2011) observed after forward pruning bunch appears on 5th leaf stage in Thompson Seedless and its clones. To achieve the bold berries with more than 18 mm diameter for export market, the cultural practices are to be followed correctly from the parrot green colour stage up to the harvest of bunch. A field experiment was therefore conducted to study the effect of cluster clipping and berry thinning on yield and quality of Thompson Seedless grapes grafted on Dogridge rootstock. The clusters were retained (75, 65 and 50 per cent) by clipping the remaining portion by following standard practices of thinning of rachis of the bunch. Maximum bunch weight (381.0 g) was recorded when 75 per cent total length of cluster was retained and berries in a bunch were thinned as compared to the control where no cluster clipping and berry thinning was performed. Berry diameter was increased from 16.94 mm in control to 21.32 mm in 50 per cent cluster retained and berry thinning treatment. The maximum berry weight was recorded in 75 per cent cluster retention and berry thinning treatment as compared to other treatments studied. The highest TSS of 19.86 °Brix was recorded in 50 per cent cluster retention treatment compared to the control. Significantly highest yield/vine was recorded in 75 per cent cluster retention and berry thinning treatment. Considering bunch and berry characters, the cluster retention up to 75 per cent and berry thinning as per the package of practices seems to be ideal in Thompson Seedless grapes for better quality production.

Patil *et al.* (2006) opined that among the different weed management treatments, the weed control efficiency during the entire crop growth period was higher with weed free check (100 per cent) followed by three sprays of post-emergent glyphosate at 3 kg a.i./ha at 90 days interval (ranging from 66.86–88.68 per cent), pre-emergent application of oxyfluorfen at 1.0 kg a.i./ha + two sprays of post-emergent application of paraquat at 2.0 kg a.i./ha at 90 and 180 days (63.50 to 88.23 per cent) and pre-emergent application of oxyfluorfen at 1.0 kg a.i./ha + one spray of glyphosate at 3 kg a.i./ha after 90 days (62.62–83.99 per cent). Higher number of bunches (36.17/vine) were recorded with weed free check followed by three sprays of post-emergent glyphosate at 3 kg a.i./ha at 90 days interval (35.17/vine) and pre-emergent application of atrazine at 2.0 kg a.i./ha + one spray of glyphosate at 3 kg a.i./ha after 90 days (33.83/vine) and were statistically on par with each other. Bunch weight was significantly increased by weed management treatments. Maximum bunch weight (280.83 g) was recorded with pre-emergent application of oxyfluorfen at 1.0 kg a.i./ ha + one spray of glyphosate at 3 kg a.i./ha after 90 days followed by pre-emergent application of diuron at 2.0 kg a.i./ha + one spray of glyphosate at 3 kg a.i./ha after 90 days (270.33 g) and pre-emergent application of oxyfluorfen at 1.0 kg a.i./ha + two sprays of post-emergent application of paraquat at 2.0 kg a.i./ha at 90 and 180 days (269.33 g) and these treatments were statistically on par with each other and significantly superior over rest of the treatments. Higher yield per vine (9.25 kg/vine) was recorded with weed free check followed by three sprays of post-emergent glyphosate at 3 kg a.i./ha at 90 days interval (8.88 kg/vine) and pre-emergent application of oxyfluorfen at 1.0 kg a.i./ha + one spray of glyphosate at 3 kg a.i./ha after 90 days (8.54 kg/vine) and was statistically on par with each other. Similar trend was observed with regard to yield per hectare.

Khilari (2006) reported that in India, grapes are grown in Maharashtra, Karnataka, Andhra Pradesh, Punjab and Haryana; however, out of these states, Maharashtra contributes to more than about half the area and ranks first in respect of area and production in the country. This subtropical area has unique features of viticulture and is distinguished by evergreen vines, two pruning and tremendous mildew pressure. Frequent vineyard surveys carried out in the Maharashtra state indicated that there are certain disorders observed during maturity and near harvest causing considerable losses in yield and deteriorations in quality of grapes. In absence of scientific trials conducted in the Universities or Research Centers, the observations made either in the vineyards or elsewhere on grower's fields or on research farms, in respect of some of the important disorders, are mentioned below.

Shikhamany *et al.* (2006) opined that an ideal training system is the one, which allows light incidence above the compensation point on the entire leaf area of the vine. Horizontal canopy with dense foliage has more parasite leaves than the diagonal or vertical canopies with the same foliage density. Leaf area Index (LAI) defines as leaf area per unit area of land is the best index of foliage density. Any canopy that has yield response to highest level of LAI among a set of canopies is the ideal one and the training system supporting that canopy is the ideal system. With this criterion, Thompson seedless vines grafted on Dogridge rootstock were trained to six training systems with varying number and orientation of cordons, to bring about variation in LAI at NRC for grapes, Pune. Observations were recorded on 4- and 5-year-old vines. Correlations were worked out between LAI as the independent variable and brix yield as the dependant variable. Efficiency of canopy was ranked based on the magnitude of the optimum LAI and the corresponding yield. Canopy of vines trained to four horizontal cordons was the most efficient with an optimum LAI of 2.71

corresponding to 2.58 kg of brix yield/vine (6 cm^2), followed by the ones trained to two diagonal and eight horizontal cordons.

Pakhale *et al.* (2007) reported that an Antioxidant potency of Indian grape cultivars varying in their skin color, seed and polyphenol content (Bangalore blue, Pandhari sahebi, Sharad seedless and Thompson seedless) and their components (whole grapes, pulp with skin and seeds) was examined as 2,2'-diphenyl-1-picrylhydrazyl (DPPH) radical scavenging activity using electron spin resonance spectrometry. The total polyphenols in Indian grapes ranged between 3-51 per cent. Extracted polyphenols caused a concentration dependent and significant loss in DPPH radical signal, similar to known antioxidants-Vitamin C, catechin and procyanidin B3 used as references. Among seedless cultivars, polyphenols from Sharad was more potent as antioxidant than Thompson, showing IC50 values of 1250 +/- 30 and 2650 +/- 125 microg/ml, respectively. The inhibitory effect of polyphenols from seedless grape cultivars was as effective as that of seeded variety. The results indicate that polyphenols extracted from Indian grapes/components (with/without seeds) exhibited free radical scavenging activity and their chemopreventive properties need to be exploited by in vivo model system.

Satisha and Adsule (2006) stated that the grape rootstocks are playing a major role in Indian viticulture to combat adverse effects of water scarcity, soil and water salinity. Apart from Dog ridge, which is the leading rootstock in grape growing regions of India, some other rootstocks are also gaining importance for both table and wine grape varieties. Rootstocks vary in their ability to root, based on biochemical constituents of mother vines and also IBA (Indole butyric acid) concentration used for treatment of cuttings. An experiment to standardize IBA concentration for better rooting of grape rootstocks revealed maximum success per cent in *Vitis longii*, SO-4 and 99 R using 1000 ppm of IBA, while it was maximum using 750–1000 ppm of IBA in Dog ridge, Salt Creek, 110 R and 1103 P rootstocks and Thompson Seedless recorded highest success per cent using 500 ppm. Early sprouting was noticed in *Vitis longii*, SO-4, 99 R when treated with 2000 ppm; Dog ridge, Salt Creek, 110 R and 1103 P treated with 1000 ppm and Thompson Seedless with 500 ppm. Cuttings taken from mother vines with moderate C/N ratio and less phenol content recorded maximum success per cent with maximum root length. This suggests rooting ability of different rootstock depends on biochemical content of mother vines and also concentration of IBA used for treatment of cuttings.

Suresh and Ethiraj (1987) reported that the effect of grape maturity on wine quality of four cultivars was studied for two years. As expected, °Brix values increased and total acidity decreased during maturation. In general, tannin and pH of the wines increased while total acidity decreased with maturity. Sensory evaluation indicated that the wine quality was highest at the early or intermediate maturity.

Bruce *et al.* (2011) reported that the ability of two electronic nose systems (conducting polymer and surface acoustic wave-based) to differentiate volatiles of grapes and wines treated with an aqueous ethanol spray (5 per cent v/v) at veraison was evaluated. Ethanol spray induced fruit ethylene production immediately post treatment, which then declined progressively. The electronic nose evaluations of grape volatiles were compared with Cabernet franc and Merlot physicochemistry and with wine gas chromatographic and aroma sensory data. Canonical discriminant and principal component analysis found that both electronic nose systems and the physicochemical measures (Brix, TA, pH, color intensity and hue, total phenols, glycosides, and berry weight) were able to discriminate between ethanol-treated and untreated grapes and wines for both cultivars. Grape physicochemical treatment differences were due mainly to variations in hue, phenolic-free glycosides, and total phenols. Aroma sensory evaluations using a consumer panel differentiated between ethanol treatments and controls for Merlot, but not for Cabernet franc wines.

Zhan *et al.* (2011) reported that the berry fresh weight and composition are under the control of complex interactions among genotype, environmental factors, and viticulture practice, which all affect not only the mean value but also the ranges of variation in berry traits. Both mean values and variation range in berry composition play a role in berry quality and, subsequently, wine typicity. This review examines recent ecophysiological, genetic, and molecular knowledge to provide better understanding of the mechanisms that influence variability in berry weight and composition. We specifically reviewed the variation range in berry weight and composition (including sugars, organic acids, and anthocyanins) among *Vitis* genotypes, the environmental and viticulture practices that cause variability for a given cultivar, the genetic clues underlying the genotypic variation, and the putative genes controlling berry weight and composition. Despite numerous studies comparing differences in the mean value of a berry trait among different environment conditions and viticulture practices, very few studies have explored the level of variation in response to those factors. Present genetic and molecular studies are mainly focused on identifying genes involved in the control of berry weight and composition, with few considerations of environmental factors that affect their expression. In the future, more effort should be directed toward integration of genetic and molecular work with ecophysiological approaches in an effort to gain novel insights into the cause of variability in grape fresh weight and composition.

Ahmad *et al.* (2008) stated that the ability of an electronic nose to classify cabernet sauvignon (*Vitis vinifera* L.) fruit based on maturity levels was investigated over two seasons. Maturity of samples collected 18, 19, and 20 weeks post-bloom was evaluated by measuring berry weight, pH, Brix, titratable acidity, total phenols, color intensity, hue, total anthocyanins, and total and phenol-free glycosides. Results were compared, using discriminate and canonical discriminate analysis, with analysis of headspace volatiles via a hand-held electronic nose. The electronic nose was able to determine differences among the three sample groups in both seasons. Additionally, in one season electronic nose measurements were compared to chemical analyses of samples collected from east and west sides of north-south oriented vineyard rows. Results demonstrated the ability of the electronic nose to distinguish fruit from vine canopy sides. Field measurements demonstrated the potential for the electronic nose as a rapid, non-destructive tool for evaluating grape maturity.

Mansfield *et al.* (2003) stated that *the Brettanomyces bruxellensis* and lactic acid bacteria are common micoroorganisms capable of modifying wine aroma and flavor. The activities of β-glucosidase against *p*-nitrophenyl-β-D-glucopyranoside was determined in a model system for 14 strains of *Brettanomyces bruxellensis* yeast and 9 strains of lactic acid bacteria (*Oenococcus oeni*). All *Brettanomyces* strains and 7 *Oenococcus* strains exhibited enymatic activity against this substrate. *B. bruxellensis* ß-glucosidase activity was primarily intracellular; *O. oeni* showed some extracellular activity. Strains showing activity greater than 1000 nmole mL^{-1} g dry cell $mass^{-1}$ 24 hr^{-1} for *Brettanomyces*, or 100 $nmole^{-1}$ mL^{-1} g dry cell $mass^{-1}$ 24 hr^{-1} for *Oenococcus*, were evaluated for their effect on Viognier grape glycosides. Neither genus was active on Viognier grape glycosides.

Zoecklein and Fugelsang (2003) observed that the replicated, sterile-filtered Pinot noir (*Vitis vinifera* L.) wines were individually inoculated with one of eight strains of *Brettanomyces bruxellensis* (Ave, M, 213, 216, Vin 1, Vin 3, Vin 4, and Vin 5) at initial viable cell numbers 50 cfu/mL. In two separate studies, population changes were monitored for 23 months or until cell densities declined to 30 cfu/mL. Significant variation was noted in both growth rate and peak population densities among strains. The concentration of select volatile components was monitored using solid phase micro-extraction, GC/MS. Significant increases in the concentration of 4-ethyl phenol occurred after accumulated cell populations reached 25 $\times 10^4$ cfu/mL. *Brettanomyces*-inoculated wines were found to have detectable concentrations of ethyl-2-methylbutyrate, isoamyl alcohol, ethyldecanoate, isovaleric acid, guaiacol, 2-phenylethanol, 4-ethylguaiacol, and 4-ethylphenol, with significant differences in their concentrations among strains. Duo-trio testing demonstrated sensory differences between the control and all inoculated wines and between wines inoculated with strains Ave and Vin 5, M and 216, and M and Vin 4. Quantitative descriptive analysis suggested that all *Brettanomyces* wines had a lower perception of fruity descriptors than did uninoculated control wines.

Whiton and Zoecklein (2002) opined that the method for the rapid determination of ethyl carbamate in wine by headspace solid-phase microextraction with detection by gas chromatography/mass spectrometry has been developed. The analysis parameters of fiber coating, extraction time, and sample temperature have been optimized. The optimized method of headspace sampling with a Carbowas/divinylbenzene fiber for 30 minutes has been found to be reproducible and linear from 10 to 80 µg/L. The method is simple, requires little operator effort, and can be automated.

Whiton and Zoecklein (2002) stated that the accuracy of two methods for determination of glycosyl-glucose (GG) in grapes and wine has been tested using several glycoside standards containing alkyl, benzenoid, and phenolic aglycones. The total GG method using C-18 sorbent material was found to exhibit poor recoveries of benzenoid glycosides. The recoveries for the phenol-free GG method were satisfactory for alkyl and benzenoid glycosides. In addition, an adaption of the phenol-free GG method to 96-well microplate format has been demonstrated.

Hartmann *et al.* (2002) observed the effect of wine matrix ingredients and conditions on the headspace sampling of 3-alkyl-2-methoxypyrazines was investigated with solid-phase microextraction (SPME) and capillary gas chromatography, using a nitrogen phosphorus detector. Changes in the recovery of

3-ethyl-, isopropyl-, *sec*-butyl-, and isobutyl-2-methoxypyrazines from the static headspace of synthetic wine matrices spiked with 5 μg/L of each analyte were investigated and reported as a function of SPME fiber type, extraction time, and temperature. The influence of pH, ethanol, phenolics, and oak was studied. Divinylbenzene/carboxen/polydimethylsiloxane (PDMS) SPME fibers at an extraction temperature of 50°C for 30 minutes with 30 per cent (w/v) added sodium chloride resulted in the highest analyte recoveries. Although PDMS (100 μm) SPME fibers at an extraction temperature of 35°C for 30 minutes with 30 per cent (w/v) added sodium chloride resulted in lower analyte recoveries, the fiber remained functional for 50 to 75 analyses after other coatings deteriorated. Changing the sample ethanol concentration from 0 to 20 per cent (v/v) resulted in an exponential decrease in the recovered analytes. Below pH 2, there was extensive loss of the analytes in the headspace. No measurable impact on alkyl-methoxypyrazine headspace concentrations was observed with exposures to selected phenolics and to oak.

Chinnathambi and Michael (1981) reported that the vegetative and reproductive anatomy of the grapevine is discussed with emphasis on recent interpretations based on scanning electron microscopy. The terminology of flowering in *Vitis* is defined and a developmental code, comprising Stages 0 to 11, is proposed for the events leading to formation of flowers. In brief, flowering in the grapevine involves three main steps: 1) Formation of Anlagen or uncommitted primordia (Stages 0 to 1); 2) Differentiation of Anlagen to form inflorescence primordia (Stages 2 to 7); and 3) Differentiation of flowers (Stages 8 to 11). The litreature on the factors that influence flowering in grapes is reviewed under three major headings: 1) Biochemical changes in apices during inflorescence primordia formation; 2) Effects of environmental factors, including temperature, light intensity, photoperiod, water stress and mineral nutrition and 3) Role of phytohormones. Anlagen develop into inflorescences, tendrils or shoots depending on the environment and hormonal factors. A hypothetical scheme for the hormonal control of Anlagen, tendril and inflorescence formation is proposed. It is suggested that flowering in grapes is controlled by the gibberellin:cytokinin balance. Formation of the inflorescence axis (the Anlage) is gibberellin controlled, but subsequent differentiation into flowers is regulated by cytokinin.

Vasconcelos *et al*. (2009) reported that the flowering of *Vitis vinifera* spreads over two seasons. Tendrils and inflorescences have a common origin known as anlage or uncommitted primordia. The fate of the uncommitted primordia depends on the cytokinin-gibberellin balance, with cytokinins promoting transition to flowering and gibberellins inhibiting it. High temperature and high light are induction stimuli for flowering. Neither photoperiod nor vernalization is very relevant for flowering induction. Inflorescence primordia development in latent buds stops after the formation of secondary and tertiary branches, approximately one month before shoot periderm formation. Buds resume growth after dormancy, with further branching of inflorescences before differentiation of individual flowers. Warm weather at budburst favors further inflorescence differentiation, resulting in more clusters per shoot, while cool weather favors differentiation of more flowers per clusters and fewer clusters per shoot. Environment and cultural practices influence flowering, either directly or indirectly via their impact on photosynthesis and nutrient availability. Cultural practices encouraging light penetration into the canopy favor flower initiation, while practices resulting in shading have a detrimental impact. Flower formation occurs through a series of sequential steps under hormone-mediated genetic control. The first genetic change involves the switch from the vegetative to the floral state, in response to different environmental and developmental signals, through the activity of floral-meristem identity genes. Second, the floral meristem is patterned into the whorls of organ primordia through the activity of floral-organ identity genes. Third, the floral-organ identity genes activate downstream effectors that specify the various tissues which constitute the different floral structures. The flowers are hermaphroditic and most are self-pollinated but cross-pollination also occurs. Fertilization is hindered by cool rainy weather and favored by warm dry weather.

Carmona *et al.* (2008) stated that the grapevine reproductive cycle has a number of unique features. Inflorescences develop from lateral meristems (anlagen) in latent buds during spring and summer and enter a dormant state at a very immature stage before completing development and producing flowers and berries the following spring. Lateral meristems are unique structures derived from the shoot apical meristem and can either develop into an inflorescence or a tendril. How the grapevine plant controls these processes at the molecular level is not understood, but some progress has been made by isolating and studying the expression of flowering genes in wild-type and mutant grapevine plants. Interestingly, a number of flowering genes are also expressed during berry development. This paper reviews the current understanding of the genetic control of grapevine flowering and the impact of viticulture management treatments and environmental variables on yield. While the availability of the draft genome sequence of grapevine will greatly assist future molecular genetic studies, a number of issues are identified that need to be addressed–particularly rapid methods for confirming gene function and linking genes to biological processes and traits. Understanding the key interactions between environmental factors and genetic mechanisms controlling the induction and development of inflorescences, flowers, and berries is also an important area that requires increased emphasis, especially given the large seasonal fluctuations in yield experienced by the crop and the increasing concern about the effect of climate change on existing wine-producing regions.

Zamboni *et al.* (2010) reported that the analysis of grapevine (*Vitis vinifera*) berries at the transcriptomic, proteomic, and metabolomic levels can provide great insight into the molecular events underlying berry development and postharvest drying (withering). However, the large and very different data sets produced by such investigations are difficult to integrate. Here, we report the identification of putative stage-specific biomarkers for berry development and withering and, to our knowledge, the first integrated systems-level study of these processes. Transcriptomic, proteomic, and metabolomic data were integrated using two different strategies, one hypothesis free and the other hypothesis driven. A multistep hypothesis-free approach was applied to data from four developmental stages and three withering intervals, with integration achieved using a hierarchical clustering strategy based on the multivariate bidirectional orthogonal projections to latent structures technique. This identified stage-specific functional networks of linked transcripts, proteins, and metabolites, providing important insights into the key molecular processes that determine the quality characteristics of wine. The hypothesis-driven approach was used to integrate data from three withering intervals, starting with subdata sets of transcripts, proteins, and metabolites. We identified transcripts and proteins that were modulated during withering as well as specific classes of metabolites that accumulated at the same time and used these to select subdata sets of variables. The multivariate bidirectional orthogonal projections to latent structures technique was then used to integrate the subdata sets, identifying variables representing selected molecular processes that take place specifically during berry withering. The impact of this holistic approach on our knowledge of grapevine berry development and withering is discussed.

Bottcher *et al.* (2010) stated that in fleshy fruit, levels of indole-3-acetic acid (IAA), the most abundant auxin, decline towards the onset of ripening. The application of auxins to immature fruit can delay the ripening processes. However, the mechanisms by which the decrease in endogenous IAA concentrations and the maintenance of low auxin levels in maturing fruit are achieved remain elusive. The transcript of a *GH3* gene (*GH3-1*), encoding for an IAA-amido synthetase which conjugates IAA to amino acids, was detected in grape berries (*Vitis vinifera* L.). *GH3-1* expression increased at the onset of ripening (veraison), suggesting that it might be involved in the establishment and maintenance of low IAA concentrations in ripening berries. Furthermore, this grapevine *GH3* gene, responded positively to the combined application of abscisic acid and sucrose and to ethylene, linking it to the control of ripening processes. Levels of IAA-aspartic acid (IAA-Asp), an *in vitro* product of recombinant GH3-1, rose after veraison and remained high during the following weeks of the ripening phase when levels of free IAA were low. A similar pattern of changes in free IAA and IAA-Asp levels was detected in developing tomatoes (*Solanum lycopersicum* Mill.), where low concentrations of IAA and an increase in IAA-Asp concentrations coincided with the onset of ripening in this climacteric fruit. Since IAA-Asp might be involved in IAA degradation, the GH3 catalysed formation of this conjugate at, and after, the onset of ripening could represent a common IAA inactivation mechanism in climacteric and non-climacteric fruit which enables ripening.

Molitor *et al.* (2012) opined that the manual horizontal cluster division may constitute an efficient tool to optimize wine composition. To test this hypothesis and to determine the optimal timing of this viticultural practice, the impact of cluster division on cluster morphology, bunch rot disease severity, and harvest parameters such as yield and total soluble solids was examined on the white *Vitis vinifera* L. cultivars Pinot gris and Riesling in the Luxembourgish winegrowing region in 2010 and 2011. Treatments were applied at five different times between prebloom (BBCH 57) and veraison (BBCH 81). In both varieties, all postbloom cluster division treatments were able to loosen the cluster structure and reduce the bunch rot disease severity. The later cluster division took place, the more pronounced were the effects on cluster structure. The density index was a suitable measure of the cluster structure and the predisposition to bunch rot infestation. Cluster division resulted in yield reductions (up to 30 per cent), which increased with time after flowering. Postbloom cluster division may help to optimize wine quality by improving grape maturity due to crop load reduction, reducing fungal contamination, and improving grape composition through the possibility of elongating the ripening period. Postbloom horizontal cluster division can be recommended as an efficient crop cultural tool for premium wine production under cool-climate conditions.

Francisco *et al.* (2011) stated that the *Vitis vinifera* pollen dispersal was studied using volumetric air samplers situated within a vineyard to measure airborne pollen concentration. The pollen dispersal period of Cabernet Sauvignon was seven days, reaching daily mean concentrations of about 90 grains/m^3, with peak pollen dispersal near midday. Daily and hourly variations in concentration were positively correlated with temperature but negatively with relative humidity. Concentration was greatest within 1.5 m of the plants and up to 1 m aboveground, after which it diminished.

Mundankar and Karibasappa (2006) reported that at this Centre, the National Repository of Grape Germplasm presently contains 112 indigenous and 288 exotic accessions of grapes. Information on different aspects of these accessions like notification details, distinguishing morphological characters, area of adaptation, yield potential, bunch and berry characteristics, resistance to different biotic and abiotic stresses are required by scientist, managers, administrators, extension personnel and farmers. A user friendly Information System for Grape Germplasm in India has been developed, which makes storage and deification of data easy and enables quick and accurate retrieval of data. The software provides the facility to store and edit data on 113 characteristics as per the UPOV, IPGR and NBPGR descriptors for grapevines. Presently, the database contains data for 138 accessions on 42 characters that have been evaluated under tropical Indian agro climatic conditions.

Venkataram and Singh (1991) observed that the hybrids of the crosses Anab-e-Shahi, Angur Kalan, Bangalore Blue, Black Champa, Convent Large Black, Thompson seedless and Queen of the Vineyard were evaluated with petiole nutrient contents and resistance to diseases. Hybrids derived from Anab-e-Shahi x Queen of the Vineyard (9/3), Angur Kalan x Anab-e-Shahi (5/4), and Angur Kalan x Black Champa crosses (26/8) were found to perform better. Dry matter content of the petioles appears to be a suitable index for higher yield. Higher chlorophyll 'b' content (light harvesting complex) was related to high yield in Bangalore blue. Use of Black Champa as seed parent and Queen of the Vineyard as pollen parent is good for imparting high vigour to the progeny. Duration for ripening depended on genetic constitution rather than on climatic factors. Glucose and fructose contents were high in Thompson seedless. Thompson seedless was very efficient in N uptake than Bangalore Blue. Hybrids 9/3 and 26/9 possessed higher P content which was related to high fertility. Ca and Mg contents had no relationship with yield whereas Fe, Zn and Mn showed some positive relationship. Seedless berries were obtained in hybrids derived from Black Champa and Thompson seedless. Hybrid 21/28 showed tolerance to downey mildew while Hybrid 30/14, tolerance to anthracnose and hybrid 21/28 to powdery mildew.

Hipparagi and Singh (1991) reported that the shoot growth and berry development pattern in different genotypes of grape was studied in five promising hybrids with five parental cultivars in a replicated trial. The five cultivars used as female or male parents were; Anab-e-Shahi, Angur Kalan, Black Champa, Thompson

Seedless and Queen of the Vineyard. Hybrids were studied against their parents in respect of vine growth parameters, yield attributes and quality components. Shoot length and leaf number per shoot at 90 and 105 days after budbreak during the fruiting season were positively correlated with total soluble solids. Fresh weight of berries at 30, 45, 60, 75 and 90 days after fruit set was positively correlated with Brix-yield/15 berries and 50 berries weight at harvest. All the five hybrids had more shoot length and leaf number than their parents. The study has shown that Anab-e-Shahi hybrids E-5/12 and E-9/3 and Black Champa hybrids E-29/4 were improvements over their respective parents with respect to yield and quality attributes, while Black Champa hybrid E-29/4 were improvements over its parents with respect to yield and quality attributes, while Black Champa hybrid E-29/5 and Angur Kalan hybrid F-5/4 were improvements over their respective parents with respect to quality.

Manjunath and Shikhamany (1991) stated that the influence of different concentrations of cyanamide (0, 1, 2 and 3 per cent a.i.) at different dates of pruning (5th October, 26th October and 16th November, 1989) on budbreak and the consequential effect of variation in budbreak on different dates shoot growth, bunch development and berry quality was studied in Thompson Seedless grape under the mild tropical climatic conditions of South interior Karnataka. Budbreak at all nodal positions, except the terminal two nodes was influenced by the dates of pruning as well as the cyanamide concentration. Irrespective of the concentration, cyanamide treatment increased the budbreak, which in turn was associated with increased yield, brix-yield/cane and increase in the proportion of water berries. Cyanamide treatment hastened budbreak by about 6 days in vines pruned on 5th October and about 4 days in vines pruned on 26th October for 50 per cent budbreak to occur. Vines pruned on 16th November and not treated with cyanamide did not show even 50 per cent budbreak. Mean shoot length was reduced by delayed pruning and cyanamide treatment. Total shoot length/cane decreased with delay pruning but was unaffected by cyanamide treatments. Reduced total shoot length/cane was associated with reduced yield and brix-yield/cane and mean bunch and berry weights. Vines pruned on 5th October resulted in higher yield than those pruned later. Higher yield was mediated through higher weight of berries and bunch and increase in the number of clusters/cane. Cyanamide treatment did not increase the yield/cane. Proportion of water berries in a bunch decreased with delayed pruning. This however, was not affected by cyanamide treatment. Cyanamide treatment did not have any influence on the quality parameters. This study has thus shown that cyanamide can be used to hasten and increase budbreak in Thompson Seedless vines pruned as late as 16th November under the agroclimatic conditions of Bangalore. Cyanamide at 1 per cent a.i. was found to be the optimum. For achieving increase in yield through budbreak by cyanamide, fruitfulness of buds on the cane should be improved.

Raghavendra and Sahijram (2005) reported on investigations on "Development of seedless grape hybrids through embryo rescue and their characterization using molecular markers" was carried out at the Indian Institute of Horticultural Research, Bangalore, from Aug 2004-Aug 2005. The main objectives of investigation were to carryout controlled crosses between cultivar Thompson Seedless and downy mildew resistant male parents, in ovulo/ex ovulo embryo culture and molecular characterization to confirm hybridity using molecular markers. Crossing of cultivar Thompson seedless (TS) with downy mildew resistant male parents would be helpful in obtaining hybrids having characters desirable from both the grower's and the consumer's point of view. Using conventional breeding procedures, it is not possible to obtain hybrid progeny form seedless x seedless crosses. Besides, using the traditional method even seeded x seedless cross may produce only up to 10 per cent seedless plants. Most seedless varieties are stenospermocarpic. Through *in vitro* methods it is possible to rescue the developing embryo before it aborts and seedless hybrid progeny can be obtained. In ovulo embryo culture prior to embryo abortion provides an attractive alternative to conventional method by allowing the recovery of hybrid from the otherwise abortive ovules of seedless x seedless or seedless x seeded crosses. Molecular characterization to confirm the hybridity was done with ISSR marker. Each cross was found to have best time of ovule explanting at different week post pollination indicating that pollen donor parent and genotype also contribute to ovule and embryo formation rates. Two hundred and forty four hybrid plantlets were recovered from 1166 (20.96 per cent) embryos cultured which were excised from 1402 ovules cultured. The utility of ISSR marker in determining hybridity of the progenies of crosses TS X SV 18402 and TS X SV 12309 has been successfully demonstrated. By employing embryo rescue technique hybrid plants have been obtained from various crosses involving Thompson Seedless' as female parent and 6 different downy mildew resistant male parents.

Investigations were carried out by Usha *et al.,*(2005) with grape cv. Perlette to assess the influence of arbuscular mycorrhiza (AM) on breaking bud dormancy, and hastening flowering, berry set and berry ripening. Eighteen year old vines of Perlette, having uniform vigour and growth, planted at 2.5 x 2.5 m and trained on a telephone system, were inoculated with three strains of AM namely *Glomus mossae, Glomus deserticola* and *Gigaspora calospora,* when the root system was active in February. Treatments with FYM and inorganic NPK were maintained as checks. The grapevines inoculated with *Glomus deserticola* took significantly fewer days for bud sprouting, inflorescence emergence, 50 per cent flowering, berry set and ripening compared with all other treatments. The observations are discussed in relation to observed changes in the soil pH and leaf nutrient status. Hastening of bud sprouting, flowering, berry set and ripening in AM treated plants helped in harvesting the fruits before the onset of rains and reduced the damage due to monsoons. It is suggested that AM strain *Glomus deserticola* can potentially be included in the fertilizer management practice as an alternative to chemical fertilizers for providing the plant nutrient requirement.

11

Pest and Disease Control

1. Quality grape production and export.
2. Keep an eye on important disease and their symptoms.
3. Keep an eye on harmful insects.
4. At what stage which disease and what insect-strike.
5. Suitable climate means what precautionary measures one has to take.
6. Take prophylactic measures than curative ones. After the treatment, 100 per cent control is not possible. Therefore take control measures and timely control of pest and diseases is very important.

In-Time Crop Care

Whatever variety of grape you grow it should be free from pests and diseases. Countries may not purchase disease-affected grapes. In Karnataka to get disease-free grape cultivation, it is a towering task and a costly affair in which only in case of ripened grape condition you won't find the pest and disease. In our place the climate is often changing. As soon as the October rain stops and flower emerges there is severe cold temperature which the bunch has to tolerate and in a very severe hot temperature the fruit ripens. In the last 8-10 years situation, you will see timely flowering till harvest. Sometimes, cold temperature, fog, cyclonic rains and hailstorm damages grape cultivation.

Such situation may not arise at once but sometimes insect and disease problem may arise, while sometimes it may go simultaneously. After October pruning the crop requires 120 days for harvest.

Four-Point Disease-Free Programme

1. Good sanitation.
2. Use of standard fungicide and insecticide.
3. A comprehensive pest and disease control programme.
4. Nutrition management.

Maintaining a Clean Garden

At the time of October pruning keep the garden clean, particularly the disease-affected leaves which may be collected and burnt before actual October pruning.

Before October pruning the insect-affected branches are cut but disease-affected leaves are left as it is on the vine. These leaves are removed during October pruning. On such leaves powdery mildew, downy mildew, Botrytis, anthracnose infection may affect plants as well as soil, and immediately on newly emerged leaves, before October pruning one may notice anthracnose. Therefore, all these leaves may be collected and burnt outside the garden.

A Comprehensive Pest and Disease Control Programme

Immediately after October pruning, anthracnose will be noticed on newly emerging, soft and succulent branches along with flea beetle. To control this after October pruning apply bordeaux paste and water soluble. Carbaryl dust, Cofidor 0.3 ml/litre may be applied in the garden.

On vines when there is new sprout and bunch emergence, keep application of fungicide/insecticide on new leaves continuously. Till the berries attain a pea size, pest and disease problem will be severe. Under such situation keep an eye on the climate and analyse the possibility of disease which may be identified and take immediate control measures. Climate means temperature, humidity, precipitation, (fog, rainfall) etc.

There is a co-relation between temperature, humidity and diseases. Understand the following:

Sl.No.	*Stage*	*Period*	*Climate*	*Disease/Insect Possibility*
1	October pruning and emergence of new sprout.	October	Hot and humid climate with rains.	Anthracnose, Flea beetle
2	New bunch, new sprouts, bunch flowering, insects.	November	Cold climate, fog, cloudy weather	Downy mildew, Anthracnose.
3	Fruit set leaves and cane growth, bunch growth, berry development	December	Severe cold wave clouds, hailstorm	Anthracnose and powdery mildew, DM Lepidopteron insects
4	Berry development stops, ripening starts, then Harvesting	January February	Cold air hailstorm fluctuation of temperature	Berry drop mealy bug, bunch berry drop.

It means after October pruning two months are very crucial, very delicate and an equally dangerous period.

Sl. No.	Disease	Temperature	Relative Humidity (per cent)
1.	Anthracnose	30-32ºC	60-100 per cent
2.	Downy mildew	12-14 ºC	60-90 per cent
3.	Powdery mildew	20-24 ºC	40-90 per cent
4.	Phomopsis	23-26 ºC	90-100 per cent
5.	Diplodia	25-30 ºC	80-100 per cent

In our place the minimum temperature during November and December will be between 12-20℃ whereas the maximum temperature is noticed between 24-32℃. In the atmosphere the relative humidity is 50-75 per cent. After watering orchard and following rainfall and mist, the spraying percentage may increase. The above diseases may spread if there is no proper phytosanitation. Temperature and RH go hand in hand and such situation is most congenial for disease spread. If temperature or RH is increasing then take up foliar spray of fungicide. During this period give foliar spray or dusting of cane. When RH increases then take up dusting. If foliar spray is difficult, the quantity required per hectare is less then use good nozzle pump. The per hectare water requirement is 500 lit. If it is more than 500 lit. Then RH increases. When the fungicide spray does not fall on the leaves, there will be fungal infection which will be very quick. While taking up fungicide foliar spray one has to keep in mind that foliar spray should reach young sprout tips, leaves and bunches. For that excess leaves should be removed. Tie the canes to the wire for proper spray, so that the leaves can harvest sunrays and good air.

Benefits of Disease Control

By keeping orchard clean, the possibility of disease occurrence gets very much reduced. If you don't do it then immediately you will find the diseases on leaves and bunches. If such a situation arises, then immediately take control measures. For this purpose you have to identify the diseases and decide which chemicals you have to spray. If control is not possible with one spray then repeat the spray but change the chemical. For control of one disease there are several chemicals available, in which contact and systemic types are available. Use both of them.

Disease Controlling Fungicides

Sl.No.	Diseases	Contact Fungicides	Systemic Fungicides
1.	Anthracnose (Karpa)	1) Bordeaux mixture 2) Copper oxy chloride	1) Benlate 2) Bavistin 3) Topsin-M 4) Score 5) Bayeor
2.	Downy mildew	1) Bordeaux mixture 2) Copper oxy chloride 3) Phytolexin-84	1) Ridomil 2) Metalyxil 3) Alliette 4) Curzat

Contd...

Table–*Contd...*

Sl.No.	*Diseases*	*Contact Fungicides*	*Systemic Fungicides*
3.	Powdry mildew (Buri)	1) Wet. Sulf 2) Mycosole	1) Beletan 2) Rubigan 3) Contof
4.	Phomopsis	1) Bordeaux mixture 2) Captan	1) Quintal 2) Benomyl
5.	Diplodia	1) Bordeaux mixture 2) Copper oxy chloride	1) Benomyl
6.	Bacterial blight (Xanthomonas)	1) Ziram (3) CoC 2) Captan	1) Benlate

Note: To understand trade name, common name or chemical name details are given (Package of practice list to be included)

Nutrition Management

In grape garden, if nutrition management is not proper, then the disease problem may increase in gardens where nitrogen application is more and the diseases increase because excess nitrogen may bring more sprouting which are prone to anthracnose disease.

In places where magnesium deficiency is observed, in such gardens downy mildew disease may cause more damage. Application of potash brings resistance against diseases not only in grape but in other crop also. We can conclude in this way, excess and unnecessary application of nitrogen should be avoided. Magnesium and potash use helps in overcoming many diseases. If the disease resistance increases in plants then disease control may become easy.

Overall Care

After October pruning, in 79-90 days, the ripening starts, in another 30 days the harvesting gets completed. During 75-90 days the disease may cause partial or total damage. In this period four important points (cleanliness, healthy leaves, controlling disease, foliar spray) may totally help in control of diseases. After this period stop spraying of any fungicide.

While taking up quality grape production cleanliness go for, exact control measures and healthy leaves –the three factors decide and controls 75 per cent disease and 25 per cent disease can be taken care. If initially not controlled it may increase, which may reflect on the quality of the fruits and leaves. The fruit bunches are the most affected ones.

October Post-Pruning Diseases and Control Measures

After October Pruning, new sprouts and disease emerges simultaneously.

1. Very soft growth or bunch - Anthracnose
2. Fruit set - Downy Mildew

3. Bunch development - Xanthomonas
4. Berry development - Powdery Mildew
5. Berry development - Botrytis
6. Berry ripening - Botrytis

Major Disease Symptoms

Downy Mildew

Initially on leaves small sized light-greenish with yellow spots are observed on leaves. In a very short time in a humid climate on lower side of the leaves spores develop, then that part will become whitish in colour. Later on white patches light yellowish colour develops. Then slowly the diameter of patches increases but in severe cases the leaves may fall. Apart from leaves the young sprouts also show whitish patch, then the growth stops or the affected sprout gets dried or the leaves may fall.

Control Measures

Downy Mildew (*Plasmopora viticola*) is a fungus. The disease initially spreads by spores on vines by turn. The left over inoculum on light green leaves may act as a source in bunches of wine while the spores may germinate in congenial condition and these spores on newly emerged shoots get rooted. If the quantum of rain is more, then the disease may spread very rapidly. The atmosphere, relative humidity up to 85 per cent and more, if persists, plays an important role, in atmospheric temperature. If the temperature is prevailing between 10° to 30°C, then only the disease may increase or decrease. But 90-100 per cent of relative humidity with 24°C to 26°C temperature, the spread of the disease will be very fast.

Timely Disease Control

1. After October pruning, when new growth starts	Carbendizim
2. After 12-15 days	Carbendizim
3. After 4-5 days	Ridomil
4. After 12-15 days	Ridomil
5. 3-4 days	Streptocyclin
6. 8-10 days	Beletan
7. 8-10 days	Microsule
8. 8-10 days	Beletan/Rubigan
9. 10-12 days	Benlate/Benomyl
10. 10-12 days	Benlate/Benomyl
Totally in 85 to 95 days	Take up 10 spray

Anthracnose (*Gloeosporium ampelophagum*) (Karpa)

First symptoms will appear on soft shoot and young leaves. On leaves around 5 mm diameter spots are visible or thick spots reddish in colour, at later stage the centre of the spots becomes brownish in colour and that centre position may dry and fall off. The young leaves are the first victims of the disease. The sunken spots on canes are observed the during rainy season. The disease is also noticed on petioles with high relative humidity causing red spots on the diseased spot.

Control Measures

Congenial atmosphere is most suitable for its spread, because the mycelium may remain on the canes after the pruning, and when the rain starts within 24 hours mycelium starts developing, then they enter the leaves and the symptoms start appearing for which 32°C for 4 days is sufficient alongwith the rains. Relative humidity (90 per cent) which is most suitable for its spread may result in a major setback to farmers or loss.

Powdery Mildew (Buri) (*Uncinula necator*)

On every soft and green vine the affected portion may become whitish in colour. Later the colour may change to brownish. The margin shows more infection; when the infection increases it spreads on entire leaves; at later stages the powdery growth on both lower and upper region is observed. If the climate is cold the margin may dry up. On the tip of new sprouts you will find powdery growth. At flower primordia stage, powdery mildew infection may cause more damage.

Control Measures

Uncinula necator is a fungus on vine greenish portion and it affects most of the fungal mycelium and spores and one season to another season they persist on vines and canes. Already infected buds sprouts in a suitable climate, therefore, spread of disease starts which may affect the newly emerged leaves. This disease spreads when the temperature is 20° to 24°C but even a temperature range of 6°C to 32°C is bad. The disease under suitable climate may complete its cycle in 5-6 days. Its relative humidity plays a destructive role with the spread of this powdery mildew disease. Under relative humidity from 70 per cent to 80 per cent, the disease may spread very fast. But high relative humidity and very dim light with very low sunshine may also cause its spread.

Dead Arm (*Phomopsis viticola*)

This disease is caused by a fungus and occurs most frequently in table grape area the fungus lives throughout winter in disease-affected cane arms, spurs and petioles left on the vines; the infection occurs during the winter when the shoots are young. The disease spreads from the old disease area to young tissues by late spring rains. The angular spots often coalesce to form large brown areas with numerous dark spots. The diseases may cause canker in the older shoots and canes. The spread of the disease area on the cluster stem causes bunch rot with the fungus growing back into the wood of the arm and gradually kills the arm.

Control Measures

This disease is caused by (*Phomopsis viticola*) the fungal Mycelium and spores persists on the canes and at relative humidity of 90 to 100 per cent spread of the diseases starts on newly emerged leaves or it enters through air-injured portion. Later the vines get infected with the onset of rains through high wind. While taking up pruning the hands of labours who are engaged in pruning the vines, which are free from this disease, also get affected. After the entry of pathogen, symptoms start being visible within a month. The disease requires 100 per cent relative humidity with 23° to 26°C temperature.

Diplodia Cane Blight

This disease causes blighted area of green canes that are grey to brown with dark brown streaks projecting at the advanced margin of the blight canker. Grey areas are often speckled with numerous black dots.

Wounds caused by girdling and pruning cuts on grapevine trunks, are sometimes infected by *Diplodia natalensis.* Cankers develop under the bark, then the bark dries, cracks and peels away. A temperature of 30-35°C is most congenial. April pruning onwards, new sprouts emerges if a rain occurs or the relative humidity reaches 80 per cent then development may take place.

Alternaria

(*Alternaria alternata*) is a fungal disease and at the initial stage on leaves small yellowish spots are observed. Later they enlarge, and the leaf lamina may show leaf spots. If the disease is severe then leaf drop is noticed; during rainy seasons this disease is severe.

Leaf Spots (*Cercospora viticola*)

This is also a fungal disease, which you can observe on the young shoots which is light in colour and spots are small in size. The spots on older leaves at margin, when there is severity, may spread to other leaves of the vine, then slowly the colour of spots become black. In an atmosphere where the relative humidity is 85 per cent and above with a temperature of 25 to 26°C may help in further spread immediately.

Bacterial Blight (*Xanthomonas compestris* pv. *Vitis vinifera*)

This a bacterial disease; its first attack will be on young emerging leaves and young shoots. Small eruptions on young leaves at the lower surface on entire lamina will be observed and the infection is observed on veins then slowly the size of spots get enlarged to 0.5 mm to 2.5 mm where there are minute black spots or pointed spots or streaks. Slowly the spot on canes and on leaves petiole are observed. After pruning within 2-3 weeks, it enters the bottom of canes and enters. When the infected canes or branches are opened, inside we can see black spots in colour or discolouration.

Control Measures

This is caused by *Xanthomonas compestris* pv.*Vitis vinifera.* It enters from a injured portion and spreads to the entire plant system, which may remain for 200 days in the plant. The innoculum may move along with the metabolites or plasma in the entire plant system. So also insects, thunder storm, hail storm and a little temperature of 20 to 24°C may help in spread of the disease very fast.

Insect Pest of Grape

Thrips (*Rhipipherothrips cruentatus*)

Thrips are the most trouble some insect pest of grape. They cause white patches on leaves of grape, curling of leaves and yellowing of blossom. There will be poor set with infested vine. They are also responsible for scab formentation of berries. This can be controlled by spraying 1 mda chloropid 0.3 ml/litre Dimethoate 1.70 ml/litre oxy demeton 1 ml/litre.

Flea Beetle (*Scelodonta strigicollis*)

This also a major pest in grape. The grubs feed on the upper surface of the leaves and tender shoots and also attack roots. This can be controlled by spraying Imda chloropid 0.30 ml/lit, oxy demeton methyl 1 ml/litre, phophomidon 0.5 ml/lit.

Vine Girdler (*Stethenias grisator*)

These beetles are found during night. They make a girdle on the arms or on the main trunk in the initial stage and completely cut the vine. This can be controlled by swabing the bark by carbaryl 20 g/litre of water.

Mealy Bug (*Ferrisiana virgata*)

Both nymph and aduls suck cell sap from the plant parts and berries. Infested portion are generally coated with a sticky honey dew on which sooty mould develop. If shoots are affected, it results in malformation of leaves and growing shoot tips, mealy bugs are effectively controlled by release of lady bird *Cryptolaemus montrouzieri, pullus alegans, Scymnus cocivora* on a large scale. A minimum of 600 beetles has to be released for an acre. Some time sticky substance like 'Tack Trap' or 'Bird tangle foot' may be applied either side of the shoot at the junction of the bunds Mealy bug can be controlled by spraying Applaud 1.25 ml + fish oil soap 5 gm/litre drencling with 1mda chloropid 2-3 ml/lit./plant. Spraying *Verticillium lacanii* 5g/litre also help in controlling of mealy bug, or *Verticillium lacanii* should be mixed with vermi wash and applied through drip to overcome mealy bug problem if the infestationis very severe then remove the bark of the vine and brush with tooth brush to clean the mealy bug, some time paste the tape one feet above the ground and on tape greese can be applied, but don't apply greese on main trunk otherwise the vine may die.

Mites

They suck the sap of the leaves, in severe case the leaves may dry. To control this spray Dicofol 2ml/litre or Ethion 2 ml/lit or Wettable sulphur 2 gm/litre of water.

Nematodes (*Meloidogyne incognita*)

The affected vine show slow decline in vigour with inadequate sprouting after pruning. Apply Neem cake 500 gm/vi.

Daane *et al.* (2006) reported that the vine mealybug is a severe, new vineyard pest in California. Typical treatment relies on application(s) of organophosphate or carbamate insecticides. We investigated the use of less-disruptive insecticides, releases of natural enemies and mating disruption as alternative control strategies. Results show a systemic application of imidacloprid or a foliar application of buprofezin reduces crop damage. Inoculative release of a parasite (*Anagyrus pseudococci*) also reduced crop damage; currently, this natural enemy is not commercially available. Similarly, mating disruption using a synthetic sex pheromone can suppress mealybug populations, but the sprayable formulation is not yet registered in the United States. There is great potential in both augmentation and mating disruption, which are being further developed for commercial use.

Nicholls *et al.* (2006) presented the results of studies in organic vineyards in Mendocino and Sonoma counties, California, in an effort to systematize the emerging lessons from our experience on vineyard biodiversity enhancement for ecologically-based pest management. In the Mendocino study, a vegetational corridor connected to a riparian forest channeled insect biodiversity from surrounding habitats into the vineyard, thus overcoming the restricted spatial limits to which the positive influence of adjacent vegetation on vineyard pest dynamics is usually confined. In addition, summer cover crops substantially enhanced biological control of leafhoppers and thrips, by breaking the virtual monoculture that vineyards become in the summer after winter cover crops dry out or are plowed under. In the Sonoma vineyard, an island of flowering shrubs and herbs provided season-long flower resources and alternate preys/hosts for natural enemies, which slowly built up in the adjacent vineyard. The island acted as a push-pull system for natural enemies, enhancing their activity but confining them mostly to the adjacent vine rows. Planting strips of summer cover crops could be a strategy to overcome the push effect of the island.

Kulkarni *et al.* (2006) stated that the chemical based, bio-intensive based and neem-based modules were tested for their bio-efficacy on insect pests of grapes. Chemical based module was found most effective treatment in reducing insect population and it was reflected in yield obtained (13.90 kg/vine) followed by Bio-intensive based–module (combination of neem formulations+bio-pesticides+chemical pesticides) (12.50 kg/vine) and only neem based modules (11.22 kg/vine and 10.20 kg/vine). All the modules were superior over untreated check (7.80 kg/vine) in reducing insect population as well as yield. Bio-intensive module consisting of Azadirachtin 1 per cent and 5 per cent formulations along with chemical insecticides and bio-pesticides was better alternative for minimizing pesticide residues in grapes. No phytotoxicity symptoms were observed in any of neem formulation treated plots. Good natural enemy activity (1–2 grub/adults/vine) of ladybird beetle was observed in neem treated plots. Compatibility studies revealed that Azadirachtin 1 per cent at concentrations up to 50 ppm had no adverse effect on growth of *Trichoderma harzianum* Tul. or *Verticillium lecanii* Zimmerman *in vitro*. Reduction in growth was noticed only at 100 ppm and higher concentrations. Studies on shelf life and berry rot studies with or without *Trichoderma harzianum* sprays revealed that the number of rotten berries was least in Azadirachtin 1 per cent at 3 ml/L + *T. harzianum* 5 ml/L (containing 1 x 106 cfu/ml) treatment followed by Azadirachtin 1 per cent at 3 ml/L treatment and Azadirachtin 1 per cent @2 ml/L + *T. harzianum*. No difference was found in number of fallen berries on any day.

Kulkarni *et al*. (2006) stated that the survey of insect pests at National Research Center for Grape, Pune during April 2004–March 2005 at weekly interval revealed that fleabeetle population was observed immediately after both foundation and fruit pruning. Highest population of 5.8 fleabeetle/vine was observed during first week of November, which coincided with bud sprouting. Thrips population was observed throughout the year with an average peak of 8–10/shoot of vine in November and December months, which coincided the flowering period. Similarly, jassids population was observed round the year with highest of 6.4/shoot during first week of the December. Mealybug population was distributed sporadically and the highest population of a 5–6 colonies/vine during the last week of February to the last week of March which coincided with the fruiting and harvesting season. Mite population occurred mainly during winter months, *i.e.*, second fortnight of December to second fortnight of February with average of 5–6 mites/shoot/vine. Correlation studies with weather parameters and insect incidence revealed that rainfall was significantly negatively correlated with population of sucking insects like thrips and jassids (-0.52** and -0.42**, respectively). Mealybug population was negatively correlated with both maximum (-0.47**) and minimum humidity (-0.50**).

Rather (2006) reported that among the phytophagous mites, eriophyiid gall mite *Colomerus vitis* (Pagenstecher) is very common and destructive species forming galls on the foliage and stunting the growth of berries in Kashmir Valley. Two predatory mites, *Amblydromella viniferae* (Rather) and *A. dalii* (Rather) were observed in greater numbers in the galls of this species. Vines harbouring large number of these predatory mites bore very less number of galls on the foliage. Screening of different grape varieties revealed that Sahibi, Khalili, Hussaini and Anab-e-Shahi are resistant to vine gall mite in the State. Himrod proved to be very susceptible variety. Predatory mites, *viz. Euseius vignus* Rishi and Rather and *E. insanus* Khan and Chaudhri dominated both in vineyards and surrounding vegetation. Both these species along with *Agistemus herbarius* (Kuznestov and Wainstein) were seen actively feeding upon eggs and nymphs of spider mites. *Lasioseius* sp. (Ascidae) and *Paracheyletia pyriformis* (Banks) (Cheyletidae) were also seen associated with phytophagous mites, though in low numbers. Surrounding vegetation if managed properly can improve the biological control of grapevine mite pests by providing refuge, hibernating sites and alternative food sources to predatory mites.

Chandra Sekhar *et al*. (2006) observed that the studies on biology of red spider mite, *Tetranychus urticae* Koch (Acariformes: Tetranychidae) on Thompson Seedless grapevine indicated that the duration of different developmental stages of *T. urticae* were 3.68 days for eggs, 2.06 days for larvae, 0.74 days for I quiescent, 1.89 days for protonymph, 0.79 for II quiescent, 2.02 days for deutonymph and 0.84 days for III quiescent stages, respectively. The fecundity was 73.20 eggs/female, while pre-oviposition, oviposition and post-oviposition periods lasted for 1.96 days, 10.53 days and 2.48 days, respectively. The total developmental period from egg to adult was recorded as 12.03 days and longevity of male was 8.60 days and that of female was 14.60 days. The seasonal incidence of *T. urticae* in relation to abiotic factors and also chemical management of mite was studied on Thompson Seedless from November 2004 to April 2005 at an experimental plot, Maheshwaram, Ranga Reddy district,

Andhra Pradesh, India. Maximum and minimum temperatures were found to be positively correlated with the population build up of the mite. Highest population of 22.11 to 25.45 mites/cm^2 leaf area was observed between March and April 2005 when the temperature ranged between 35 and 37.3°C. Chemical management studies showed that abamectin at 2.70 g. a.i ha^{-1} was significantly superior over all other treatments with 92.65 per cent population reduction over control and also recorded highest fruit yield 26757 kg ha^{-1}. The other treatments in descending order of their efficacy were fenazaquin 125 g. a.i. ha^{-1} (90.82 per cent), difenthiuron 300 g. a.i. ha^{-1} (89.29 per cent), spiromecifen 120 g. a.i.ha^{-1} (87.31 per cent), milbemectin 5 g. a.i. ha^{-1} (86.48 per cent) and dicofol 231 a.i. ha^{-1} (79.82 per cent).

Patil *et al*. (2009) stated that we compared hematologic parameters and liver and kidney function tests in occupationally exposed pesticide sprayers of grape gardens (n = 60) and normal healthy participants (n = 30), 20-45 years of age, in Western Maharashtra (India). Venous blood samples were collected from both groups. Compared with control participants, sprayers showed the following-significantly increased serum C reactive protein (117.6 per cent), liver function marker enzymes-serum aspartate transaminase (57 per cent), alanine transaminase (37.4 per cent), alkaline phosphatase (16.9 per cent), serum bilirubin (41.8 per cent), creatinine (18.4 per cent), blood glucose (11.4 per cent), and urea (13. per cent); and decreased acetyl cholinesterase activity (30.9 per cent) and serum cholesterol (12.12 per cent). Serum total protein, globulin, and the A/G ratio were not significantly altered, however, but serum albumin decreased slightly (3.3 per cent, $P < .05$). Compared with the control group, hematologic parameters significantly decreased in sprayers-hemoglobin (6.9 per cent), hematocrit (3 per cent), mean corpuscular volume (3.8 per cent), mean corpuscular hemoglobin (4.5 per cent), mean corpuscular hemoglobin concentration (5.8 per cent), and red blood cell count (7.7 per cent), whereas the white blood cell count increased (18.2 per cent). The decreased serum acetyl cholinesterase value indicates a high degree of pesticides absorption, which leads to impairment of liver and kidney functions and slightly altered hematologic parameters in the occupationally exposed pesticides sprayers of grape gardens studied here.

Kehrli *et al*. (2011) stated that in Europe, the polyphagous planthopper *Hyalesthes obsoletus* Signoret (Hemiptera, Cixiidae) is assumed to be the most important vector of the grapevine yellows disease bois noir, which is caused by phytoplasmas of the stolbur 16Sr-XII-A group. For a better understanding of the epidemiology of bois noir in Switzerland, distribution and host plant preferences of *H. obsoletus* were studied in the field as well as in the laboratory. A national survey revealed that *H. obsoletus* is present in vineyards of southern, western and northern Switzerland; however, no specimens were caught in the east of Switzerland, where evidence for the disease is lacking. Even though field bindweed (*Convolvulus arvensis* L.) was much more abundant in vineyards than stinging nettle (*Urtica dioica* L.), *H. obsoletus* adults were captured almost exclusively on the latter. A second, more targeted field study confirmed *H. obsoletus'* clear preference for *U. dioica*. From June to September 2008, adults of *H. obsoletus* were primarily captured on *U. dioica*. However, few individuals were also caught on *C. arvensis*, hedge bindweed (*Calystegia sepium* L.) and four other dicotyledons. Emergence traps set up above *U. dioica* and *C. arvensis* highlighted that significantly more nymphs emerged from stinging nettle than field bindweed. Besides, a clear association between the distribution of *U. dioica* patches and the incidences of bois noir infected vines could be observed in a commercial vineyard. In conclusion, there is good evidence that *H. obsoletus* is also the most important insect vector of bois noir in Switzerland and that the insect prefers to feed and develop on *U. dioica*. Stinging nettle therefore plays a major role in the epidemiology of bois noir in Swiss vineyards.

Van Helden (2011) opined that the flavescence dorée is caused by a phytoplasme spread by the leafhopper *Scaphoideus titanus* and is present in the south of France. Leafhopper control is based on up to 3 obligatory insecticide sprayings (T1; T2; T3) using neurotoxic insecticide compounds over large areas (Imposed Sprayed Area = ISA). Spraying dates are defined according to the observation of the first larvae (T0) at T1= T0+4 weeks, T2 = T0+6 weeks (T1 and T2 aim the larvae) and T3 = T0+10 weeks (aiming at adults). Such schedules and area definition are very efficient but lack scientific basis on insect dynamics and migration and encounter opposition from farmers. Over the last 5 years progress was made to reduce pesticide use while maintaining sufficient vector control. A project was launched in the Saint Emilion region after two plots were found contaminated in 2006. 13000ha of vineyards entered the ISA. Farmers and regional Plant protection service organised into a small association to test a more 'integrated' approach for vector control. Spraying zones were limited to a buffer zone of 2 Km around the contaminated plots. Farmers between 500m and 2 Km were allowed to reduce sprayings to only 1 single application at T0 + 5, while adult trapping was used to monitor overall results. Traps were yellow delta traps (Piège Tri-_nglué) that are highly efficient. In most of this zone a single insecticide spray reduced trap catches to almost nonexistent, considered as a proof of sufficient control of the vector. In some areas trap catches were considered too important (>3 adults on a single trap in a single week) and a second spraying (T3) was imposed. This was probably due to some farmers either not spraying at all or organic farmers that only were allowed Rotenone, that is rather inefficient. This approach was repeated in 2008. Total cost of trapping and monitoring of larvae (traps, technicians, monitoring) is around 3 to 4 Euros per ha. These costs were largely compensated by the reduction in spraying (64 and 72 per cent reduction in 2007 and 2008 compared to the traditional approach). Moreover, the combination of reduced spraying and vector monitoring seemed well accepted by the farmers. In addition all plots were monitored for plants showing symptoms of phytoplasm, around 1200 samples were analyzed first year. Over 80 per cent of the samples appeared to be 'Bois Noir' phytoplasm but some Flavescence Dorée samples were found in both 2007 (2 samples) and 2008 (6 samples). Scouting and analysis costs were around 25 Euro per ha. In other areas where spraying frequency was reduced simultaneously but without vector monitoring, a strong increase in FD symptoms was observed, together with high insect populations. This probably reveals that many farmers do not apply the single imposed spraying. The trapping seems a useful tool both to show efficient vector control and to incite farmers to respect the minimal spraying frequency.

Linder *et al.* (2011) reported that the leafhopper *Scaphoideus titanus* (Hemiptera: Cicadellidae) is the principal vector of *flavescence dorée*, a severe grapevine disease spreading in Europe. The project PROPSCAPH united four European research institutes in order to improve risk management schemes against the spread of flavescence dorée's vector on propagation material. Amongst others, the impact of the standard hot water treatment against the disease on the development of the *S. titanus* eggs was studied. In autumn 2008, one-year-old and two-year-old cuttings were collected in *S. titanus* infested vineyards in France, Italy and Switzerland and sent to our institute. After storage in a cold chamber until March 2009, half of the cuttings were exposed for 45 minutes to a hot water treatment at 50°C. Thereafter, treated and untreated cuttings were placed in cages and the hatching of *S. titanus* nymphs was assessed. At least 10-times more nymphs hatched from twoyear-old cuttings than from the one-year- old cuttings. This confirmed the ability of *S. titanus* to lay eggs on propagation material. However, hot water treatments significantly reduced the number of hatched nymphs, *e.g.* the exposure to hot water killed about 80 per cent of deposited eggs. In conclusion, hot water treatments do not only have a major impact on *flavescence dorée*, they also kill most of its vector's eggs. Thus, the hot water treatment of rootstocks and scions before grafting is strongly recommended in order to reduce the risk of *S. titanus'* propagation. Before commercialization grafts should be hot-water treated once again, since the chemical protection of rootling is not always 100 per cent effective in the nursery fields. All these measures together will ensure the propagation of plant material free of *flavescence dorée* and they will minimise the risk of diseases and vector's spread.

Diez-Navajas and Ortiz-Barredo (2011) observed that the integrated production Regulation for grapevine in the Basque Country was published in 2004 (BOPV 221/ 2004, official bulletin). A previous 3-year study was carried out to evaluate the real situation of growers to assume this regulation in their exploitations. Several deficiencies were reported, including technical and management aspects. This work is focused in Rioja Alavesa, an area with 12,869ha of vineyard in the Northwest of Spain producing 81,325,000 wine grape kilo. Integrated disease control interest is rising, although the number of fungicides applications in the last two years against downy mildew and other fungal diseases reached up to 12 treatments, being usual 5 or 6 per season. These data prompted us to implement the strategy of fungal disease control. We installed 12 monitoring weather stations in different trials, according to different environment and geographical characteristics. The stations were equipped with sensors of temperature, relative humidity, leaf wetness, soil temperature, radiation sensors, rain gauge, and an anemometer. Registered data can be discharged and recuperated from internet, at the same time that the risk assessment for downy and powdery mildew and botrytis diseases. In parallel, one sticky glass trap to capture spores by impact was installed near each station and two more traps in each field. Pheromone traps to capture *Lobesia botrana* were also distributed in the fields close to the spore traps, as the abundance of this insect gives an idea of the possible future botrytis damages and spread. After these assays we would be able to validate the stations for each studied zone and assure that the risk assessment made by climate parameters follows a real model, taking into account the disease incidence in the field as well as the spore and vector populations.

Sentenac *et al.* (2011) reported that the control of mealybugs by growers currently relies on the use of neurotoxic insecticides. The possible use of the predator *Chrysoperla lucasina* (Lacroix) is the unique alternative method. We propose to evaluate the biological activity *in situ* of the bred *Ericydnus sipylus* (Walker), major parasitoid of *Heliococcus bohemicus* Sulc. The releases of parasitoids were carried out in 2007 and 2008.

During the 2007 season, the first three releases of *E. sipylus* did not appear to have any effect on the population of mealybugs. However the fourth and fifth release made later in the season did show significant differences between the experimental plots. The biological control showed positive results in 2008 and 2009, as the number of *H. bohemicus* in the "release" blocks was significantly lower than those of the "control" blocks. A reduction of 50 to 60 per cent of the number of *H. bohemicus* was found for the 2007 winter generation and from 75 to 85 per cent for the 2008 summer generation. Despite of stopping the release of *E. sipylus* rates of 80 per cent were found in the 2008 winter generation, and 80 per cent in early 2009 summer generation. If the difference of the population of *Pseudococcidae* between the experimental plots was due to the introduction of the parasitoid *E. sipylus*, the lower level of the population found in summer 2008 and 2009 in the "control" blocks was probably due to: a) the presence of other parasitoids like *Anagyrus szodensis* or predators like *Exochomus quadripustulatus* (Linné), *Hippodamia variegata* Goeze, *Harmonia axyridis* (Pallas), *Coccinella septempunctata* Linné, *Nephus quadrimaculatus* (Herbst) and *Chrysoperla carnea* (Stephens) b) the dispersion and activity of *E. sipylus* beyond the plots where it was introduced.

Naud *et al.* (2011) reported that the political roadmap about pesticides in France is to reduce quantities by 2, "if possible", before 2018. Research is needed to design and evaluate new pest management solutions. A decision workflow system, name GrapeMilDeWS was designed, at the plot scale, to handle grapevine powdery and downy mildews. GrapeMilDeWS stipulates throughout the season if and when fungicide sprayings should take place. GrapeMilDeWS has been experimented on a network of plots in different French wine regions. We give some results about the number of sprayings generated by GrapeMilDeWS and the crop protection performance obtained. We discuss methodological aspects such as partnership and data required to test and check such decision system.

Hein *et al.* (2011) stated that the using organic nanofibers as dispensers for pheromones and kairomones in plant protection for disrupting insect chemical communication is a novel approach aiming at popularizing this technique in organic and integrated plant production. Expected advantages of the nanofibers are highly controlled spatiotemporal release rates of pheromones/kairomones, improved climatic stability, and mechanized application. Dispenser types used so far show deficiencies in one or more of these requirements. Mechanical application of pheromone dispensers is a new approach to reduce the costs of manual labour. Therefore the environmentally compatible, highly specific and efficient technique of mating disruption may become an alternative to the use of synthetic pesticides in integrated pest management. The electrospun nanofibers are highly elastic, which prevents break-off of smaller pieces, and polymers used are biocompatible. Due to the tiny scale of nanofibers the mass input both for pheromones and for polymeric nanofibers is extremely low. Major environmental benefits are high control specificity, very low concentrations of residues and minimal risk towards development of resistant insect strains. We tested organic polymer fibers, made from the biodegradable polymer Ecoflex® (BASF), treated with *Lobesia botrana* (Lepidoptera: Tortricidae) sex pheromone, (E,Z)-7,9-dodecadienyl acetate, in cage tests placed inside vineyards. With this experimental setup it is possible to show the applicability of mating disruption dispensers for the use in plant protection against insect pests in vineyards. We used Isonet LE dispensers (Shin-Etsu) as a positive control. The Ecoflex® nanofiber-pheromonedispensers show a mating disruption effect which is comparable to the efficacy of the Isonet LE dispensers for at least three weeks. After that, the mating disruption effect observed tapers off to a level where it is insufficient for plant protection purposes. But, in principle, our first prototype of a nanofiber pheromone dispenser has proved its efficacy for mating disruption purposes. Modifications of the fibers, currently under development, open up the possibility of extending the disruption effect to a period of several months' duration, enough to cover the entire flight period of *Lobesia*.

Besides mating disruption techniques and progress in monitoring techniques (*e.g.* the use of food traps against females), biological control may reveal itself very efficient at controlling rape moth populations (Thiery *et al.*, 2011). Parasitoids active to control grape moths are known for long in vineyards; few of them were already described in the middle of the 19th century in French vineyards and their efficiency was already recognized especially against the diapausing and the first spring generations of the moths. Rather numerous attempts to release egg parasitoids have been done in different European countries using different species of trichogrammas. The results obtained varied a lot and could not yet clearly promote the use of this technique in vineyards. We believe that a biological control based on larval parasitoids could efficiently be developed as a valuable alternative to chemical control. In the present paper, we focus on larval parasitoids among which ichneumonids and chalcidoids (Hymenoptera) dominate, and present results obtained in different French vineyards (Bordeaux vineyard, Perpignan and Montpellier area, Côtes du Rhône and Alsace). We discuss factors that may favour or reduce their efficiency as biocontrol agents.

Lucchi and Santini (2011) stated that the life cycle and the ecology of the grapevine moth (GM) *Lobesia botrana* (Den. and Schiff) feeding on *Daphne gnidium* L. (to date considered the original host for *L. botrana*) were investigated in a Tuscan Natural Park in the last 2 years. The studied environment was located 2 km away from the beach and hosted numerous shrubs of *D. gnidium* on which *L. botrana* has been feeding forever, in the absence of *Vitis vinifera*. In the mentioned environment the moth seasonal flights were monitored and the larval feeding behaviour along the whole season was defined.

Up to 2007, two species of grapevine leafminers were known in Italy as well as in Europe: *Phyllocnistis vitegenella* Clemens and *Holocacista rivillei* (Stainton). They are usually considered minor pests and do not require specific control measures. However, an increasing importance of these pests has been observed in north-eastern Italy in recent growing seasons (Duso *et al.*,2011). An additional leafminer species belonging to the genus *Antispila* was also detected in 2007 in northeastern Italy. The origin of this species, which is new for Europe, is still unknown. The phenology, pest status, and natural control of these leafminer species have been investigated in a number of vineyards located in the Veneto and Trentino regions. The implications of leafminer occurrence for integrated pest management (IPM) on grapevines are discussed.

Hommay *et al.* (2011) stated that the several releases of 20,000 and 40,000 individuals of *Trichogramma cacoeciae* were made at a central point in a vineyard in order to follow the dispersal of trichogrammes and the distribution of parasitism on *Ephestia kuehniella* egg-cards. The dispersal of trichogrammes essentially occurred along the release row and over the nearest rows, since the vine rows formed plant screen that channelled their dispersion. The daily captures of trichogrammes increased with the number of trichogrammes released and decreased over time. It increased with minimal temperature and solar radiation, but did not depend on other meteorological factors. The number of parasitised egg-cards decreased with the distance from the release point and increased with the amount of wind received at the control points. The doubling of the number of trichogrammes released influenced parasitism by increasing both the discovery rate and the exploitation rate of egg-cards.

Cocco *et al.* (2011) opined that a mating disruption experiment against the vine mealybug, *Planococcus ficus*, was carried out during 2008 in 3 commercial vineyards in northwestern Sardinia. Within each vineyard, 2 adjacent plots of variable size (0.2, 0.4, and 0.5ha, respectively) were established as mating disruption and control plot. Plastic dispensers (Suterra Inc., USA) with 100mg each of the synthetic sex pheromone were deployed in mid-May, before the first flight of adult males, at a rate of 625 dispensers per hectare. The effectiveness of the mating disruption technique was evaluated by comparing the number of males captured in pheromone traps, the leaf population density, the percentage of ovipositing females, and the crop damage. Male flights were monitored with traps baited with 0.01mg of the sex pheromone. The percentage of females with ovisacs and their fecundity were estimated by collecting adult females after each peak male flight and rearing them individually under laboratory conditions. The mealybug density was assessed from June to September by counting every 2 weeks nymph and female mealybugs on basal leaves of 20-30 vines per plot. The crop damage was evaluated at harvest by assessing the percentage of damaged bunches on the same vines. In all the vineyards, the number of males captured in mating disruption plots was lower by far than in control plots, while the percentage of females with ovisac was significantly different only in the first generation. Females collected from the field after the first male flight produced more eggs than those collected after the second and third flight, with no significant difference between treatments. The density of mealybugs on leaves in mating disruption plots was lower than that in control plots, but the difference was not significant. Also, the percentage of damaged bunches at harvest did not differ between treatments. Mating disruption experiments to control the vine mealybug produced mixed results. This could be due to the limited size of experimental plots and/or the suspected parthenogenetic reproduction of Sardinian populations of *P. ficus*.

Heliococcus bohemicus is an increasing pest in vineyard in the north-east of France. A biological control was initiated in 2007 and 2008 (Fleish *et al.*, 2011). A study in laboratory, to measure the two antagonist biological characteristics, the mealybug and one of its parasitoids *Ericydnus sipylus*, was undertaken. The description of the mealybug life cycle is led under the climatic conditions: 25°C and 60 per cent R. H. It appears that *H. bohemicus* has three larvae instars. Females did not lay off eggs and the larvae hatched after 23.5 days on average. The development of first and second instars lasted near 34 days and for the third instars it was near 14 days. The biological characteristics of the parasitoid were studied at 21°C and 50 per cent R. H. Under these conditions the pre-imaginal development lasted 35 days and the length of adult emergence was 57 days on average. The fertility is equal to 15 individuals. Others parameters are studied and are explained in the article. The method of the mass rearing is developed. The pest insect is developed on potato tubers turned green by light.

Bagnoli (2011) stated that in the context of the Euphresco-Propscaph project, a survey on *Scaphoideus titanus* egg distribution on grapevine has been under way since January 2009. Samples of one and two to four-year old vine wood of the "Trebbiano Toscano" variety from an untreated vineyard in Latium (Central Italy) and samples of one and two-year old of the "Kober 5BB" rootstock from an abandoned field in Veneto (North Italy) have been analysed. The vine samples, collected during the winter pruning, were stocked at 4°C till the beginning of the experimentation, when the same materials were cut into pieces of different lengths and put into rearing boxes (cm 29x20x12) maintained at 24°C, 75 per cent RH and LD 16:8 photoperiod to obtain the first instar *S. titanus* larvae. In the middle of May in two Latium untreated vineyards, several original capture devices were put on the plants, in particular around the trunk, the cordon, the cane of one year old and the buds of the vines to survey the emergence of *S. titanus* larvae. Results confirmed that *S. titanus* females prefer to oviposit in the bark of two or more year old wood (0.09-0.31 larvae per cm^2), even if very few specimens were obtained from the one-year old wood (0.0002- 0.0036 larvae per cm^2) as well. The field experiments with the capture devices evidenced for the first time that even the bark of the trunk is a preferential site of the cicadellid oviposition as well.

The leafhopper *Scaphoideus titanus* is the vector of the Flavescence dorée. In this study, we investigated (Chuche *et al.*, 2011) the aggregative behaviour of the larvae. We conclude from different experiments that larval aggregation occurs at the plant scale and the age and colour of the food source could be factors cueing an aggregation. These aggregation patterns should be studied in more details in order to gain knowledge in the epidemiology of Flavescence dorée and eventually to develop control strategies based on inter-individual epideictic regulation.

Mazzoni *et al.* (2011) reported that the *Hyalesthes obsoletus* Signoret is a vector of the stolbur phytoplasma (group 6SrXIIA), causal agent of the grapevine disease 'Bois Noir'. In plant hoppers the intraspecific communication is based on substrate-borne vibrational signals. The aim of this study was to investigate the mating behaviour of *H. obsoletus*. Vibrational signals were detected from nettle cuttings by laser vibrometer. Both male and female can start the mating process by the emission of calling signals.

The male calling signal is a single short syllable (Male Syllable 1) whereas the female's is a train of pulses of variable length. Initially, the male/female duet is based on the alternating emission of calling signals (recognition phase). In a second step, the male starts emitting long syllable trains (male courtship signal) in which the female occasionally inserts single pulses (courtship phase). The male courtship signal is made of two different types of syllables (Male Syllable 2 and 3), the switch from one to another often triggered by the emission of a female pulse. When partners are in close vicinity the male emits a specific pre-copula signal that precedes the copula (pre-copula phase). Rivalry tests, either pairs of males or trios, showed a strong competition between males, expressed in the form of chorusing, rivalry signals and aggressiveness.

Huth *et al.,*(2011) state that the common earwig *Forficula auricularia* has been classified as a beneficial predator in vineyards. Amongst others the insect feeds on grape pests like different tortricids. In recent years within many regions of the viticultural area of the Palatinate the individual densities increased to an extremely high level. Earwigs may cause direct damages such as contamination of the grapes with faeces, eroded berries and transfer of pathogens. The chemical agent 2- ethyl-1,4-benzoquinone, released from the abdominal glands while earwigs are menaced and likewise contained in faeces, may have a negative influence on the wine quality. All these facts constitute a deterioration of grape quality for wine-growers. The high number of earwigs in the grapes after vintage causes a negative image by consumers. This study was carried out to investigate possible relations between the population dynamics of earwigs and specific environmental conditions in vineyards. The main focus of the research project was focused on chemical, ecological and biological strategies to reduce the population densities. Another important point of survey was to study the life cycle of earwigs especially in vineyards. For sampling purposes in the cluster zone a special life trap made of bamboo tubes has been developed. This type of trap showed the highest catch rate of the four trapping types tested. For monitoring the ground dwelling earwigs pitfall traps were used.

Kampimodromus aberrans is an important predatory mite in European vineyards and fruit orchards (Tirello *et al.,* 2011). Pesticides can exert profound effects on *K. aberrans* populations that are influenced by the pesticide experience of predatory mites. Resistance to organophosphates and dithiocarbamates has been suggested for some Italian populations of *K. aberrans* but this phenomenon has not been shown in the laboratory. The effects of three different pesticides (chlorpyriphos-ethyl, methoxyfenozide and flufenoxuron) on *K. aberrans* were investigated in field and laboratory conditions. In field tests, predatory mites were released at the beginning of February on defined plots. *Kampimodromus aberrans* densities were evaluated on insecticide treated plots and on control plots by examining leaf samples in the laboratory from June to August. The lethal and sub-lethal effects of the same pesticides were tested in the laboratory using a residual bioassay. Field experiments showed some effects of the three pesticides on predatory mite populations. Results of bioassays were partially correlated with those obtained in the field. Implications for IPM are discussed.

Lorenzon *et al.* (2011) stated that the *Kampimodromus aberrans, Typhlodromus pyri* and *Amblyseius andersoni,* important predatory mites in European vineyards, were reared in the laboratory on five potential food sources: the herbivores *Panonychus ulmi* and *Eotetranychus carpini*, the non-specialized feeder *Tydeus caudatus*, the pollen of *Typha latifolia*, and the mycelium of grape downy mildew *Plasmopara viticola*. Experiments were conducted using grape leaves as a substrate. The developmental times, survival and fecundity of predatory mites were determined on each food type. None of these species developed on *T. caudatus*. Developmental times of *K. aberrans* and *T. pyri* were faster on pollen than on spider mites or *P. viticola* mycelium. The fecundity of *K. aberrans* and *T. pyri* was higher on pollen than on other foods. *Amblyseius andersoni* developed faster on pollen than on *P. ulmi* or *P. viticola* mycelium, and showed a low fecundity on *P. viticola* mycelium.

The leafhopper *Empoasca vitis* (Göthe) (Homoptera: Cicadellidae) is a serious pest in European vineyards. The assessment of economic damage caused by *E. vitis* was associated with no definite results probably because of a different susceptibility and sensibility of grape cultivars to the pest. *Empoasca vitis* abundance and symptoms associated with the activity of this pest were evaluated during three years on 10 grape cultivars grown in collections located in northeastern Italy (Fornasiero *et al.*, 2011). *Empoasca vitis* densities were estimated from May-June to September by sampling leaves every 7-10 days. At vintage time, the percentage of leaf surface with symptoms caused by *E. vitis* feeding was estimated. Carménère, Sauvignon Blanc and Tocai Friulano showed to be very susceptible and very sensible, whereas Cabernet Sauvignon, Chardonnay, Verduzzo Friulano and Pinot Gris showed an opposite behaviour. However, such correspondence was influenced by meteorological conditions and plant stress factors. These results may be useful for establishing economic thresholds and sampling plans for *E. vitis* in vineyards.

Linder *et al.* (2011) reported that the multicoloured Asian ladybeetle *Harmonia axyridis* Pallas is now established in Central Europe and is likely to spread soon over the whole continent. In the United States, *H. axyridis* moves to vineyard shortly before harvest. When disturbed or crushed during vinification, the ladybeetles release alkylmethoxypyrazines, which can strongly affect the quality of white and red wines. In Europe, wine growers are beginning to worry about the impact of *H. axyridis* on the quality of their processed wines. In this study we artificially contaminated harvested grapes of the variety 'Chasselas' and 'Pinot noir", the two main varieties of Switzerland, with three different densities of *H. axyridis* and determined the impact of crushed ladybeetles on the taste of processed wines. The addition of ladybeetles did neither affect fermentation nor the basic chemical properties of processed wines. However, at the organoleptic level the contamination of grapes had a significant effect. Contaminated 'Chasselas' wines were of lower fruitiness, fineness, structure, acidity, equilibrium and exhibited a "rancid oil" odour. Moreover, the quality of contaminated 'Pinot noir' was rated low, in particular, its fruitiness, spiciness, fineness, structure, softness, robustness as well as intensity and quality of tannins was poor. Even though *H. axyridis* is rarely present in vineyards of Western Switzerland, our results emphasis that there is a risk of wine contamination. Thus, the evolution of the multicoloured Asian ladybeetle in European vineyards should be carefully monitored in order to anticipate future problems.

As per the Lucchi *et al.* (2011) the larval feeding behaviour of the grapevine moth *Lobesia botrana* (GM) on grapevine leaf and shoot tissues in the absence of grapes is described. GM larvae were obtained from eggs laid on the grapevine leaves in a field cage covering two rows of a Chardonnay vineyard in which all the bunches were previously removed. Larvae developed till pupation feeding only on leaves and shoots and showed the characteristic leaf rolling habit of the Tortricidae. All of them were able to pupate and to give adults that mated in laboratory conditions. Females laid fertile eggs and gave rise to a new generation of larvae. Pupal weight, number of eggs laid, and longevity of leaf/shoot fed insects were compared with those of the insects reared on berries and artificial diet.

Laboratory and field tests were performed by Mazzoni *et al.* (2011) to ascertain a possible role of visual cues for *Scaphoideus titanus* during the location of a host plant. In laboratory, two choice colour tests were made in a Y-tube to compare the attractiveness of blue (peak wavelength: 475nm), green (521nm), yellow (573nm) and red (650nm). The observed preference order was yellow > red > green > blue for males and red > yellow > green > blue for females. In field, nine groups of three coloured (yellow, red, blue) sticky traps were positioned in an organic vineyard and replaced once a week, for two summers (2008 and 2009). In both seasons the yellow traps captured significantly more individuals (either males or females) and had the highest sex rate (percentage of males on the total captures) followed by red and blue. Our investigation suggests that, despite a partial different response between laboratory and field tests, *S. titanus* adults are influenced by visual cues present in their habitat. The possible reasons of the different responses observed in lab and field are discussed.

Pozzebon *et al.* (2011) stated that the green leafhopper *Empoasca vitis* Goethe (Homoptera: Cicadellidae) is an important pest of grapes in Europe. Its feeding activity may cause a reduction in photosynthesis, mesophyll conductance and transpiration rate with implication for yields. For these reasons insecticides are widely used to control this pest with potential effects on beneficial arthropods. Field experiments were conducted to evaluate the impact of a number of insecticides (chlorpyriphos, flufenoxuron, thiamethoxam, and indoxacarb) on *E. vitis* populations. Pesticide side-effects on the predatory mite *Typhlodromus pyri* were assessed. Thiamethoxam was the most effective towards *E. vitis*. Pest population levels were slightly reduced by chlorpyriphos, suggesting that this strain was resistant to organophosphates. Indoxacarb and flufenoxuron were associated to intermediate effects. The predatory mite populations were affected especially by thiamethoxam.

Mating disruption is an effective and sustainable integrated pest management technique that is now established in a broad range of cropping systems (Briand *et al.*, 2011). For example, it is a popular alternative to insecticides for controlling the grape berry moth *Eupoecilia ambiguella* Hb. (Lepidoptera, Tortricidae), a main pest of European vineyards. A downside of this environmental friendly pest control technique is the challenging assessment of its efficiency in the field. In order to obtain reliable results, newly processed pheromone dispensers have to be tested at a large spatial scale where pest densities, crop variety, cultural practices, microclimate etc. can vary considerably among trial sites. The realisation of statistically sound results therefore asks for numerous independent replicates, which are time-consuming, space-demanding and expensive. The aim of this work was to facilitate the evaluation of the efficiency of mating disruption. We tested the suitability of small insect enclosure field cages for a pre-evaluation of the impact of sex pheromones on insect mating. Our enclosure field cages consisted of a cubic metal frame of 35cm side length that was covered with gauze tissue. Cages were installed in the centre of pheromone treated and untreated vineyards. Thereafter, various numbers of grape berry moth couples were exposed for 1, 2 or 3 days. The impact of pheromones was assessed by counting the number of mated females, *e.g.* the number of dissected spermatophores. Our data showed that mating was significant lower in pheromone treated compared to untreated vineyards. Moreover, there were significant differences among pheromone dispensers tested and mating disruption generally increased with the amount of pheromone diffused. Overall, the exposure of eight couples for one night seemed to be optimal for comparing different control schemes. Our study showed that small insect enclosure field cages offer a fast, simple, cheap and reliable method to measure the impact of pheromones on insect mating. Data obtained in such trials may therefore help predicting the value/use of setting-up large-scale field trials. In conclusion, small insect enclosure field cages may facilitate the development and assessment of semi chemical lures.

Mori *et al.* (2011) stated that the nettle (*Urtica dioica* L.) is the most important host plant for the phytoplasma associated with the Bois noir (BN) disease of the grapevine and for its vector, *Hyalesthes obsoletus* Signoret. A higher abundance of the vector inside the vineyards is favoured by the nettle growing in edge rows and surrounding ditches. Foliar applications of insecticides appeared not an effective measure for vector and disease control, whereas nettle control could be useful to reduce BN incidence in vineyards. In particular the efficacy of nettle control by selective herbicide applications in early autumn and early spring was demonstrated. Many authors have suggested that the nettle control applied during *H. obsoletus* flight could have negative effects, because the vector adults could be forced to migrate onto grapevines. The aim of this research was to evaluate if nettle cutting during the adult flight favours the colonization of the vineyards by the vector. For this purpose, three different weed managements (no cut, one cut during adult flight, frequent cuts) were applied on the herbaceous vegetation of a ditch bordering a vineyard. In this ditch, nettle was one of the principal components. The flight dynamic of *H. obsoletus* was recorded by yellow sticky traps placed on herbaceous vegetation along the ditch and at different positions inside the vineyard. Results showed that nettle cutting induced a reduction in vector presence in the ditch, but favoured its dispersion inside the vineyard. The implications of these observations on disease control are discussed.

Chuche *et al.* (2011) observed that the leafhopper *Scaphoideus titanus* is the vector of the *Flavescence dorée* and transmits the phytoplasma causing the disease while feeding. To better understand the transmission process, we developed for this leafhopper the electropenetrograpy technique which allows studying the feeding behaviour. We used the direct current method (DC system) to study the fundamental processes of the *S. titanus* feeding behaviour. We present the first results of the characterization of the electrical patterns obtained with this insect.

Thiery (2011) stated that the viticulture has an impressive range of progress for the future years concerning the use of integrated pest management. This can be illustrated by insect pest management for which several tools have been developed or could be developed rather soon for an integrated protection management. The present talk aims at presenting through the experience gained in on insect pest or vectors management gaps and potential progress that can be achieved. The European grapevine moth *Lobesia botrana* can serve as a case study to analyse the different possibilities that can be used in IPM. Few existing examples of IPM compatible techniques will be presented: use of behaviour modifying chemicals in push-pull strategies, monitoring techniques, mathematical models, biotechnical insecticides (*Bt*) and natural enemies. The different advantages and limits are discussed, and we attempt to identify gaps where scientific or development efforts are required to progress.

Steel and Greer (2006) stated that climatic conditions of temperature, light (quantity and quality) humidity and rainfall have major impacts on vine performance, crop yield and berry quality when vines are exposed to extremes in these various climate parameters. A major climatic contrast between Eastern Australian and inland central vineyards is the day temperatures are significantly higher at the inland vineyards (35–42°C) whereas relative humidities are higher at the eastern vineyards (30 cf. 80 per cent RH). This difference in climate causes a high predisposition of grapes to non-Botrytis bunch rots, most notably bitter rot (*Greeneria uvicola*) and ripe rot (*Colletotrichum acutatum*), in the wetter eastern sites and little to no infection on vines in the inland vineyards. By contrast, high light exposures, coupled with extremely hot (>35°C) temperatures at inland vineyards cause severe skin damage in the form of sunburn. Sunburn damage does, however, increase the incidence of Botrytis latent infections in grapes. In both cases bunch rots and skin damage causes a loss of berry quality and has severe impacts on wine quality. Management practices include maintaining open canopies for Botrytis control; however, this has the downside of increasing non-Botrytis bunch rot infections. A greater effect of this practice on west compared to east-facing rows suggests combinations of sunburn and/or heat stress or even higher UV-B (280–320 nm) may explain the exacerbation in non-Botrytis bunch rot incidence. While exposure to UV-C (<280 nm) elicits formation antifungal metabolites in vine tissues, this effect is not seen with UV-B which may in fact encourage bunch rots. Future climate changes that lead to increased wetness in the inland wine grape regions of Australia will be detrimental in bringing higher incidences of bunch rots and reducing the competitive advantage in grape growing that these regions currently enjoy.

Rawal *et al.* (2006) a statistical models were developed to optimize the role of ecological variables and simultaneously predict downy mildew incidence in grapes (cv. Anab-e-Shahi) at Indian Institute of Horticultural Research, Bangalore, India. Further, as a measure of goodness-of-fit, the coefficient of determination (R^2) was used to evaluate the empirical models developed. The initial results indicated only 39 per cent of the variability in per cent disease incidence was reflected collectively by preceding week's ecological variables *viz.*, minimum temperature (X_1), maximum temperature (X_2), relative humidity at 07.30 h (X_3) and at 13.30 h (X_4), evaporation (X_5), amount of rainfall (X_7) and wind speed (X_6). This may be due to the reason that per cent disease incidence was in the range zero to as high as 90 per cent during the period of study. To eliminate such a high variability, new models were developed separately for severity less than 10 per cent and severity exceeding 40 per cent. Correlation analysis showed that both maximum and minimum temperatures negatively influenced, whereas humidity levels both at morning and evening h were positively influenced towards the absence of disease incidence. The results indicated that if the disease severity was less than 10 per cent, about 65 per cent of the variability in per cent disease incidence followed the equation of $y = -12.29 - 0.065X_1 - 0.04X_2 + 0.01X_3 + 0.156X_4 - 0.53X_5 - 0.31X_6 + 0.03X_7$. Correlation analysis in case of high disease incidence showed that relative humidity, maximum temperatures, wind speed and number of rainy days jointly influenced the disease. In the case of high severity about 60.3 per cent variability in per cent disease incidence followed the equation of $y = 88.23 - 4.17X_1 + 3.43X_2 - 1.77X_3 + 0.83X_4 + 3.06X_5 + 1.36X_6 - 0.41X_7$. This model would be highly useful in predicting the disease severity and thereby programming for the disease management could be worked out, which will further reduce the fungicide load on the vine.

Sawant and Sawant (2008) reported that the results of in vitro and three-year field trials (2002, 2003 and 2004) with sodium, ammonium and potassium bi-carbonates indicated that all the three bi-carbonates namely sodium, ammonium and potassium were equally effective in planta. Since excess sodium and nitrogen can be problematic in grapes, potassium bi-carbonate was selected for further studies as it was found safe for use. In in-vitro studies potassium bicarbonate reduced conidial germination gradually at concentrations above 100 ppm. Potassium bicarbonate alone was effective at 10 g/L when given as preventive sprays at 10 days interval starting from 40 days after pruning. The efficacy was on par with sprays of wettable sulphur at 2.0 g/L. When sprayed on infected leaves potassium bicarbonate at 10 g/L reduced sporulation of the pathogen up to 80 per cent. Use of a non-ionic surfactant improved the efficacy of potassium bicarbonate to control powdery mildew on bunches, probably because of better coverage. Half dose of potassium bicarbonate (5 g/L) in combination with hexaconazole (0.5 ml/L) showed better control of the disease than when these chemicals were used alone. Potassium bicarbonate was also found compatible with Myclobutanil, Flusilozole and Pencnazole and hence can be used in combination with these fungicides also.

Rawal *et al.* (2006) reported that statistical models were developed to optimize the role of ecological variables and simultaneously predict powdery mildew incidence in grapes (cv. Anab-e-Shahi) at Indian Institute of Horticultural Research, Bangalore, India. Statistical model developed showed that about 61.4 per cent of the variability in per cent disease incidence could be due to preceding week's ecological variables *viz.*, minimum temperature, maximum temperature, and relative humidity during morning and evening, evaporation, amount of rainfall and wind speed. Further, optimized model ($Y = 199.03 - 7.96x_1 - 6.23x_6 + 4.16x_5$) showed that about three abiotic factors namely, minimum temperature (x_1) and wind speed, x6 (negative influence) and evaporation, x_5 (positive influence) could, themselves, predict the incidence to the extent of 59.4 per cent. Further, as a measure of goodness-of-fit, the coefficient of determination (R^2) was used to evaluate the empirical models developed.

From this study, it could be possible to reduce the unnecessary spray of fungicides which help in avoiding the over usage of fungicides and better return to the farmers.

Mundankar *et al.* (2006) opined that a computer based expert system to assist farmers in taking day-to-day decisions to manage Powdery mildew disease in vineyard has been developed using Visual basic and MS Access. The 'NRCG-SKAI PMExpert' is a rule based decision support system. It requires daily weather data on maximum and minimum temperature, maximum and minimum relative humidity (RH) and rainfall and field data on pruning date, presence of powdery mildew disease and nutrient deficiency symptoms and name and date of the last fungicide application in the past fifteen days. It generates: a) diagnosis based on weather risk, growth stage risk, nutrient deficiencies affecting powdery mildew and previous fungicide applications; b) recommendations on the choice of fungicides and their doses for control of the disease. The software also provides detail guides for the management of powdery mildew and nutrient deficiency which supports in better understanding of the recommendations generated by the software, information on weather data equipments and proper recording of weather data ensures the correct inputs to the expert system. This paper describes and discusses the NRCG-SKAI PMExpert software for effective management of powdery mildew in Indian vineyards.

Sawant and Sawant (2008) stated that the preliminary studies indicated that *E. ampelina* (de Bary) Shear is not eliminated from lesions on mature shoots even after 48 hrs dip in carbendazim 0.05 per cent or COC 0.15 per cent solution. The hot water treatment of 50°C for 30 minutes, generally considered as safe for grapevine cuttings (Waite, 2005), was also not effective in inactivating the inoculum. In vitro studies with *Trichoderma* spp. indicated significant mycoparatism of *E. ampelina* within a few hours of coming in contact. The mycoparasite caused lysis of hyphae as well as the spores. In in planta studies, *E. ampelina* could not be isolated from stem lesions pasted with *Trichoderma* spp. containing 2×10^8 spores ml^{-1}, indicating effective parasitization of the pathogen in the lesions, thus showing potential for elimination of the stem borne inoculum. These techniques could be useful for treatment of infected planting material for raising disease free nursery for treatment of infected planting material.

Banerjee *et al.* (2008) observed that the rate of degradation of mancozeb and the residue dynamics of its primary metabolite ethylenethiourea (ETU) in table grapes is reported. Mancozeb residues dissipated following first order rate kinetics with a half-life of 15.4 and 18.2 days in the case of standard and double doses of 1.5 and 3 kg a.i/ha, respectively, when applied 4 times before harvest. The residues of the primary metabolite of mancozeb, *viz.*, ETU were detectable on the day of the spray application itself, indicating immediate conversion of free mancozeb molecules to ETU when exposed to open environment. ETU residue level increased with time up to 5 days at both the doses as a result of gradual conversion of mancozeb to this toxic metabolite. After 5 days, the residue level of ETU started decreasing, indicating further conversion of ETU residues to other relatively non-toxic metabolites. A pre-harvest interval of 20 days may be recommended for dissipation of mancozeb and ETU residues to below detectable level.

Mohapatra *et al.* (2010) stated that a mixture of trifloxystrobin and tebuconazole is excellent in controlling both powdery and downy mildew of grapes. The objective of the present work was to study the behaviour of trifloxystrobin and tebuconazole on grape berries and soil following treatment with Nativo 75 WG, a formulation containing both fungicides (trifloxystrobin 250 + tebuconazole 500 g kg^{-1}). This study was carried out for planned registration of this mixture for use on grapes in India.

Initial residue deposits of trifloxystrobin and tebuconazole on grapes were below their maximum residue limit (MRL) of 0.5 and 2 mg kg^{-1}, respectively, when Nativo 75 WG was applied at the recommended dose of 175 g product ha^{-1}. The residues dissipated gradually to 0.02 and 0.05 mg kg^{-1} by 30 days, and were below the quantifiable limit of 0.01 mg kg^{-1} at the time of harvest (60 days after the last treatment). Trifloxystrobin and tebuconazole dissipated at a pre-harvest interval (PHI) of 36 and 34 days, respectively, from the recommended treatment dose. The acid metabolite of trifloxystrobin, CGA 321 113, was not detected in grape berries at any point in time. Soil at harvest was free of any pesticide residues.

Residue levels of both trifloxystrobin and tebuconazole were below MRLs when grapes were harvested 30 days after the last of four applications of 175 g product ha^{-1} (trifloxystrobin 44 g AI ha^{-1}, tebuconazole 88 g AI ha^{-1}) under the semi-arid tropical climatic conditions of India. Copyright © 2010 Society of Chemical Industry

Craig and Wayne (2011) opined that removing basal leaves 2 weeks postbloom significantly reduced powdery mildew severity on clusters in each year of a study in a New York Chardonnay vineyard. In contrast, removing leaves 5 weeks postbloom had no effect. The effect was not significantly different whether one leaf or two above and below each cluster was removed. Shoot density of vertical shoot-positioned (VSP) vines was lower than that of Umbrella-Kniffen vines and was associated with a significant reduction in disease development in one year of the study. When VSP-training was combined with early leaf removal in the absence of fungicide sprays, it reduced mean disease severity by 32 per cent relative to untreated clusters on Umbrella-Kniffen-trained vines. However, there was no effect of training system in the second year of the study. In South Australia, doubling the volume of irrigation water supplied to vines that received a standard reduced deficit irrigation program resulted in two- and seven-fold increases in foliar powdery mildew severity in two consecutive seasons. Results highlight that viticultural practices targeted primarily at controlling vine growth and crop quantity and quality can also significantly affect the development of powdery mildew.

Zoecklein *et al.* (2000) stated that sour rot on White Riesling grapes increased Brix, glucose to fructose ratio, titratable acidity, tartaric acid, glycerol and gluconic acid while reducing berry weight. Sour rot did not influence pH, acetic acid, ethanol or laccase activity. Two classes of grape aroma components, potentially volatile terpenes (PVT) and free volatile terpenes (FVT) were not influenced by sour rot. However, quantitative differences in selected free aroma compounds between "clean" and rot degradated fruit were observed. Rot reduced free geraniol, nerol, and linalool concentrations, while increasing trans-furan linalool oxide, benzyl alcohol, 2-phenylethanol, 2-methyl-1-propanol, and 3-methyl-1-butanol.

In field trials, effects of the growth regulators Gibb 3 and Regalis® on cluster structure of grapes and the reduction of botrytis disease and sour rot disease were evaluated. Gibb 3 can only be used on Pinot-varieties. Regalis® also can be used on Rhine-Riesling. The growth regulators Gibb 3 and Regalis® were very effective in loosening the cluster structure at Pinotvarieties. Reduction of botrytis disease was compared to fungicide treatments. In most cases, the yield of grapes decreased without an increase of the sugar content. Results of Regalis® on Rhine- Riesling varied greatly without relevant reasons. No differences were found when Regalis® was applied with different amount of water and at different development stages. Gibb3 proved to have a better effect if applied with greater amounts of water. The quality of red wines increases because of higher anthocyan- and phenol-contents. White wines produced from grapes treated with Regalis® didn't show better tasting results in spite of lower yield.

Basu *et al.* (2011) stated that the grapevine varieties with tight clusters are extremely susceptible to infection by grey mould (*Botrytis cinerea*). Severe pathogen attack leads to reduced yield and affects wine quality. A couple of management practices can reduce tightness of berry clusters, among them moderate defoliation and the use of plant growth regulators. Prohexadione-Calcium (trademark Regalis®) is a gibberellin acid biosynthesis inhibitor with low toxicity and persistence in the plant. Over a period of three consecutive years the effect of prohexadione-Ca on grape cluster architecture was tested in cv. Riesling and combined with anti-*Botrytis* fungicide applications and leaf removal measures. While prohexadione-Ca treatments significantly increased the flexibility of grape clusters in 2007, no significant effects were evident in 2008 and 2009. However, even a minor reduction of index values of 0.2 – 0.3 may have an influence on the tightness of clusters and thus on the susceptibility to *B. cinerea* infections. In all three years severity of *B. cinerea* was significantly reduced in Riesling grapes when the following measures were combined: application of Regalis®, fungicide treatments and moderate defoliation. Therefore, future strategies to control grey mould should aim at a combination of different management strategies.

Van Helden *et al.* (2011) report that the nematode *Xiphinema index* is, economically, the major virus vector in viticulture, transmitting specifically the Grapevine fanleaf virus (GFLV), the most severe grapevine virus disease worldwide. A second virus (ArMV) and its vector (*X. diversicaudatum*) do exist in the Bordeaux region but seem of lower importance. Plots that have become completely infected, no longer providing sufficient yield or quality, have to be uprooted to be replaced by new healthy plants. Increased knowledge of the spatial distribution of this nematode, both horizontally and vertically, and of correlative GFLV plant infections, is essential for efficient control of the disease. Vertical distribution data showed that the highest numbers of individuals occurred at 40 to 110cm depth, corresponding to the layers where the highest densities of fine roots were observed. Horizontal distribution revealed a significant aggregative pattern with patches of 6 to 8m diameter, together with a significant neighbourhood structure of nematode densities, thus identifying the relevant sampling scale to describe the nematode distribution. Nematode patches often, but not always, correlate significantly with those of GFLV-infected grapevine plants. The nematodes are able to survive in the soil even without food for several years. Since nematodes live deep in the soil, they cannot be attained by nematicides (that are, anyway, progressively forbidden). Because of this lack of efficient control measures of the vector, a fallow period of 7 years is advised to farmers but rarely observed for economic reasons. In some cases, high reinfestation levels have occurred with short fallow periods but this does not always occur. Sampling of numerous plots in the Bordeaux region, uprooted because of virus infestations, showed that nematode numbers are unpredictable. No clear relation was found between the percentage of each virus and the population of its vector. *X. diversicaudatum* was rarely present in significant amounts, even though infestation percentage of plants with ArMV could be high. *X. index* was sometimes present in high numbers (up to 300 per litre of soil) in some samples but mostly numbers were quite low or nematodes even not detected. These results suggest that the fallow period can be adapted following nematode sampling. In the case of very low nematode numbers it could be reduced to 1.5 to 3 years whereas longer periods need to be observed when high numbers are found. Nematode sampling grid should be rather dense, and sampling should ideally be done twice, just after uprooting and before replanting. The use of 'nematicidal' antagonistic plants against *X. index* is currently being investigated in our lab.

Kast and Belyer (2011) reported that the OiDiag-2.2 is a system of two tools. The first one helps the vine growers to find the date for the start of the sprays against powdery mildew (*Erysiphe necator*). The second tool calculates an index value considering the climatical and ontogenic inputs for the risk of powdery mildew attack. Vine growers should get a table of relevant fungicides with values for the maximal time lag after the spray, which is completely covered by these fungicides considering these index values.

Viret *et al*. (2011) reported that to control the main fungal diseases in accordance with the epidemiological development of the fungi in viticulture, the use of forecasting systems is one of the major progresses. Agrometeo is an interactive platform (www.agrometeo.ch) and a general tool for agriculture, including actual and historical weather data, modules for field crops, grapevine and fruit orchards. The grapevine module contains forecasting for downy mildew and grape berry moths; leaf area adapted spraying, growth development model, pesticides index, and descriptions of the main diseases and pests. The forecasting modules use data from a weather station network covering the whole viticulture area of Switzerland. Grapevine downy mildew forecasting is done with Vitimeteo, a new expert-system for the forecast of grapevine downy mildew, designed by the research institute of Freiburg (Germany), Agroscope ACW and the company GEOsens. The software generates graphics and tables freely available for the growers on the Internet. Leaf area adapted spraying was developed in analogy to the tree row volume concept on fruit orchards. The experiments show a reduction of about 20 to 30 per cent of the use of pesticides by calculating the precise dose needed for a given leaf area. The module in www.agrometeo.ch allows a ready to use solution for the calculation of the amount of plant protection products based on the width and height of the leaf canopy and the row distance, depending on the registered amount indicated on the package. The number of Agrometeo users is in expansion confirming the grower's interest.

Caffi *et al.* (2011) a research was carried out in order to verify the long-term viability of *Plasmopara viticola* oospores. It is commonly assumed that oospores may survive across grapevines growing seasons, but no experimental evidence of this assumption has been provided to date. Grapevine leaves were collected from untreated plants in Emilia-Romagna in fall 2004 to 2008. Fragments of these leaves were kept in soil under natural conditions. In spring 2005 to 2009, part of the leaf fragments collected in the previous autumn were used to evaluate oospore germination using the floating disk bioassay; in spring 2009, the bioassay was also performed using the leaf residue collected in fall 2005 to 2007. Results demonstrated that *P. viticola* oospores remained viable for the whole period considered in this work, *i.e.*, about 65 months (from fall 2004 to spring 2009).

Rossi *et al.* (2011) opined that either by themselves or in combination with mycelium in the dormant buds, ascospores produced in chasmothecia (the sexual fruiting bodies) of *Erysiphe necator* are an important source of primary inoculum for grapevine powdery mildew disease. In northern Italy, *E. necator* overwinters mainly as the sexual stage (*i.e.*, mycelia are not usually present in dormant buds), but no data are available on the abundance of chasmothecia in the vineyards. Therefore, the dispersal of chasmothecia was studied in commercial vineyards of northern Italy, from 2005 to 2007 (16 combinations of 10 vineyards x 3 years); the vines were not sprayed with fungicide during this study. The dispersed chasmothecia were collected on filter papers in funnels that were placed on the trunk of affected grapevines from mid-August to the end of leaf fall. Each filter paper was examined microscopically, and the chasmothecia were counted. The first chasmothecia of the season were dispersed between late August and early October; cumulative numbers of the chasmothecia dispersed subsequently increased (with different dynamics depending on the vineyard and year) but dispersal stopped at the end of leaf fall. Over all three years and 16 vineyards, chasmothecia averaged $3/cm^2$ of trap surface, with a maximum of $15/cm^2$. The numbers of mature chasmothecia that dispersed roughly depended on the powdery mildew severity on leaves: high numbers of chasmothecia were associated with disease severity ≥ 80 per cent. According to estimates based on the current data, vine bark in vineyards trained with a Guyot, Geneva double curtain, or spurred cordon pruning system could contain till 18, 19, or 44 million chasmothecia/ha, respectively.

Calonnec *et al.* (2011) reported that the simulations are performed to explore the sensitivity of epidemics of powdery mildew of grapevine to variation of parameters related to the pathogen, the plant growth or the crop management. For early inoculation, the three parameters linked to the sporulation, the dispersion process (cid) and the plant vigour (vig) are the most discriminating. The parameter of ontogenic resistance is less discriminating, and the height of shoot topping and the distance between buds are not discriminating. We also examined the relationship between vine vigour and disease variables at key periods in the epidemic process. An increase of the parameter of vigour from 0.2 to 1 amounted to a higher number of leaves at flowering (Nflo) and a higher rate of leaves emergence (RLE). The RLE was correlated with the rate of diseased leaves emergence and with the severity of the disease at shoot topping, whereas Nflo was correlated to the diseased leaves area at day 240.

The percentage of young leaves during first sporulation event (s1s2) was correlated to the rate of shoot development (RDS). These two variables from host allowed to discriminate the years. The duration and dynamic of infectious tissue can considerably vary function on the development of secondary shoots therefore on the climatic conditions and vigour.

Schnee *et al*. (2011) reported that the current control of grape powdery mildew requires news strategies, able to limit the pathogen development and requesting a reduced number of chemical sprayings. This study proposes to exploit host plasticity in order to investigate the effect of a plant growth changing on either the epidemic process, or organs susceptibility. An experimental plot combining two cultivars, three rootstocks and two opposite crop management for creating a vigour differential was surveyed regularly during the vegetative growth and different variables were measured to characterise host growth and pathogen. Shoots coming from vinestocks localized in the different vigour area were sampled for organ susceptibility assessment. Two groups of vigour were obtained, not based on the established treatment but on the individual growth potential, that differ mainly by the number of secondary leaves. The vigorous group presents the higher level of disease, increasing with the appearance of secondary leaves. A low vigour modifies the susceptibility of the tissues, by an earlier appearance of ontogenic resistance phenomenon. The relevance of vigour control and the concomitant physiologist mechanisms are discussed as integrative strategy in the grape powdery mildew management.

Gessler and Jermini (2011) reported that the blackrot caused by *Guignardia bidwelli* is causing damages in recuperated vineyards in southern Switzerland planted with the cultivar Isabella. The source of inoculum is attributed to abandoned vineyards. Heavy and prolonged rains may also favor infection by *Plasmopara viticola* on Isabella especially if a heavy load of primary inoculum can be expected. A 6 year experiment was made to test and confirm minimal necessary fungicide applications and timing to control Black rot and downy mildew. Infection conditions during the period between flowering and up to six week after, have led to heavy losses reaching 100 per cent without adequate control. A combination of the two pathogens was registered, with *P. viticola* bunch symptoms without visible sporulation. We recommend therefore using fungicides or fungicide combinations with an effect on both pathogens. Mostly two well timed sprays with an appropriate fungicide active against both pathogens are sufficient to control fully the diseases on bunches and avoid loss. Under particular rainy situation a third applications gave 100 per cent control. Timing should be before rain events leading to prolonged leaf wetness during the time period between flowering and six week afterwards. Winegrowers in the area are not professional winegrowers or farmers, having little specific knowledge; therefore they will apply fungicides only at particular date (Saturday) following the instruction of the manufactures. Applications are made by knapsack sprayers and vineyards are usually terraced irregularly often as pergola. Therefore we developed an easy applicable scheme to determine application timing and an instruction system to determine fungicide concentration so to avoid over and under dosaging. Here we present the most relevant points we teach to our winegrowers.

Jermini *et al.* (2011) opined that the ascomycete *Guignardia bidwellii* is the casual agent of the black rot of grape. Native to North America introduced to Europe in the early 19th century and today regularly observed in France, Italy, Germany and southern Switzerland. The black rot is not generally considered as an important disease but it is able to induce heavy crop losses particularly in regions characterised by weather conditions favourable to *Plasmopara viticola*. Present studies focused on the genetic structure of the pathogen population, to get a better understanding of the epidemiology and life cycle of the disease. We successfully extracted and amplified DNA from mummified berries and developed six SSR markers for genotyping of *G. bidwellii* isolates. The SSR markers were used to study the pathogen populations of two Merlot vineyards in France and a mixed variety vineyard (Merlot, Regent, Isabella, Solaris, Bianca and Chambourcin) in southern Switzerland. All three populations were related and shared common SSR alleles. Gene diversity was similar for the two French populations, and the Swiss population was nearly clonal in a genotype frequently observed in France. Only a few private genotypes were found, therefore the populations differed mainly in genotype frequency. Based on our data we challenge the importance of sexual reproduction for the propagation of the fungus. The low genotype and allele diversity observed could be explained either by a generally low genotype polymorphism in Europe or by a low SSR marker polymorphism. More research is therefore needed to assess SSR marker polymorphism on an American population, to develop additional SSR markers for a higher genotype resolution and to understand an eventually difference in the genotype frequency between leaf and berry infections.

Legler *et al.* (2011) observed that grapevine powdery mildew is a key disease all across Europe and is currently controlled almost exclusively using fungicides. To develop a new biofungicide product against this disease, caused by *Erysiphe necator*, a large culture collection of *Ampelomyces* strains collected worldwide from different powdery mildew species was screened to assess for the first time the level of variability of different characteristics of these biocontrol agents known to occur as natural mycoparasites of powdery mildews in the field. This preliminary screening revealed a considerable diversity in many characteristics of different *Ampelomyces* strains including both culture patterns and mycoparasitic activities against the asexual and sexual stages of *E. necator*. The screening protocol developed in this work could be used in further studies on *Ampelomyces* in order to develop new effective commercial biofungicide products against powdery mildew infections of grapevine and other crops.

Fermaud *et al.* (2011) suggested that the water activity (Aw) is a physical property of a humid solid, which is of prime importance in food microbiology. This parameter is an assessment of available water, unbound to nutrient molecules, which allows damaging bacteria, yeasts and fungi to grow in a food product. As most moulds which do not grow below an Aw level of *ca.* 0.8, *Botrytis cinerea* was unable to grow on artificial media at an Aw of 0.93 and below. In 2008, in an experimental vineyard (*Vitis vinifera* L. cv. Merlot noir) near Bordeaux, *Botrytis* bunch rot development was assessed and the evolution of water activity was investigated at the surface of healthy grape berries during fruit development. Our results showed that Aw decreased steadily from the berry herbaceous stage to maturity and was correlated negatively and significantly ($P = 0.01$) with disease incidence. Furthermore, an experiment under controlled conditions showed the significant effect of relative humidity (RH) on the water activity level at the berry surface. When healthy berries at two different development stages ("herbaceous" and "beginning of colour change") remained at 80 per cent RH for 6 days, the Aw values decreased significantly ($P = 0.05$). On the basis of the close relationship between water activity and Botrytis bunch rot development, the potential of Aw for use as a new disease indicator is discussed as well as the influence of climatic conditions on water activity measured at the grape berry surface.

Kortekamp and Kockerling (2011) stated even though "Esca" may be as old as viticulture itself, new attention has been directed to this disease in recent years due to the dramatic increase in economic importance. A sudden wilting of Esca-affected grapevines, followed by the death of the entire plant, occurs in many grapevine-growing areas of the world. Esca is a complex disease that comprises several symptoms caused by a set of fungal pathogens. Unfortunately, infected plants can often not be cured, since foliar or wood treatments using fungicides do not lead to an effective control of the respective pathogens in the infested wood. Given the large number of wounds made to propagation material during the various nursery stages and also made as a result of repeated pruning, hygiene and wound protection is of particular importance. Therefore, the effect of various chemical and biological treatments on Esca-associated fungi, such as *Phaeoacremonium aleophilum* (*Pal*), *Phaeomoniella chlamydospora* (*Pch*), and *Fomitiporia mediterranea* (*Fmed*), and on the disease progress of artificially infected plants was investigated *in vitro* and in field and greenhouse experiments. For this purpose, several putative contamination pathways and different test systems have been evaluated to verify the ports of entry for fungal spores and the efficiency of fungicides to control the disease. In the year 2009, vineyards of the DLR research station and of local grape growers were inspected and the impact of Esca in these vineyards was evaluated. As published by other authors, the number of affected vines increased with their age. Furthermore, there are no clear differences in susceptibility and none of the varieties planted in the Palatinate showed any suitable resistance. Plant material of diseased vines was collected and tested for the presence of Esca-inducing pathogens. Mainly *Phaeomoniella chlamydospora* (Pch), *Botryosphaeria* species, *Cylindrocarpon* sp., *Trichoderma* sp. and mould fungi such as *Alternaria* sp. and *Penicillium* sp. were frequently found. All type of wounds created artificially represented suitable ports of entry for all Esca-associated pathogens. Especially, infections with *Cylindrocarpon destructans* significantly reduced viability of the scions and had severe effects on plant growth as tested with infected cuttings. Wound infections caused by polluted water that had been used to soak propagation material led to a reduced growth of cions, especially when contaminated with *Cylindrocarpon* spores. Growth tests *in vitro* showed that most fungicides tested exhibited antifungal capacities. The *Trichoderma*-based product Trichostar® was an effective biological agent in the field, preventing infection of wounds when applied one day before inoculation, whereas most fungicides were not able to sufficiently prevent colonisation of wounds by fungi.

The aim of researches performed by our group (Lecomte *et al.*, 2011) is to determine which internal and external factors have a key influence on grapevine wood diseases, especially on Esca. The relationship between Esca leaf symptoms, the size of inner necrosis, the fungal endophytes associated with the disease and the whole microflora that can be detected from the grapevine wood, is discussed. Recent studies on leaf symptom development showed that summer temperature conditions likely play a major role in the expression of Esca symptoms. Cultural factors, like plant vigor or type of soils, also have presumably a significant influence. Several evidences suggest that other abiotic factors linked to the training systems (*e.g.* forms with very short cordons) or to pruning decisions (*e.g.* winter hand pruning characteristic of the modern grape-growing training systems) have also to be taken into account, alone or combined with other factors. Whole data indicate that Esca is a multi-factor disease and that many complex pathological scenarios could explain the grapevine trunk disease expression. A control strategy, mainly based on cultural measures to avoid a too early development of inner necrosis, is proposed.

Vijay and Venugopalan (2011) made an attempt to develop nonlinear stochastic models for disease forecasting in Grapes. Grape (*Vitis* spp.) is an important crop for the farmers for getting higher returns and with consumer for delicacy and as a medicinal fruit. Downy mildew is one of the most destructive vine diseases known leading to total crop losses. Thus, the remunerative and successful cultivation of grapes has been hampered. To avoid this, by understanding prevailing weather conditions which influence the onset, initiation and progression of disease, judicious need based control measures are the need of the hour. To this end, the present investigation was carried out to understand the role of weather factors on downy mildew incidence in Grapes (cv. Anab-e-Shahi) and disease progression over time epoch by developing suitable statistical models. Efforts were made to develop models individually for backward and fore pruning periods, resulting in meaningful interpretation to the researchers. Also, an attempt was made to investigate statistical considerations involved in the error structure and subsequent methodologies to be followed, while developing non-linear models. Using the nonlinear models developed, an index was also developed to compute quantitative information about the biological parameters concerning intrinsic infection rate and maximum mildew severity over time-epoch. Statistical models developed for backward pruning data (May-June) showed that maximum temperature, Evaporation and relative humidity at 7.30 hrs, observed with a time lag of one week, collectively explain about 89.4 per cent of the variation in downy mildew incidence. Statistical models developed for fore pruning data (September-October) showed that minimum temperature, relative humidity at 7.30 hrs and 13.30 hrs, observed with a time lag of one week, collectively explain 88 per cent of the variation in weekly downy mildew incidence. Logistic and Gompertz nonlinear stochastic statistical models developed expressed the disease progression to the extent of 97-99 per cent. These models when used to compute quantitative information about the biological parameters concerning intrinsic infection rate and maximum mildew severity over time-epoch showed that, in general, for backward and fore pruning data, the rate of disease severity was maximum during the fourth-fifth week and fifth- sixth week after pruning, respectively. Hence, appropriate management strategies for controlling the disease should be oriented within the period identified in the investigation, separately for backward and fore pruning. Resultant nonlinear models were used to compute the Area under Disease Progressive Curve (AUDPC). A perusal indicates that the values obtained by logistic and Gompertz are ranged from 48 to 84 and 25 to 65 respectively for backward pruning data. However, for the fore pruning data the results showed that AUDPC values were higher as it ranged from 78 to 86 and 61 to 65 respectively. These results indicate that the downy mildew rate of progression in Fore pruning is much severe than in backward pruning. SAS programming codes were generated for model building. The message arising out of this present investigation is that proper prophylactic measures, if taken by considering the model resulted significant weather factors along with knowledge about disease progression over time as depicted by nonlinear models, separately for backward and fore pruning, not only results in an efficient and economic management strategies for controlling downy mildew incidence in grapes (cv. Anab-e-Shahi) but also considerably reduce crop yield loss thereby providing better return to the farmers.

Pliny was the first to refer this disease from Italy in the 1st century of Christian Era. De Bary described this disease from Germany in 1874. The disease was first reported by Burrill (1886) from Central Illinois, Indiana from USA. It was reported to be the most destructive disease in Rhodesia (present Zimbabwe) resulting in total failure of grape cultivation. Similarly, heavy losses due to the anthracnose of grapevine epidemic on an entire province of Argentina were reported in 1929. Jenkins and Bittencourt (1943) in a survey on the history of vine anthracnose reported that the disease engaged the attention of viticulturists in Brazil, Argentina, Paraguay, Chile and Uruguay as early as 1880 -1890. Anderson (1956) opined that the disease was present for centuries in Europe which was introduced into the United States through cuttings or young vines, hardly a fair bargain for the exchange of downy mildew and black rot into Europe from USA.

The disease has been reported as one of the oldest vine diseases in South Africa, where several hundred vine yards have been uprooted owing to its ravages. It caused 80-100 per cent loss in America during the epidemic year of 1950-51, whereas 20-50 per cent loss has been reported in middle Asia. Crop loss as high as 80 per cent have been reported from Russia (Winkler, 1965)

In India, it was first recorded near Pune as a minor disease by Butler in 1903. Later on, this disease was reported from Madras (Tamil Nadu), Uttar Pradesh and Mysore (Karnataka). Losses due to this disease (10-15 per cent) have been reported from Punjab and Haryana (Bedi *et al.*, 1969a). Anthracnose disease of grape was in severe form during rainy season in almost all the districts of Punjab (Jeyarajan *et al.*, 1970).

Anthracnose has been reported from the entire grape growing areas of India and assumed serious proportion in the states of Karnataka and Andhra Pradesh (Sridhar and Sohi, 1976). Gurme and Kore (1977) reported that during 1975 rainy season, the incidence of grape anthracnose was found to an extent of 50 per cent at the Agriculture Farm, Parbhani, Maharashtra. Gupta *et al.* (1981) observed that the pathogen *Sphaceloma ampelinum* affected all the aerial parts of plants but serious loss resulted when inflorescence and berries were attacked. The disease has also been reported from Japan, Spain, Persia, New South Wales, China, Arizona, Italy and Central Asia (Shanmughvelu, 1989).

A gradual decline in yield and growth of cultivar, Thompson seedless of grape vine (*Vitis vinifera* L.) due to this disease has been reported from Punjab and Haryana (Bedi *et al.*, 1969a; Bakshi *et al.*, 1970). Anthracnose of grape was widespread and appeared in severe form during rainy season in almost all the districts of Punjab. In Tamil Nadu, the severe incidence of anthracnose was observed during high moisture, rainy conditions (Jeyarajan *et al.*, 1970)

Suhag and Grover's (1972b) survey found that every vineyard of Haryana was infected with anthracnose. Most of the commercial cultivars of grapevine showed mild to severe infection. The cultivars which were known to resist invasion by the pathogen under field conditions were showing susceptibility under artificial conditions of infection. In Haryana, on an average the diseased vine caused an approximate loss of over Rs. 4,000 per hectare per year. Sridhar and Sohi (1976) reported that anthracnose appeared in severe form in all grape growing areas of India and assumed serious proportion in the state of Karnataka and Andhra Pradesh. During 1975 rainy season, the incidence of grape anthracnose was found to be around 50 per cent at the Agricultural College Farm, Parbhani (Kore and Gurme, 1978).

Kumar and Thind (1992) surveyed different vineyards in Punjab and Haryana during March to September in 1990. They observed that the disease was less severe during early growing season but increased with onset of rainy season and was maximum between July to September. Pampanagouda (2000) in his survey on anthracnose of grape over northern Karnataka reported the highest incidence in Raichur followed by Koppal and Bellary districts while moderate to severe incidence was recorded in Bijapur district. The author concurred with the observation that the anthracnose was severe in *kharif* (back/April pruning) season than *rabi* (fore/October pruning) season.

Contrary to the above observations, heavy infection of Anab-e-Shahi variety has been reported immediately after October pruning because of continuous rainfall in Hyderabad during 1962 (Govinda Rao and Dakshinmurti, 1964).

Anderson (1956) described the symptoms as, the fungus attacks all the aerial parts of the plant in the green stage but is most often observed on the new shoots and fruit. On the leaves the first evidence is the appearance of small, irregular, dark brown spots. These later turn gray in the centre with a border of dark brown tissue. The central tissue often falls out later in the season. On shoots and tendrils small isolated light brown spots appear, at first circular in outline but later elongating into elliptical, slightly sunken areas. The tissue at the border of these lesions becomes slightly raised, forming a dark coloured rim, while the central area turns to ashy gray. Lesions similar to those on the shoot are often found on the veins and petioles of the leaves. The lesions may be so numerous as to give rough, scarred appearance to shoot. Sometimes in midsummer during wet weather, small pink spots may be observed in the depressed centre of the lesions that contain the spore masses of the causal agent. In severe infection of the disease on shoots restricts the growth and sometimes dwarf, light green leaves with down ward curling may be observed and rarely defoliation may occur. The disease on berries produces typical "Bird's eye spot" having gray centre surrounded by reddish brown zone. Initially the spots on berries start as a small dark red spots with slight depression and there is no shrinkage of the berry that remains firm.

Suhag (1971) described the symptoms of anthracnose caused by *Sphaceloma ampelinum*. All aerial parts of vine particularly foliage and canes but rarely berries were affected. It was often most destructive on the new vegetative growth from the late June till September.

Chohan (1974) described the symptoms of anthracnose where shoots and tendrils were highly susceptible to the fungus which produced usually circular, light brown spots, may become elongated and have slight sunken areas. The border of these lesions had slightly raised areas and forms a dark coloured rim, while area in the centre became ash grey. On the leaves, spots are small, irregular and dark brown. On the berries, the spots have reddish brown zone with grey centre and may enlarge and make the fruit misshapened. Similar symptoms were described by Prasad and Nirvan (1965), Suhag and Grover (1977a), Pathak (1980), Chadha and Shikhmany (1999), Singh (2000) and Shanmughvelu (2003).

De Bary (1874) described the disease and the conidial form of the pathogen which he named as *Sphaceloma ampelinum* de Bary. Most of the early studies of this pathogen were by Viala and Pacottet (1906) in France who established the polymorphic nature of the pathogen and named it as *Manginia ampelina* Viala and Pacottet. Brown (1922) reported the isolation of a *Gloeosporium* species from grape but he did not characterize it. The name *Gloeosporium ampelophagum* (Pass) Sacc. was retained for the imperfect state of the pathogen till the discovery of the perfect state by Shear in 1929, who named it as *Elsinoe ampelina* (de Bary) Shear. The perfect state was found in the cankers of previous year's growth or on pruned canes, which had been left on the ground. Such cankers produced small, black, inconspicuous ascocarps, in which globular asci were embedded and each ascus contained hyaline, triseptate with 4-5 x 15-16µm sized ascospores. In addition another fungus *C. gloeosporioides* (*Colletotrichum* state of *Glomerella cingulata* (Stonem) Spauld and Schrenk, as a causal agent of grape anthracnose has been established in the recent years (Porwal and Sharma, 1974 and Kumar *et al.*, 1994). They have observed that the pathogen is more frequently isolated during rainy seasons in India.

Suhag *et al*. (1982) identified anthracnose isolates from 20 grape cultivars into three groups SA1, SA2 and SA3. Of these, SA3 seemed to be the most virulent and aggressive. Kumar *et al*. (1995) tested 62 isolates of *Gloeosporium ampelophagum* by detached leaf technique as well as potted plant technique. Based on the time taken for initiation of first symptom and per cent disease intensity on leaves, the isolates were graded into three distinct groups; GA-I, GA-II and GA-III. Among these, GA-I appeared to be more pathogenic and aggressive/virulent with shorter incubation period.

Kore and Gurme (1971) studied the growth and sporulation of the fungus, *S. ampelinum* on various solid and liquid media *viz*. Potato dextrose, Richards, Asthana and Hawker's, Czapek's, Waksman, Kirchoff's, Brown's and Oat meal agar. Among the solid media, significantly superior growth and sporulation was observed on Potato dextrose agar followed by Waksman's agar and Oat meal agar. The acervuli were more in number during advanced stage of growth on oat meal agar than all other media tried. Among liquid cultures, significantly high dry weight of mycelium and abundant sporulation were observed on Richard's, Kirchoff's and Potato dextrose media. Asthana and Hawker's and Brown's media proved to be the poorest for growth and sporulation of the fungus.

Suhag (1971) studied growth and sporulation of the fungus *S. ampelinum* on various solid media and reported that Potato dextrose agar media was better than other media. Of the four solid media tested, maximum mycelial growth and sporulation of *S. ampelinum* was in Yeast extract agar and least in Czapek's agar. Among three liquid media, mycelial growth of *S. ampelinum* was the maximum in Czapek's medium, while the least growth was seen in Potato dextrose agar medium (Cheema *et al.*, 1978).

Cheema *et al.* (1981) standardized the preparations of cv. Perlette of grapes leaf extract medium. The number of spores produced in leaf extract agar was quite high. The leaf extract medium supported sporulation of all the isolates. Zhang *et al.* (1991) grew the fungus *S. ampelinum* on several common culture media and identified Sabouraud's agar and Potato dextrose agar as the best. Among the four solid media tested, maximum sporulation of *S.ampelinum* was in Czapek's agar media and least in Leonian agar and Potato dextrose agar (Ramaswamy and Anilkumar 1995).

Pampanagouda (2000) recorded maximum growth of *S. ampelinum* on potato dextrose agar followed by Richards' agar while least growth in Sach's agar whereas Richards' medium was found to be significantly superior for mycelial growth in liquid cultures and least in the Sach's medium.

Kumar *et al.* (1994) also made a noteworthy mention that, isolations carried out during Aug-Sept yielded mostly *Colletotrichum gloeosporioides* cultures, which were fast growing with shorter incubation period and generally weak pathogens. Similarly, nine to eleven morphological groupings have been reported from Punjab by Sabourry (1978) and Cheema *et al.* (1978) respectively and the isolates were mostly fast growing with shorter incubation (7-days) period.

Jameel Akhtar and Singh (2007) studied variability in *Colletotrichum capsici*, causing anthracnose of chilli. They found out variability amongst the five selected isolates on different media, in terms of growth, shape, colour, texture and zonation. The results indicated that the growth of isolates on PDA ranged between slow to fast with black and wavy margins having thin texture and poor zonation while on Richards' medium the growth of isolate was intermediate with circular, smooth white colonies having thick texture without zonation. The differences were sharp on Czapek's medium producing circular, smooth margins; the pigmentation varied between greenish, white and grayish with clear zonation. Similarly the isolates exhibited marked variation in NPA and Leonian media producing black to greenish black, thick textured colonies. The PDA and Czapek's media were found to support the maximum growth producing unbranched, aseptate conidiophores, the conidia in mass/group gave pinkish appearance while individually they were hyaline.

In comparison over traditional techniques/methods RAPD (Random amplified polymorphic DNA) and other molecular approaches like RFLP (Restriction fragment length polymorphism) (Abang *et al.*, 2002) and AFLP (Amplified fragment length polymorphism) (O'Neill *et al.*, 1997) and sequence analysis (Balesdent *et al.*, 1998, Zhang *et al.*, 1998) have resolved many taxonomic problems in fungi. RAPD is a very widely used technique, which is effective in differentiating between isolates of many different fungal genera (Sreenivasaprasad *et al.*, 1996, Sicard *et al.*, 1997).

Tan *et al.* (1996) investigated the genetic differences among the citrus scab pathogens *Elsinoe fawcetii, E. australis* from South America and *Sphaceloma fawcetii* var. *scabiosa* from Australia that were indistinguishable morphologically but showed pathological differences. Two pathotypes of *Elsinoe fawcetii* from Florida and two pathotypes from Australia were distinguished on the basis of host range. Restriction analysis of the amplified internal transcribed spacer (ITS) of ribosomal DNA with several endonucleases and sequence analysis of the ITS readily differentiated *E. australis* from *E. fawcetii*, and *Sphaceloma fawcetii* var. *scabiosa*. RAPD analysis indicated that *E. fawcetii*, isolates from Australia and Florida were more closely related to each other than to *E. australis* isolates from Argentina. However, all the Australian isolates were separable from all Florida isolates by their RAPD profiles. There was good correspondence between the RAPD profiles and the two pathotypes identified in Australia. The authors opined that molecular analysis will be a rapid, useful tool in identifying exotic types of citrus scab on shipments of fruit and help reduce introduction of these types into new areas.

In the RAPD comparison of avocado and mango isolates of *Colletotrichum gloeosporioides* from various subtropical fruit crops and production areas in South Africa revealed that the only RAPD primers producing polymorphic bands were OPCO2, OPAO2, OPCO8, OPB12, and OPE14. OPCO2 was the most discriminatory and could distinguish isolates according to the host and symptoms from which isolations were made. Differences could be observed when the RAPD band patterns of each production area were compared based on cluster analysis confirmation. Differences were also observed when representative band profiles from different hosts, *viz*, avocado, mango, papaya and citrus were compared. Cluster analysis showed that mango and citrus isolates were more closely related, while, papaya isolates more distant, but more closely related to mango and citrus isolates than avocado isolates (Freeman *et al.*, 1998).

Wang *et al.* (1998) carried out work on RAPD markers identification linked to anthracnose resistant gene in wild grapes native to China. Shivakumar (2006) found out molecular variability through RAPD using 13 primers in *Colletotrichum gloeosporioides*.

Persistent use of carbendazim as well as of other chemicals in the management of anthracnose on grapes has led to the development of resistance phenomenon among the pathogens. The failure of fungicides is mainly due to association of high fitness among the resistant strains (Bollen and Zaayen, 1975, Delp, 1980, Dekker and Georgeopoulos, 1982, Delp, 1989, Kumar and Thind, 1992 Wuest *et al.*, 1974, Mohan Chander and Thind, 1995, and Hemanthakumar *et al.*, 1998 and 2001a).

Kumar and Thind (1992) in their monitoring programme of grape vineyards in certain parts of Punjab during 1990s that were continuously using carbendazim for anthracnose control reported the reduced sensitivity in *Gloeosporium ampelophagum* to carbendazim (Bavistin).

Mohan Chander and Thind (1995) found out reduced efficacy of carbendazim due to frequent use of these fungicides. Testing 52 isolates of *G. ampelophagum* from different vineyards in Punjab revealed the presence of 13 isolates with highly reduced sensitivity to Bavistin with MIC values ranging from 40 to >100µg/ml and most of the resistant isolates were equally pathogenic and produced symptoms on leaves treated with Bavistin up to 1000 µg/ml. The resistance was found to be quite persistent in nature. There was cross-resistance to other formulations of carbendazim but the resistant isolates appeared equally sensitive to multisite inhibitors like Captaf, Indofil M-45 and Bordeaux mixture. Alternation of Bavistin with Indofil M-45 or Bordeaux mixture proved highly effective in controlling the disease in a resistance-affected vineyard.

Kumar and Thind (1992) found that many anthracnose strains/isolates on grapes (*G.ampelophagum*) developed resistance to carbendazim concentration as high as 100 µg/ml, at the same time there was existence of isolates highly sensitive to carbendazim at concentration as low as 1.0µg/ml.

Deokate *et al.* (2002) tested 210 *G. ampelophagum* isolates for their sensitivity to carbendazim on Czapek-Dox agar following poisoned food technique, of which 184 isolates were sensitive and 26 were resistant with a MIC (Minimum Inhibitory Concentration) ranging between 4.76 to 4.93µg/ml.

Thind *et al.* (1998) tested four triazoles and a contact fungicide against carbendazim resistant and sensitive isolates of *G. ampelophagum*. Growth inhibition assays showed that, difenconazole possessed maximum inhibitory action against both types of isolates followed by tebuconazole and fluazinam. Using detached leaf technique difenconazole @25 µg/ml was again proved to be the most effective against both the isolates. Tebuconazole, flusilazole and triadimefon provided good control at 50-100 µg/ml. Fluazinam, a contact fungicide arrested the disease completely at 500 µg/ml.

Mohan Chander and Thind (1995) opined that *G. ampelophagum* being a slow growing fungus, requires nearly a month for detection of resistance in terms of colony size using fungicide-amended medium. Germ tube length can thus be taken as an important criterion for detecting resistant strains in a conidial population of *G. ampelophagum* when tested at discriminatory dose of Bavistin (5.0 µg/ml) as all sensitive isolates are eliminated at this dose. This method seems to be quick and reliable as the results are obtained within 72-96h and that the frequency distribution of resistant types can easily be ascertained in a population.

Further they observed that spore germination in both resistant and in sensitive did not seem to be much affected by Bavistin as evidenced by the reaction of sensitive and resistant isolates tested at lower concentrations of carbendazim. However, germination was markedly affected between these two groups at 50 µg/ml (R=42.8 per cent, S=0 per cent) and 100 µg/ml (R=24.0 per cent, S=0 per cent). The germ tube morphology appeared to be influenced. The germ tubes were normal and continued to grow at higher concentrations of Bavistin in case of resistant isolate. Contrarily, in the sensitive isolate there was no growth of germ tubes above 0.1 µg/ml of Bavistin and these appeared stunted and deformed.

Ramaswamy and Anilkumar(1995) studied spore production and germination of carbendazim resistant isolates (Gm3, Gm5 and Gm10) of *Gloeosporium ampelophagum* and compared with wild sensitive isolate (Ga). Carbendazim resistant isolates were capable of sporulating in presence of 25 µg/ml of copper oxychloride unlike sensitive isolate while they were highly sensitive to thiram. The resistant isolates were capable of germinating in presence of kitazin up to 25 µg/ml while sensitive isolate failed to germinate even at 5 µg/ml. Similarly, the resistant isolates were found capable of germinating in mancozeb even at 50 µg/ml concentration while sensitive isolates failed to germinate at this concentration.

Hemanthakumar *et al.* (2001a) worked on the cross resistance in the carbendazim resistant isolates of *G. ampelophagum* to other fungicides. They found out that the highly resistant isolate showed some resistance to thiophanate methyl up to 1500 ppm whereas sensitive isolate could grew only in one ppm concentration. Similarly the carbendazim resistant isolates could grow in concentration of 750 ppm of IBP (kitazin) as against 500 ppm concentration tolerated by the sensitive isolate. The carbendazim resistant and susceptible isolates did not show any difference in their resistance or susceptibility to hexaconazole, mancozeb, copper oxychloride, Bordeaux mixture, dithionon and chlorothalonil. They further observed that cabendazim resistant isolate showed greater tolerance to thiophanate methyl and IBP compared to susceptible isolates showing existence of possible cross resistance in CR isolates.

Alternate use of risky fungicides with compounds having different mode of action has been advocated in a number of diseases confronted with resistance problem (Delp, 1980). Deokate *et al.* (2002) recorded that use of carbendazim alternately with benomyl, dinocap, fenarimol, thiophanate methyl, triadimefon and tridemorph reduced the growth of CR (carbendazim resistant) isolates significantly on agar plates as well as on grape berries. This reduction was more prominent with thiophanate methyl and fenarimol. Similarly carbendazim used in mixture with benomyl, fenarimol, thiophanate methyl and tridemorph reduced the growth of the CR isolates.

The pathogen over-winters as dormant mycelium and sclerotia in the cankers that remain deep-seated on the branches and are able to cause fresh infections in spring on young new shoots by producing conidia under favourable weather conditions (Govinda Rao and Dakshinamurti, 1964 and Suhag and Grover, 1972b).

Suhag and Gover (1973) recorded that in India the perfect stage of anthracnose pathogen has not been reported so far and the most common name accepted for the pathogen is *Sphaceloma ampelinum*. The mycelium surviving in the cane forms conidia. Cankers on uncut vines or on pruned twigs can produce conidia. The conidia can penetrate the unwounded shoots, leaves, petioles, tendrils, stems and fruits. According to Suhag and Grover (1977b) cankers as well as infected plant debris are the main source of infection that withstands the extreme summer and winter conditions.

Ellis and Erincik (2005) mentioned that the fungus over winters by sclerotia laid down initially in late summer or autumn as host tissue harden and may remain viable for 3-5 years. Peres *et al.* (2005) working on life cycles of *Colletotrichum acutatum*, a causal agent of strawberry anthracnose commented that the pathogen appeared to compete poorly as a saprophyte but could survive in soil and on an inanimate surface for various periods. The authors found out that the pathogen survived in air-dry soil for more than one year, however at field capacity the conidia survived less than two weeks.

Bottomley (1937) and Duplessis (1940) observed that anthracnose of grape was increased due to rainfall. Mohanraj (1970) reported that high temperature and prolonged high humidity were necessary for successful infection. Increase in spore germination and germ tube elongation of *Sphaceloma ampelinum* at higher temperature was found parallel with increased diseased incidence at higher temperature.

During 1962 in Hyderabad, Anab-e-Shahi variety of grape has been reported with a heavy infection of the disease immediately after October pruning because of continuous rainfall (Govindarao and Dakshinamurthy, 1964). Prasad and Nirvan (1965) made observations that rain and dew is highly favourable for the spread of the spores and their germination. The distribution and spread of the spores depend mainly on the splashing effect of the rains. Spores were detected in wind borne splash drops up to seven meters. Sporulation in spring is favoured by wet, humid weather; however it may occur at temperatures as low as 2°C. Brook (1973) reported that sufficient wetting of the lesion is required to initiate spore production. Wetness periods for infection range from 7-10h at 12°C (in spring) to 3-4 h at 21°C (summer) in Auckland.

The spore distribution mainly depends upon the washing rains. Spores can be detected in wind-borne splash drops for a distance up to 7 m (Chandrasekhara Rao, 1992). Conidia are spread by rain splash from sclerotia on lesions of stems, petioles or old bunch stalks (Ellis and Erincik, 2005).

Tanaka *et al.* (1974) reported that optimum conditions for growth and sporulation are 28°C. Suhag and Kaushik (1982) observed the production of lesions after 30-42 hours of incubation and reported the optimum temperature for spore germination between 30-35°C.

Pearson and Goheen (1988) observed that the optimal temperatures for disease development are 24-26°C, conidia can germinate and infect at a range of 2-32°C with incubation period varying from 13-14 days. Temperature and moisture are the main environmental factors affecting disease development.

Anthracnose is more damaging during years of heavy rainfall. A minimum leaf wetness period of 3 hours is required for successful infection at 25°C. In India anthracnose becomes more severe and damaging during monsoon period. Prasad and Nirvan (1965) reported a positive correlation between heavy rainfall and the disease incidence. Singh and Joshi (1971) found that due to high temperature and low relative humidity during summer, the vine death due to disease was negligible, but continuous rainfall during July-September provided the most favourable conditions for the disease development. After November as rain subsided, there was decrease in relative humidity and temperature that created unfavourable conditions for the progress of epidemic of anthracnose on grapes. Prasad and Nirvan (1965) observed the early appearance of infection on new growth in rainy season. Continuous rains and cloudy weather were the most conducive for infection and rapid spread on young leaves. The disease was favoured by high humidity above 70 per cent and the temperature range between 20°C-30°C.

In the continuous drizzling or rains above 48 hours the disease appeared in an epiphytotic form (Sridhar and Sohi 1976). Kore and Gurme (1978) found that the range of temperature for the grape anthracnose fungus growth was from 9to 35°C and maximum growth and sporulation were obtained at 29°C and no growth was seen at 40°C. Further, 100, 90 and 80 per cent relative humidity were found to be the best for growth and sporulation of the fungus. However, growth and sporulation decreased as relative humidity levels decreased. Light treatment study revealed that continuous darkness was found to be the best source for growth and sporulation followed by diffused and continuous light.

Suhag and Grover (1977a) at Hissar, in their epidemiological studies recorded the severity of the disease in warm and wet weather. The authors opined that relative humidity and precipitation were more important than temperature for disease development.

Rao and Satyanarayana (1991) carried out step down regression analysis with meteorological parameters and disease which indicated that rainfall (49.9 mm/wk) distributed over 3.16 days/wk with 39 h sunshine/wk (cloudy weather) favoured anthracnose build up and severity on Anab-e-Shahi cultivar of grapes.

Mohan Chander and Thind (1995) made observations that rainfall seemed to influence the disease epidemiology significantly. The disease was generally more severe during July-August when the rains were abundant. Positive correlation between rainfall and anthracnose was observed.

Mohanraj (1970) found out efficacy of Captan, Cuman L and Difolatan as well as mycosystatin among the antibiotics. Suhag and Grover (1972) recorded in laboratory tests that the cankers were effectively checked by the use of ferrous sulphate, brestan, dodine and hinosan whereas miltox, maneb, captan, thiram and ziram inhibited mycelial growth of the pathogen temporarily.

Sastry *et al.* (1989) recorded the efficacy of Bavistin, Thiram, and Difolatan inhibiting the mycelial growth of the pathogen. Pampanagouda (2000) reported thiophanate methyl, benomyl among the systemic and Bordeaux mixture as well as mancozeb among the nonsystemic fungicides was found to inhibit the pathogen (*S. ampelinum*) growth significantly *in vitro.*

Gupta *et al.* (1981) reported the inhibition of spore germination in *Colletotrichum capsici* due to phytinoids of onion, garlic and neem. Michealgrainge and Saleem Ahmed (1988) found inhibition of growth in *Sphaceloma ampelinum* due to onion bulb extract attributed to presence of alkaloids. More than 50 terpenoids have been isolated from neem tree. The fungicidal properties of neem and its products has been reported by Jones *et al.* (1989) and Parveen and Alam (1993).

Sawant and Sawant (2006) have recorded significant mycoparasitism of *Trichoderma harzianum* and *T. viride* on anthracnose fungus that caused lysis of the hyphae and the spores *in vitro*.

Taupeau (1986) observed that in Europe, where the disease was first signalled, it has practically disappeared due to regular treatment with Bordeaux mixture aimed at controlling downy mildew. Ramakrishnan and Sundaram (1955) recommended that, for effective control of anthracnose, removal and destruction of the diseased pruned shoots is the foremost requirement followed by dormant fungicidal sprays such as iron sulphate ($FeSO_4$) and Bordeaux mixture (5:5:50).

Prasad and Nirvan (1965) reported the efficacy of Fycol-8E (oil based copper formulation) sprayed five times at fortnightly interval against anthracnose of grapes in Uttar Pradesh. Similarly Bedi *et al.* (1969) reported the efficacy of blitox and aureofungin in controlling the leaf and twig infection in Punjab. Many such reports indicate the effectiveness of different fungicides from time to time that include; difolatan, thiophanate methyl, Benomyl, Bavistin and Thiram, baycor, Daconil, Dithane flowable, foltaf, etc among the non-systemic (Sridhar and Sohi, 1976, Rawal and Muniyappa, 1981, Upadhyaya and Yadav, 1982 and Rawal, 1987).

Thind and Bedi (1988) obtained highly effective control of the disease with sprays of Bavistin, Agrozim and Topsin-M at 0.1 per cent dosage. In Australia, sprays of dithionon, captafol, chlorothalonil, dichlofluanid, captan, fluazinam were also found to be promising against this disease (Magarey *et al.*, 1993).

In the recent period many new formulations such as RIL-F004, RIL 006/C, Samarth (a combi-product of hexaconazole and adjuvants) and Shield SC (Shitole *et al.*, 2003 and 2005, and Jahagirdar *et al.*, 2004a&b) have been tested and found to be highly effective. To counteract this resistance phenomenon among the anthracnose isolates alternate use of risky fungicides like benzimidazoles with dithiocarbamates or copper based fungicides has been advocated (Mohan Chander and Thind, 1998). Hemanthakumar *et al.* (2001b) reported the superiority of hexaconazole alone and in combination with carbendazim in countering the carbendazim resistance. Many researchers have attributed the superiority of onion extract as well as the fungicidal properties of neem (*Azadirachta indica*) and its products in managing the anthracnose on grapes due to presence of alkaloid and terpenoids respectively (Singh and Pande, 1966, Ghewande, 1989, Jones *et al.*, 1989, Ganpathy and Narayanswamy, 1990, Parveen and Alam, 1993 and Pampanagouda, 2000).

Most of the *vinifera* varieties have been found to be susceptible to moderately tolerant. The disease has been found to be serious in variety Pachadrakshi (Ramakrishnan and Sundaram, 1955). It has been reported in severe form on Anab-e-Shahi and Kali Sahebi in Hyderabad region of Andhra Pradesh, Maharashtra and Coimbatore of Tamil Nadu (Muthuswamy and Abdul Khader, 1960 and Olmo, 1970).

Bakshi *et al.* (1970) recorded the Delight variety of grapes as tolerant and comparable in terms of quality and taste with that of Thompson Seedless. Goyal *et al.* (1971) observed that the varieties Bharat Early and Hussain remained completely free from infection while Phakdi recorded maximum infection and minimum incidence was recorded in White Muscat.

Mathur *et al.* (1973) screened 67 varieties during 1966 and 1967 of which 21 varieties were found to be resistant. In Karnataka, the disease has been reported to be serious on Anab-e-Shahi and Thompson Seedless, Phakdi, Bhokri and Selection 7 (Shelke *et al.,* 1997). Gurme and Kore (1977) screened 57 varieties in Maharashtra (Parbhani) of which Athens, Black Champa, Bangalore Blue, Black Cornicorn, Champion, Muscat Hamburg, Kishmish Beli, and Champanel were found highly resistant.

Sandhu *et al.* (1981) evaluated 179 F1 grape hybrids of three different crosses, *viz.* Karachi Gulabi X Thompson Seedless, Karachi Gulabi X Himrod and Karachi Gulabi X Perlette for resistance against anthracnose of grapes. The first two crosses yielded three and one resistant hybrid vines respectively while the last named cross failed to produce any resistant F1s.

Datar and Ashtaputre (1985) screened 80 varieties on leaf stage infection and recorded Himrod as highly resistant genotype among the 13 resistant varieties identified.

Prasad *et al.* (1992) evaluated grape hybrids and found Bangalore Blue and its derivative E-21/28 (B'lre Blue X Convent Large Black) as highly resistant against anthracnose.

Shelke *et al.* (1997) screened five varieties *viz,* Kishmish Charni, Thompson Seedless, Tas-A-Ganesh, Sonaka and Manikchaman against anthracnose under natural epiphytotic conditions. All the genotypes were found moderately susceptible. Thind *et al.* (1997) evaluated grape cultivars Beauty Seedless, Cardinal, Delight, Himrod, Perlette, and Thompson Seedless of which Perlette and Beauty Seedless as the most susceptible.

Jayashankar *et al.* (2000) made *in vitro* selection of proembryonic masses grapevine (*Vitis vinifera* L.) 'Chardonnay' (clone 02Ch) by exposing to culture filtrates of *E. ampelina*. For 4-5 cycles of recurrent selections with medium containing 40 per cent fungal culture filtrate, putative resistant lines RC1 and RC2 respectively were established. The selected lines inhibited the growth of *E. ampelina* and *Fusarium oxysporum* (isolated from watermelon) in a dual culture assay and reduced the growth of mycelium on a conditioned medium test. The SDS-PAGE analysis revealed that an extracellular protein is involved in the resistance. Plants regenerated from selected lines exhibited resistance to infection by *E. ampelina* in both green house tests and detached leaf bioassays.

Dhanumjayaraol *et al.* (2006) reported that the higher total acidity enhances resistance. Higher total sugars and reducing sugars enhance susceptibility to powdery mildew disease in grape. High varietal variation was found to disease intensity. Leaf area, leaf lamina thicknesses have no role and stomatal density has role in disease intensity.

Mundy *et al.* (2012) stated that *Botrytis cinerea* is a fungus responsible for considerable damage to a wide range of crops worldwide, including grapes. Botrytis bunch rot caused by *B. cinerea* is the major disease problem that must be managed by the New Zealand wine industry each season. However, the fungus is not easily managed as it can be both necrotrophic and saprophytic, with a range of overwintering inoculum sources. New Zealand grape growers have asked whether it is necessary to remove tendrils at the time of pruning in order to minimise botrytis bunch rot infection at harvest. This review provides a summary of the information currently available on the importance of tendrils in the epidemiology of botrytis bunch rot under New Zealand conditions. Gaps in knowledge and areas for further investigation are also identified.

Mundy *et al.* (2012) reported that the Botrytis bunch rot is a disease that requires management under New Zealand conditions in order to prevent financial losses for wine grape producers. A survey was conducted to investigate the potential spore production of tendrils in 16 vineyards from the Marlborough and Hawke's Bay wine growing regions. Significant differences in spore production potential of tendrils were detected between sampling before flowering and after fruit set. Significant regional differences were also detected, with tendrils from Marlborough producing fewer spores. Under ideal laboratory conditions for incubation of the fungus, spore production by tendrils was low at the start of the season and was reduced further following application of the industry standard fungicide programme. Based on these experiments and a review of other data it is recommended that tendril removal is not required at pruning as part of vineyard disease management.

Hill *et al.* (2010) opined that the Botrytis bunch rot (botrytis) can reduce grape yield and wine quality. Standardised assessment methods are needed to allow greater precision in wine making and to allow the use of disease management decision support models in grape production. This study developed a botrytis disease assessment key to assist the accuracy of visual disease assessment. Associated computer training software, designed to improve the accuracy of disease assessments, was also developed and tested. The mean absolute error in estimates of percentage botrytis severity was significantly ($P<0.05$) lower after eight assessors used the training software. Vineyard sample size required for estimation of disease severity was determined and a disease incidence-severity relationship for estimating severity from incidence was developed.

Kim *et al.* (2010) reported that the risk of the appearance or the intensification of a crop disease can be assessed using information about the weather, the pathogen or the crop. Weather data for use in disease risk prediction can be obtained from measurements at a nearby weather station. While weather measurements can represent accurate weather conditions at the site where the weather station is located, these data are representative only of a small area near the station. To obtain weather information over a larger area, spatial interpolation and remote sensing can be used to estimate the likely weather conditions in other locations. It is crucial to obtain weather data at an appropriate temporal resolution (*e.g.* daily or hourly) for a given disease in order to predict the disease. A weather database system is being constructed to provide high-quality climatic data (*e.g.* daily temperature, humidity and rainfall), which can be used to quantify the link between weather conditions and disease outbreaks.

Elmer *et al.* (2009) stated that the bunch rot of grapes caused by *Botrytis cinerea* is responsible for crop losses exceeding 30 per cent under favourable disease conditions.

The potential for greater losses exists due to a market-driven, industry-wide shift towards 'nil pesticide residues' spray programmes. Plant and Food Research is developing nil residue biological and natural products in partnership with New Zealand Winegrowers, Technology New Zealand, the Foundation for Research, Science and Technology, and Botry-Zen Limited. The biological product BOTRY-Zen®, based upon *Ulocladium oudemansii* (Patent PCT/NZ01/00111), is applied early season for control of *Botrytis*, and ARMOUR-Zen®, a chitosan-based natural product formulation, is applied mid- and late season, and complements BOTRY-Zen. Two developmental products, NP2 (PCT/NZ2005/000167) and BCA-L1, have provided excellent control of *Botrytis* during mid- to late season. NP2 is a plant-based natural product suitable for use mid- season, while BCA-L1 colonises the berry surface, thereby preventing *Botrytis* infection preharvest. These field trial results demonstrate that nil residues, full season biological control of *Botrytis* is achievable, with crop losses being similar to those under current fungicide-based spray programmes.

Mundy (2008) stated that nitrogen is an important macro-nutrient for grapevines, influencing a number of metabolic pathways and plant structures. This review provides a summary of some direct and indirect effects of nitrogen fertilization on bunch rot incidence in the field. Direct changes due to increased nitrogen nutrition include greater thickness of the wax on berry surfaces and increased concentrations of the biochemical defence compounds produced in them. Indirect changes that result from increased nitrogen include development of tighter bunches and denser canopies. As integrated management systems become more complex, a better understanding of different components such as nitrogen nutrition is required, to allow vineyard managers to make informed decisions on how potential changes to current practices may influence disease and berry quality parameters.

Mundy and Beresford (2007) reported that the percentage of detached grape berries inoculated in the laboratory that became infected by *Botrytis cinerea* during ripening was correlated with increasing sugar concentration and increasing yeast assimilable nitrogen (YAN) concentration. These factors could not be separated to provide a causal relationship between YAN and per cent berries with disease due to the close relationship that existed between sugar concentration, YAN, vine development and time. Wounding increased the percentage of berries with disease, particularly when the berries were immature. These findings do not support the hypothesis that the inherent susceptibility of grape berries to botrytis infection increases with increasing YAN. Field observations by growers that nitrogen fertilizer increases bunch rot may be due to nitrogen causing increased canopy density, which in turn causes a microclimate more conducive to the development of botrytis bunch rot.

Beresford *et al.* (2006) stated that the epidemic descriptors that quantified the location in time and rate of disease increase of botrytis bunch rot were developed using percentage disease severity data. Data were obtained from fungicide evaluation trials in three regions of New Zealand and from Tasmania in Australia. Mean percentage severity versus time was logit transformed and linear regression was used to predict the date at which 5 per cent severity was reached, the daily severity increase rate at 5 per cent severity, the severity on the date of harvest and the daily severity increase rate at harvest. These descriptors will be of general use for the critical comparison of bunch rot epidemics between sites and between seasons. They will be used in quantitative analyses to identify factors that drive botrytis bunch rot epidemics in order to develop botrytis prediction models.

Henshall *et al.* (2005) observed that the wetness duration measured by flat plate sensors inside and outside a grape canopy was recorded from December-March. Sensors outside the canopy generally recorded longer wetness duration than sensors inside the canopy. For days with rain, short wetness durations detected by outside sensors were not detected by inside sensors because of sheltering by the canopy. When wetness arose solely from dew, duration inside was much shorter than outside for prolonged wet periods. Wetness was used to calculate infection periods according to two botrytis bunch rot risk models. Agreement between sensors was worse inside the canopy than outside, although on occasions when rainfall exceeded 10 mm there was greater uniformity between sensors. For region-wide disease risk monitoring, wetness duration measured outside leaf canopies at standard meteorological sites would give a "worst-case" estimate of disease risk. Regression equations are presented that allow estimation of inside wetness duration from outside wetness duration for rainy and non-rainy days.

Agnew *et al.* (2004) reported that the standard practice for grape disease management in Marlborough 10 years ago was to spray for diseases using calendar based spray programmes. Communication of grape disease management research has led to a change of practice for many grape growers. The key tool developed by the research programme was targeted spraying for diseases. This involved field monitoring for diseases and the use of decision support software, which contained the Bacchus *Botrytis cinerea* risk model. Adoption of target based spraying resulted in a reduction in fungicide usage of up to 50 per cent. The research involved industry consultation, field trials on vineyards and communication of information to participants. The involvement of grape growers as advocates from the outset ensured that there was industry ownership of the outcomes. A change in practice by the Marlborough grape industry has resulted in financial savings for growers and environmental benefits for the wider community through reduced fungicide applications to wine grapes.

Obanor *et al.* (2004) opined that *the Botrytis cinerea* can be detected in juice pre- and postharvest using the monoclonal antibody BC-12.CA4 in plate trapped antigen-enzyme linked immunosorbent assays (PTA-ELISA). The amount of antigen predicted by PTA-ELISA was found to increase as grape juice samples were progressively diluted. This effect was also observed in *B. cinerea* contaminated boysenberry and blackcurrant juice samples. Detection of *B. cinerea* infections by PTA-ELISA was improved by incubation of grapes at 20°C for 48 h and was similar to visual assessments of infected grapes. PTA-ELISA is faster (1-3 days) than visual *B. cinerea* assessments (10-14 days). PTA-ELISA measurements can be standardised, removing human bias in determining *B. cinerea* infection levels.

Mundy and Agnew (2002) reported that the field trials were established in four vineyards in January 1999 to evaluate the effects of four mulch mixtures containing vineyard and winery wastes, on different soil and plant parameters. The effect of mulch treatment on numbers of colony forming units of soil fungi was assessed using soil dilution plating. During 3 years (1999-2002) the numbers of soil fungi tended to be higher in soil sampled from under mulch-treated plots than in soil from non-mulched plots. The effect of mulches on incidence and severity of botrytis bunch rot (caused by *Botrytis cinerea*) was also investigated. Incidence of the disease was not increased by the mulch treatments, and in some instances bunch rot incidence was less on grapes from mulched plots than un-mulched plots.

12
Processing

Raisin

The grapes meant for raisin making should preferably be seedless, white, with thin skin and TSS above 24 Brix. Select fully ripe bunches and remove damaged and diseased berries. Selected bunches are dipped in a boiling NaOH (Caustic Soda) solution of 0.3 per cent for 3 Sec. followed by cold water washing. Treated bunches are spread on perforated trays, kept in an air tight sulphur box and fumigated by burning sulphur at the rate of 3g/kg of fresh grape for 3 hours. Sulphited bunches may either be dried in a hot air drier at 60-65°C temperature or they may be allowed to dry under shade at room temperature till the raisins get about 15 per cent moisture content. Soon after drying raisins should be separated from dried rachies quickly and thoroughly washed in lukewarm water, air dried and kept for moisture equalization for a period of one month in an air tight container. Afterwards, raisins are packed in 400 gauge polythene bags and stored at cool temperature (3-5°C). IIHR variety 'Arkavati' has been found to be better for raisin making.

Seeded white grapes with high total soluble solids (TSS) may be used for making 'Munnaca' while coloured grapes like Beauty Seedless can be used for making black raisins to be used in confectionery.

Soda Oil Dipped Raisin

Seedless grapes are dipped in a sodium solution having a thin film layer of olive on the surface; the dipped grapes are dried in trays in direct sunlight. The raisins are medium to dark brown in colour, fairly tender, slightly oily but not sticky. The grapes are dipped in warm solution containing 3 kg cooking soda and 100g sodium hydroxide in 100 ltr of water for 3 minutes on which1 ltr of olive oil is floated. After removing from the dipping solution the grapes are dried in sun.

In another method grapes are immersed in a water solution containing 4.5 per cent potassium carbonated, 0.5 per cent sodium carbonate and 1 per cent completely emulsified olive oil for about five minutes. The grapes are spread in trays and turned in trays after two days.

The Australian method involves dipping of grape in alkaline emulsion of commercial dipping oil which is mixture of about 70 per cent ethyl esters with free oleic acid emulsifiers.

1. Standard cold dip emulsion:

Potassium carbonates	2.4 kg
Dipping oil	1.5 ltr
Water	100 ltr

2. Standard cold dip

Eemulsion (Two third)	1.7 kg
Dipping oil	1.0 ltr
Water	100 ltr

Trimmed and washed bunches in perforated plastic crates are dipped in water containing the commercial dipping oil (a mixture of potassium carbonate and ethyl oleate) The required quality of dipping oil is poured in water in a cement tank to raise the pH of the solution to 10. The pH is tested with litmus paper, grape trays are allowed in the dipping solution for five minutes and taken out draining the solution completely, the grapes are spread on fine mesh (1 cm x 1 cm) nylon nets stretched in tiers, of nylon nets with a clearance of 9" to 10" between two successive one and built in sheds with tin roof. Drying completes in 10 days.

Zoffoli (2006) reported that maturation of table grape (*Vitis vinifera* L.) extends from a period of almost 40 days from veraison to harvest. During this phase of fruit development occur the most significant physiological change that allows the accumulation of sugar, acid, phenolic compound and the increase of volume (Coombe and Bishop, 1980). The main deterioration factors of table grapes during storage at low temperature are decay and water loss. The non-climacteric (minimum ethylene production during ripening) response of fruit ripening and the low respiration rate determine that softening is not a relevant postharvest problem as in the majority of fruits. In the case of decay, gray mold is considered the universal disease and drying of the stem alters the appearance at the market level. These two aspects require a constant attention in the postharvest handling of table grape. Postharvest handling of table grape includes determination of optimum harvest time, harvest and cooling operations, sulfur dioxide application and water loss control.

Fourie (2006) reported that the decay on table grapes, whether pre- or post-harvest, can cause considerable financial loss for producers, as well as the trade. In South Africa and many other countries, fungi such as *Botrytis, Rhizopus, Aspergillus* and *Penicillium* as well as bacteria and yeasts are primarily accountable for pre-harvest decay. Post-harvest fungal decay is mainly caused by Botrytis cinerea; however, other pathogens, depending on the storage conditions, may also cause losses. To reduce the risk of decay development, an integrated approach of decay management should be followed. Pre-harvest control mostly relies on strategies to prevent or reduce the risk for infections and does not only entail the application of fungicides. The application of correct cultivation and vineyard practices is of utmost importance. Decay monitoring during the growing period, especially near to optimum harvest maturity, is required. Post-harvest control starts by observing and recording weather conditions prior to and during packing. Careful handling during harvesting and packing, to minimize injuries, and hence infections, is a prerequisite for reducing decay development during storage. Handling of grapes after packing and during cold storage needs to focus on limiting the likelihood of further infections and in reducing the spread of decay. In this effort, temperature management plays a vital role. Packing aids, of chemical or non-chemical nature, may also be used for decay control during storage. The use of SO_2 sheets has become an integral part of decay control on table grapes intended for medium or longer-term storage. As important is the selection of the applicable packaging materials, where internal packaging in particular, may have a profound effect on grape quality maintenance.

Sawant *et al.* (2006) studies on fungi associated with post harvest decay in table grapes meant for overseas markets were done on grapes collected from different vineyards in Maharashtra during 1997–2003. During storage at 0±0.5 and dag;C, post harvest decay was minimal and caused due to *Alternaria alternata* and *Cladosporium sphaerospermum*. Infection of both these fungi was restricted to individual berries. On shelf, *Aspergillus* spp. mainly *A. niger, Botryodiplodia theobromae,* and *Penicillium* spp. were found associated with decayed berries, while *Rhizopus stolonifer* was associated with bunch rot. *Botrytis cinerea* and *Phomopsis* were not detected in any sample.

Deshpande and Shukla (2006) stated that the grapes occupy far more land in the world than any other single fruit and account for nearly half of the total world production of all fruits. Grapes are usually harvested in the months of March–April and August–September in central parts of Madhya Pradesh and Western Maharashtra region. Grapes being perishable and high moisture fruit start decaying after about 4 days of storage when kept under ambient conditions with about 28–30°C temperature and 65 per cent –70 per cent relative humidity. There are studies on grape storage in different packages with and without application of fungicidal spray; however, the studies related to the packaging of grapes in modified atmospheric conditions using polymeric packaging is lacking. As in the case of other fruits, if the grapes are stored under MAP conditions, shelf life could be enhanced due to the arrest in respiratory activity. Different commonly available polymeric films (especially LDPE and PP) could be used for storing the grapes in small pouches by vegetable sellers and consumers/users after the produce is brought to the fruit and vegetable mandi and prior to the actual consumption. The objective of the present investigation was to study the effects of two polymeric films in packaging garden fresh grapes to extend their post harvest storage life under ambient condition. In this study, two polymeric films were used for packaging of grapes for short-term storage under ambient conditions and protected from direct light. The films used were low-density polyethylene (LDPE) 25µ, 50µ and 75µ and polypropylene (PP) 25µ and 38µ. At the end of 7th day of packaging, grapes kept under ambient condition in LDPE and PP bags had no colour change and no mould growth/fungal attack. At the end of packaging period, *i.e.*, 10th day, samples in all bags turned brownish and mould growth was visible. The physiological loss in weight from the non-packaged grapes was 5.71 per cent after 5 days and 9.87 per cent after 7 days compared to 3.07 per cent after 5 days and 5.91 per cent after 7 days storage for the grapes packaged in PP films. Sensory evaluation studies indicated acceptability of grapes packaged under ambient conditions in LDPE and PP pouches up to the period of 7 days.

Narayana Reddy *et al.* (2006) conducted a experiment with Thompson Seedless grapes to study the effect of post harvest treatment with AOX inhibitors alone and in combination with polyamines on the shelf life when stored at 4°C. Cumulative physiological loss of weight (CPLW) increased gradually during storage in grapes but the AOX-inhibitors and polyamines effectively reduced the CPLW and berry shatter during storage (Spermine + n-PG) 1.0 mM, (Spermidine + n-PG) 1.0 mM recorded higher TSS. The shelf life was higher (48.6 days) with SHAM 1.0 mM and 47 days with n-PG 1.0 mM and in the combination treatment (spermidine + SHAM) 1.0 mM gave the highest shelf life of 51 days followed by spermine 1.0 mM + SHAM 1.0 mM (45 days) and rest of the treatments were at par with each other.

Somkuwar *et al.* (2006) opined that the growing grapes on rootstocks are a common practice in India, since rootstocks helps in improving the quality of grapes through enhanced uptake of nutrients. Secondly, increase in cell wall turgidity of the grape berries by pre-harvest treatments of calcium nitrate as dipping has a major contribution in increasing the shelf life. Considering this, a trial was conducted on seven-year-old Thompson Seedless grapes grafted on four rootstocks, *viz.*, Dogridge, Salt Creek, 1613 C and St. George. Bunches were dipped in 1.0 per cent calcium nitrate solution. At harvest, the grapes from each rootstock were harvested from calcium nitrate treated and untreated vines and kept under cold storage for 30 days. After removal from cold storage, the observations were recorded in shelf for five days. To study the role of nutrients in increasing shelf life, the nutrients were analyzed from rachis of the bunch. Among the different rootstocks, Dogridge rootstock was found better for minimizing PLW in Thompson Seedless compared to other rootstocks. On the 3rd day under shelf, minimum PLW of 6.26 per cent was recorded when the berries were treated with calcium nitrate compared to untreated control. On the same day under shelf, fallen berries were minimum on Dogridge rootstock in calcium nitrate treated (4.05 per cent) compared to untreated (7.0 per cent) grapes. Maximum berry fall was recorded in Thompson Seedless grafted on St. George rootstock. The same trend was also recorded for per cent rotten berries. The pedicel browning was more on St. George rootstock having shriveled berries leading to the reduced shelf life. In treated grapes, higher calcium of 101.96 ppm and magnesium (87.70 ppm) was recorded in Dogridge rootstock compared to 98.79 ppm and 63.70 ppm in untreated, respectively.

Consumption of grapes (*Vitis vinifera* L.) is beneficial to health. In addition to the nutrients such as sugars, vitamins and minerals, grapes are also rich in phenolic compounds and other natural antioxidants (Doshi and Adsule, 2006). However, very little information is available on effect of post-harvest handling and storage temperature on changes of the phytonutrients in grapes. The present work was done to investigate the effect of post-harvest storage using restricted ventilation (1–2 per cent) polythene cluster bags/plastic boxes at ambient temperature and 2–4°C in the refrigerator, on physicochemical parameters, phenolics compounds and the antioxidant activity in grapes. A representative sample of grape bunches stored at ambient temperature without cluster bag/box was taken as control. It was found that the physiological loss in weight, total soluble solids and titrable acidity were gradually increased with storage period at ambient temperature up to 25 days and at 2–4°C up to 60 days, though the changes were relatively less in the samples stored in the bags and boxes as compared to control. The reducing sugars, total phenolics, flavonoids, procyanidin monomers and anthocyanins extracted in the aqueous methanol were significantly reduced ($p < 0.05$) in the control sample with the increase in storage period till 10 days followed by gradual increase; however, relatively less variation was observed in the case of the samples stored in bags and boxes at both the temperatures. All the samples stored at 2–4°C showed much less per cent polymeric color in the range of 8–11 per cent as well as no significant variation in the procyanidin monomers content throughout the storage period of 60 days. The antioxidant activity analyzed for various samples correlated well with the total phenolics content of the respective samples. Thus, the storage-life and the antioxidants level in grapes could be well retained at low temperature storage using restricted ventilation packaging.

Three experiments were conducted (Narayana Reddy *et al.,*) to elucidate the involvement of polyamines in the shelf life of grapes. Thompson Seedless had significantly longer shelf life followed by Kishmish Rozavis and Anab-e-Shahi. Putrescine was the major polyamine in grapes followed by spermidine and spermine. There was a general decline in the intensity of polyamine, and particularly spermine declined to undetectable levels during the storage in the varieties or treatments with shorter shelf life. Storage of Anab-e-Shahi grapes at 4°C extended the presence of spermine and resulted in increased shelf life. Undetectable levels of spermine spell the end of shelf life in grapes. Regarding various physico-chemical parameters, Thompson Seedless and Sonaka recorded significantly lowest PLW, berry shatter and highest TSS and acidity.

Whereas viticulture has been historically evolved in an empiric way, we are today confronted with a viticulture characterized more and more by technicality. The knowledge being increasingly complex and closer to interpreting the reality, viticultural zoning advances to answer more and more profound questions (Tonietto, 2006). Geographical indicators for grapes – soil, climate, adaptation of the grape varieties – are one of the major contributions of viticultural zoning, which offers a whole range of practical applications. It is growing in importance, be it on account of the technical resources which are becoming more efficient and which make possible the development of an increasingly integrated, consistent and useful zoning, be it on account of a more and more globalized market. The article points out the importance of geographical indicators for table grape producing and for winegrowing regions. It deals with viticultural production, quality and typicity of the products with regards to their organoleptic characteristics. Some aspects of the contribution of zoning are detailed: site selection, the agroviticultural and enological management, the delimitation of the viticultural territories and zoning of regions with viticultural potential, including particularities of grape production in tropical regions. Methodologies for zoning studies are presented, such as the Géoviticulture Multicriteria Climatic Classification System; moreover, the text shows the indirect contributions regarding the table grape and wine market, the sustainable territory development and the possibility of the establishment of Geographical Indications in some cases. A new concept for some tropical wines is presented: the just-in-time quality. Résumé Indicateurs Géographiques en Viticulture Mots-Clés : climat, sol, terroir, typicité, Système CCM Géoviticole, indications géographiques, marché, qualité just-in-time L'importance du zonage vitivinicole, y compris les indicateurs géographiques, est en train d'augmenter, soit en fonction des moyens techniques chaque fois plus performants, qui rendent possible le développement des zonages de plus en plus intégrées, consistants et utiles, soit en fonction d'un marché de plus en plus mondialisé. Indicateurs géographiques pour la viticulture - sol, climat, adaptation des cépages, ce sont des contributions importantes du zonage vitivinicole, qui offre une série d'usages pratiques. Quelques aspects de la contribution du zonage sont particularisés : la gestion agroviticole et œnologique, la délimitation des territoires viticoles et le zonage des régions à potentiel viticole, et également quelques particularités de la production dans les régions tropicales. Approches méthodologiques sont pressées, comme est le cas du Système CCM Géoviticole. L'article montre les contributions indirectes du zonage dans la production du raisin de table et du raisin de cuve au niveau du marché, et au niveau du développement territorial et du développement soutenible. La possibilité de valorisation des produits des régions de la zone intertropicale par les Indications Géographiques est signalée. Un nouveau concept pour quelques vins produits en zone tropicale est présenté : la qualité just-in-time.

Patil and Patil (2006) reported that the Thompson Seedless variety of table grape occupies a major area under cultivation in India. The technology for home wine making from Thompson Seedless variety of grape was standardized. The wine was prepared from three different yeasts strains and analysed for biochemical and organoleptic parameters. The biochemical characteristics like pH, TSS, titrable acidity, protein content, total sugars, reducing sugars, non-reducing sugars, alcohol, tannins and aldehydes were estimated. The sensory properties like aroma, bouquet, vinegar, flavour, astringency, sweetness, etc. were studied as per 20 point scale. The wine fermented with *Saccharomyces ellipsoideus* No. 101 strain recorded highest score of (16.5) with 8.25 per cent alcohol, 134 mg/100 ml of tannins, 95 mg/litre of aldehydes and was found most acceptable for consumption.

Twenty exotic and 3 indigenous wine varieties were evaluated for their growth, yield and fruit quality at the experimental farm of the National Research Centre for Grapes, Pune, India (Karibasappa and Adsule, 2006). Among the 14 red varieties, maximum fruit yield was recorded in Pusa Navrang (19 t ha^{-1}) followed by Convent Large Black (18.47 t ha^{-1}) and Shiraz (17.87 t ha^{-1}). Cluster to cane weight ratio varied significantly among varieties; however, cluster to cane weight ratio was minimum in Champanel (0.29), a *V. champini* red cultivar and maximum in Ugni Blanc (7.56), a white wine variety. Juice yield in hectolitre ha^{-1} varied significantly among the varieties. Ugni Blanc recorded the maximum juice yield of 208.8 hectolitres ha^{-1} (hl ha^{-1}) followed by Chenin Blanc (202.3 hl ha^{-1}), Grenache (178.9 hl ha^{-1}) and Carignane (178.7 hl ha^{-1}). Whereas, juice yield was low in cvs.: Champanel (17.9 hl ha^{-1}), Sangiovese (33.1 hl ha^{-1}), Chardonnay (40.8 hl ha^{-1}), Pinot Noir (41.3 hl ha^{-1}) and Black Prince (58.3 hl ha^{-1}). Highest TSS of the must was noticed in Pinot Noir (24.80 °Brix) followed by Merlot (23.10 °Brix) and Cabernet Sauvignon (22.20 °Brix). Must TSS was generally recorded low among white varieties except in Sauvignon Blanc (20.80 °Brix) and Vermentino (20.60 °Brix), musts of Chardonnay (18.0 °Brix), Chenin Blanc (18.50 °Brix) and Clairette (17.80 °Brix) were low in TSS, whereas Ugni Blanc (19.8 °Brix) and Garganega (19.60 °Brix) recorded moderate high TSS. Acidity of the most varied between 4.25 g L^{-1} in Pinot Noir to 6.40 g L^{-1} in Saperavi, while in white varieties it was varied between 3.39 g L^{-1} in Angur Kalan and 5.48 g L^{-1} in Garganega. Must pH varied between 3.0 (Carignane) and 4.95 (Arka Trishna) among red varieties, while it was varied between 3.07 in Garganega and 3.69 in Sauvignon Blanc. Based on juice yield and must quality parameters cvs., Shiraz, Cabernet Sauvignon, Merlot, Zinfandel and Pusa Navrang among red varieties and Sauvignon Blanc, Ugni Blanc, Vermentino and Garganega among white varieties are recommended for commercial wine production.

Somkuwar *et al.* (2006) reported that the growing grapes on rootstocks are a common practice in India, since rootstocks helps in improving the quality of grapes through enhanced uptake of nutrients. Secondly, increase in cell wall turgidity of the grape berries by pre-harvest treatments of calcium nitrate as dipping has a major contribution in increasing the shelf life. Considering this, a trial was conducted on seven-year-old Thompson Seedless grapes grafted on four rootstocks, *viz.*, Dogridge, Salt Creek, 1613 C and St. George. Bunches were dipped in 1.0 per cent calcium nitrate solution. At harvest, the grapes from each rootstock were harvested from calcium nitrate treated and untreated vines and kept under cold storage for 30 days. After removal from cold storage, the observations were recorded in shelf for five days. To study the role of nutrients in increasing shelf life, the nutrients were analyzed from rachis of the bunch. Among the different rootstocks, Dogridge rootstock was found better for minimizing PLW in Thompson Seedless compared to other rootstocks. On the 3rd day under shelf, minimum PLW of 6.26 per cent was recorded when the berries were treated with calcium nitrate compared to untreated control. On the same day under shelf, fallen berries were minimum on Dogridge rootstock in calcium nitrate treated (4.05 per cent) compared to untreated (7.0 per cent) grapes. Maximum berry fall was recorded in Thompson Seedless grafted on St. George rootstock. The same trend was also recorded for per cent rotten berries. The pedicel browning was more on St. George rootstock having shriveled berries leading to the reduced shelf life. In treated grapes, higher calcium of 101.96 ppm and magnesium (87.70 ppm) was recorded in Dogridge rootstock compared to 98.79 ppm and 63.70 ppm in untreated, respectively.

Devarajan *et al.* (2011) stated that the effect of grapevine canopy side (north versus south, and east versus west) on grape and wine volatiles of Cabernet franc was evaluated using two electronic nose systems (conducting polymer-based and surface acoustic wave-based) during two growing seasons. Data from three sampling dates for each season from both electronic noses were compared with physicochemistry and wine aroma sensory evaluations. Univariate and multivariate statistical analyses generally indicated grape physicochemistry indices were not able to differentiate ($p>0.05$) between canopy sides consistently across growing seasons and sampling dates. Both of the electronic nose systems corroborated 100 per cent discrimination of canopy sides for grapes and wine using canonical discriminant analysis. On average, the surface acoustic wave-based electronic nose was able to explain <50 per cent of variation for grapes and <60 per cent for wine using the first principal component, compared to >80 per cent for the conducting polymer-based electronic nose. Wine aroma sensory evaluation was able to differentiate canopy sides in three of four evaluations.

Gardner *et al.* (2011) reported that the cold soak is a pre-fermentation maceration process at cold temperatures, traditionally used to enhance red wine color. This study monitored changes *in Vitis vinifera* L. cv. Cabernet Sauvignon volatiles using a commercial conducting polymer electronic nose (ENose) during a five day cold soak and post-fermentation. PCA distribution of juice volatiles detected by the ENose during cold soak showed PC1 accounted for 95.7 per cent of the variation. Various volatile associations were made with specific ENose sensors. In comparison, PCA distribution of must chemistries had 52.4 per cent of the variation accounted by PC1. The PCA distribution of wine volatiles detected by GC-MS showed PC1 accounted for 97.1 per cent of the variation between control and cold soak treatments where control wine volatiles were associated with several ethyl esters, while cold soak wine volatiles were associated with diethyl succinate, isovaleric acid, benzyl alcohol, 3-methyl butanol, cis-3-hexenol, gamma-nonalactone, benzaldehyde, 2-methyl propanol, phenethyl acetate, 1-octanol, beta-damascenone, terpinene-4-ol, gamma-butyrolactone, ethyl acetate, hexanoic acid, citronellol, phenethyl alcohol, and n-butanol. Comparatively, PC1 accounted for 100 per cent of the total variance when using the ENose to measure volatile composition. Sensory evaluation did not demonstrate significant differences in aroma between control and cold soak wines. This study demonstrates differences in volatile chemistry between control and cold soak wines, as well as the ability to use a conducting polymer as a simple tool for analysis of volatiles.

Zoecklein *et al.* (2008) opined that the wine discrimination and analysis is typically done through chemical analysis and sensory evaluation by a trained panel. Both of these methods are proven to be successful in wine discrimination, but require extensive preparation, time and money. The electronic nose is an objective, rapid-analysis tool that has been used in the food industry for a number of applications. The purpose of this study was to determine if an electronic nose can accurately discriminate between Cabernet Sauvignon (*Vitis vinifera* L.) wines made from grapes that have received different pre-harvest but post-bloom spray treatments to enhance growth. Aqueous ethanol, which has been shown to impact fruit maturity, was sprayed on the grape clusters at 13 weeks post bloom in different concentrations (control, 5 per cent and 10 per cent v/v). Chemical analysis was able to accurately discriminate between the wines produced from these grapes. Triangle difference testing by a consumer panel was not able to differentiate between the different treatments. The electronic nose data was able to accurately identify the control group and the 5 per cent EtOH treatment 90 per cent of the time. Placement of the 10 per cent EtOH group was only 13 per cent correct. The results show the promising potential for an electronic nose to discriminate between control and treated wine samples.

Martin *et al.* (2008) studied to evaluated two electronic nose systems (conducting polymer and surface acoustic wave-based) to differentiate volatiles of grapes and wines treated with an aqueous ethanol spray (5 per cent v/v) at véraison. Ethanol spray induced fruit ethylene production immediately post-treatment, which then declined progressively. Electronic nose evaluations of grape volatiles were compared with Cabernet franc and Merlot physicochemistry and wine aroma sensory data. Canonical discriminant and principal component analysis found that both electronic nose systems and the physicochemical measures (Brix, TA, pH, color intensity, hue, total phenols, glycosides, and berry weight) were able to discriminate between ethanol-treated and untreated grapes and wines for both cultivars. Grape physicochemical treatment differences were due mainly to variations in hue, phenolic-free glycosides, and total phenols. Aroma sensory evaluations differentiated between ethanol treatments and control for Merlot, but not Cabernet franc wines.

Viognier grapes grown in northern Virginia, USA, and resultant wines were evaluated as a function of training system (Zoecklein *et al.*, 2008). Treatments included Vertical shoot-positioned (VSP), Smart-Dyson (SD), and Geneva double curtain (GDC), with vines of all treatments spaced 2.4 m apart in 3.0 m-wide rows. In addition to increased cluster numbers and crop yield, GDC training generally increased fruit zone sunlight interception and fruit exposure, while it decreased cane pruning weights per m of cordon, compared to SD and VSP. Crop adjustments were made between bloom and véraison in six seasons, to result in average yields of 10.5 kg/vine (GDC), 9.9 kg/vine (SD) and 6.0 kg/vine (VSP), with the lower SD canopy bearing 30 to 40 per cent less crop than the corresponding upper SD canopy. Crop loads (yield/cane pruning weight) were generally between 4 and 12; GDC crop load approached 20 in three seasons, while SD approached 14 in one. Leaf area per crop ratio was determined one season and exceeded 1.8 cm^2/kg of fruit for all systems. Fruit was harvested at similar Brix values, with differences in berry weight, pH, titrable acidity, malic and tartaric acids among treatments generally not significant. Volatile compounds were analyzed using headspace solid-phase microextraction GC-MS. Fruit showed consistent differences in linalool, α-terpineol, β-damascenone and n-hexanol concentrations among training systems. SD had the highest concentration of most free volatiles quantified in both juice and wines, while GDC frequently produced wines that had the highest concentration of phenol-free glycosides. Triangle difference sensory testing demonstrated identified differences between GDC and SD in wine aroma and flavor, and between VSP and SD in flavor. GDC wines generally had higher fruity and floral aromas compared to the other systems.

Mansfield and Zoecklein (2003) reported that the in Cabernet Sauvignon must, total and phenol-free glycosides (expressed as glycosyl-glucose) rose during fermentation while skin concentrations dropped. Wines were heated post fermentation, prior to dejuicing (rising 2 to 3°C per day from 23 to 42°C, and held for one day at 42°C), or after bottling (at 42°C for 21 days) to determine the effect on total glycosides, glycosidic fractions, and anthocyanin complexing. Pre-dejuicing thermal vinification resulted in higher total (12 per cent) and phenol-free (18 per cent) glycosides. Large polymeric pigments rose 28 per cent and small polymeric pigments rose 41 per cent. Skins had lower total glycosides (16 per cent), and no significant difference in phenol-free glycosides. Post bottling heat treatment resulted in lower total (15 per cent) and phenol-free (16 per cent) glycosides, and increased hue (25 per cent). Large polymeric pigments increased 62 per cent compared to control wines.

Gump *et al.* (2002) stated that the five methods for evaluating nitrogen status were compared using 70 Cabernet Sauvignon juice samples: nitrogen by *o*-phthalaldehyde (NOPA), arginine NOPA, enzymatic ammonia, Formol and high-performance liquid chromatography (HPLC). Parallel recovery studies using model solutions of various amino acids and ammonia, presented singly and in combination, were also conducted. The results from two fruit-processing methods were compared using immature and mature berries. NOPA measurements were significantly higher in mature, pressed whole berry-derived samples, compared with homogenized juice. Adjustment of formaldehyde pH prior to analysis was found to be critical to consistency of the Formol method. Average amino acid recoveries for the Formol titration ranged from 82 to 99 per cent. Average recovery for proline was 16.9 ± 0.4 per cent. Ammonium nitrogen was also recovered (84 ± 3 per cent) in the Formol procedure. Formol results trended significantly with NOPA. The correlation coefficient between Formol and NOPA plus NH_4^+ was 0.87, with the Formol values being higher. The average deviation between the Formol and HPLC plus NH_4^+ was 7.3 per cent, while that between the NOPA plus NH_4^+ and HPLC plus NH_4^+ was -7.3 per cent.

Zoecklein *et al.* (2000) reported that the grape glycosides are, in part, an important source of varietal wine aroma and flavor. A glycones may be aliphatic residues, monoterpenes, sesquiterpenes, norisoprenoids or phenolic compounds. Studies were undertaken to evaluate and compare methods of isolating non-phenolic or phenol-free glycosides. Phenol- free glycoside fractions were obtained by three isolation methods: the addition of polyvinylpolypyrrolidone (PVPP), isolation at pH 10.0 using C18 solid phase extraction cartridges, or isolation at pH 13.0 using Oasis™ hydrophilic-lipophilic balance (HLB) extraction cartridges. Glycoside separation using the Oasis cartridges reduced the phenol content of the phenol-free glycoside fractions in Chardonnay juice and wines by an average of 49.7 per cent compared to 33.6 per cent using C18. The differences in the phenol concentration in the phenol-free glycoside fractions with these two methods were greater with red juices and wines. For the Cabernet Sauvignon juice and wines the phenol concentration was reduced by an average 103 per cent and 73 per cent using the Oasis and C18 cartridges, respectively.

Whiton and Zoecklein (2000) stated that the headspace solid-phase microextraction (SPME) is a fast and simple sampling method for analysis of volatile compounds, but quantitation can be affected by sample matrix and sampling conditions. Sampling time and temperature are particularly important in controlling analyte response, and the effects vary for different compound classes and volatilities. Tests with model solutions containing a range of typical wine volatiles show that increasing temperature and sampling time can increase sensitivity for higher boiling polar compounds but decrease sensitivity for very volatile compounds. Sample matrix elements such as ethanol concentration can also have different effects on the responses of different compounds.

Palma *et al.* (2000) stated that supercritical fluid extraction with methanol-modified CO_2 was used to extract glycosides from grapes. An optimization design involving 12 extraction variables was applied to achieve quantitative recoveries. The most important factor was the amount of organic modifier, a consequence of the high degree of glycoside polarity. By the proposed method, the total time of analysis can be decreased relative to that required for more conventional extractions. The full method can also be automated.

Thool *et al.* (2008) reported that the grape is a seasonal crop and gets spoiled fast. Therefore all the grapes of a season must be utilized within short span of 4-5 days. Present natural process of grapes (Kishmis making) requires 12-15 days. Hence the investment of space and infrastructure are large and cannot be afforded by Indian farmers. There are also locations where the ambient conditions are good for grapes growing but adverse for Kishmis (Raisins) making which has almost twice of profit than grapes and has long shelf-life. An experimental drying chamber has been designed and fabricated which simulate the climatic condition. Also a PIC microcontroller based controller is developed to provide flexibility in varying conditions to produce all varieties of raisins (brown, golden, green). This controller has such flexibility that it can dry other fruits and vegetables like tomatoes using infrared drying.

In less than 50 years grape production in the tropics has changed from being somewhat of a horticultural curiosity to grapevines becoming the most widely grown temperate fruit in the tropics. Currently world production is over 3 million tonnes annually, with India producing more than a million tonnes and Brazil just under this figure each year. Other tropical producers include Yemen 163,000, Peru 136,000, Thailand 40,000, Colombia 19,000, Tanzania 14,000 and Venezuela 11,500 tonnes. Grapes are also produced in the tropical part of Mexico and Saudi Arabia, while many other tropical countries produce small amounts for local consumption (Possingham, 2006). Many different climates exist within the tropics resulting from variations in temperature, humidity, elevation and in the amount and distribution of the rainfall. Those best suited to growing ever-green vines and producing grapes have average minimum temperatures above 15°C and a reliable dry period of at least 3 months during which bunch formation and ripening occurs. In both the dry and wet tropics, the growth and cropping cycle of the vine can be manipulated to extend from 5 to 12 months by a combination of pruning, modifying vine water status and the judicious use of chemical regulators. In the tropics overhead trellises, which separate bunches from the vine canopy and reduce bunch rotting, are the most common form of vine training. Some elevated areas within the tropics have cold periods that induce leaf fall and in these areas the growth cycle of the vine is climate controlled and not able to be manipulated. Initially grape vines were cultivated in the tropics for the production of table grapes and this is still their main end use. With their capacity to produce more than one crop per year grapevines can provide massive yields of nutritious fresh fruit at times of the year when supplies of tropical fruits are limited. New developments in this area include the growing of an increased range of seedless varieties and the selection and breeding of new clones and varieties better suited to the tropics. A further innovation has been the use of plastic shelters and plastic-houses with open sides to, shield the vines from rain, reduce the incidence of bunch rots and slightly modify ripening times. The adoption of post harvest techniques originally developed for grapes from temperate areas has been of major importance for tropical grapes. These involve fumigating the grapes with gaseous and slow release sulphur dioxide papers combined with storage at temperatures just above zero. These treatments greatly extend the shelf life of tropical grapes and allow them to be transported to distant markets. A relatively new development has been the use of tropical grapes for the production of both wine and dried raisins. Good wine is now produced in the slightly elevated, wet tropical areas of northern and eastern Thailand, from both the Shiraz and Chenin Blanc varieties. Grapes from the larger crop that ripens in the cooler low rainfall months are of better quality and have less disease than those from the monsoon crop. In a region with a somewhat similar climate, the Deccan Plateau of South India, a range of winemaking varieties are now being grown and cropped once per year, with the grapes being used exclusively for wine production. Good wine is also made from grapes, from vines grown with irrigation and cropped twice per year, in dry tropical areas of both Venezuela (Carora) and Brazil (San Francisco Valley). As with virtually all wine grapes grown in hot areas it is essential to supplement the ferments with tartaric acid to provide an appropriate acid balance to the wine. Raisin production in tropical areas of high humidity represents perhaps the ultimate challenge for tropical grapes; however, this problem has largely been solved in South India using techniques developed in Australia for producing raisins rather than drying them on paper on the ground. In this scheme of things, the Thomson Seedless (Sultana) grapes are dried on wire netting racks with roofs that protect them from the rain, after treating them with emulsions of fatty acids (mainly ethyloleate) and potassium carbonate that hasten drying and lead to the production of light-brown coloured raisins. It is clear that tropical grape production is now a major world industry and the grapes are able to be utilized for fresh eating, wine and raisin production.

Turkey ranks fourth in the world in vineyard acreage with 567,000 hectares, and 3,700,000 metric tons being produced annually. Fresh table grapes account for 25 per cent of this production while 17.5 per cent is seedless raisin grapes and 15 per cent is seeded raisin grapes. Of the remaining production 2.5 per cent is used for wine and cognac with the remaining 40 per cent being utilized in juice, jelly, and other processed products (Turkben and Bahattin, 2001). Nearly all of the seedless raisin grapes produced in the Aegean Region are dried naturally on the ground or on wire fences after dipping into potash (K_2CO_3) and olive oil solution.

Turkey, which is one of the leading seedless raisin grape exporting countries, has many problems with drying and marketing. In this paper, the drying methods used in Turkey and their effects on marketing both in foreign and domestic markets along with optimal drying methods are discussed.

Bresksa *et al*. (2010) reported that the six raisin grape cultivars and 10 new raisin grape selections were analyzed for antioxidant activity (ABTS assay) and for total and individual phenolic compounds. Samples were freeze-dried and values are reported on a dry weight basis.Antioxidant activity across the 16 samples ranged from 7.7 to 60.9μmol Trolox/g DW, with A95-27 exhibiting the greatest activity. Total phenolic content, determined in gallic acid equivalents using the Folin-Ciocalteau assay, ranged from 316.3 to 1141.3mg gallic acid/100g DW and was strongly correlated (r =0.990) with antioxidant results. Concentrations of individual phenolics were determined by HPLC. trans-Caftaric acid was the predominant compound in all samples. A95-15 contained the lowest concentration (153.5μg/g DW) of caftaric acid, while Fiesta contained the highest concentration (598.7μg/g DW). Selections A56-66, A95-15, and A95-27 had much higher levels of catechin (86.5-209.1μg/g DW) and epicatechin (126.5-365.7μg/g DW) than the other samples.

Tripathi *et al*.,(2008) observed that during screening of twenty six essential oils against Botrytis cinerea, the essential oils of the ten plants *viz. Chenopodium ambrosioides, Eucalyptus citriodora, Eupatorium cannabinum, Lawsonia inermis, Ocimum canum, O. gratissimum, O. sanctum, Prunus persica, Zingiber cassumunar* and *Z. officinale* were found to exhibit absolute fungitoxic activity (100 per cent growth inhibition). The essential oils of *O. sanctum, P. persica* and *Z. officinale* were selected for further investigation because these oils showed lower Minimum Inhibitory Concentration (MIC) as compared to the other fungitoxic oils. The selected oils were subsequently standardized through physico-chemical and fungitoxic properties. The MIC values of *O. sanctum, P. persica* and *Z. officinale* were found to be 200, 100 and 100 ppm (mg/l) respectively. The oils showed fungistatic nature at their respective MIC. The oils were thermostable, and exhibited a wide range of fungitoxicity against 15 other post-harvest fungal pathogens. The oils had the potency to withstand high inoculum density. The antifungal potency of oils was found to be greater in comparison to some prevalent synthetic fungicides. Practical applicability of the essential oils was observed in control of grey mould of grapes caused by *B. cinerea* during storage. The *O. sanctum*- and P. persica-oil-treated grapes showed enhancement of storage life up to 5 and 4 days respectively. The storage life of *Z. officinale*-oil-treated grapes was found to be enhanced up to 6 days. The oils did not exhibit any phytotoxic effect on the fruit peel. Therefore, the oils could be recommended as a potential source of ecofriendly botanical fungicide, after long term and wide ranging trials.

Sharma and Awasthi (2003) observed that the persistence of forchlorfenuron residues in grape berries at harvest following its dip application as single or split doses to grape berry clusters and periodic dissipation of forchlorfenuron residues in grape berries following foliar spray application were studied. Periodic dissipation of forchlorfenuron residues following its fortification in soil and water were also studied. Splitting the dip application concentration of forchlorfenuron to grape berries reduced its residues in the berries at harvest, which persisted for more than 65 days from all treatments. In case of foliar application, however, the residues of forchlorfenuron in/on the grape berries persisted for 15–20 days only from three treatment concentrations of 2, 3 and 4 ml/l and dissipated with half-lives of 3.4–4.5 days. The residues of forchlorfenuron dissipated faster in soils maintained at field capacity moisture condition than in air dry soils. There was wide variation in its residue persistence in soil (DT50 ¼ 15:1^{-1} 21.3 days) depending on soil type and moisture condition. Forchlorfenuron residues persisted for more than 30 days in water and its dissipation was fastest at a water salinity level of 3.85 mmho/cm although the rate of dissipation was not significantly affected by the change in salinity level from <0.04 to 5.90 mmho/cm.

Satisha and Prakash (2005) opined that four grape varieties namely Flame Seedless, Thompson Seedless, Sharad Seedless and Tas-a-Ganesh of six month old plants were subjected to three levels of moisture stress *viz.*, control, 50 per cent stress and 100 per cent stress for 14 days. There was marginal reduction in photosynthesis and greater reduction in transpiration with increased water use efficiency (WUE) in all the varieties from control to 50 per cent stress. None of the varieties could survive beyond 3-4 days at 100 per cent stress. Among the varieties Flame Seedless had highest WUE followed by Thompson Seedless at 50 per cent stress. A nine fold increase in abscises acid content was observed in Flame Seedless at 50 per cent stress than at control while it was about 1-3 fold in other varieties. Reduction in cytokinin content was observed with increased moisture stress. Root to shoot length ratio (RSLR) and root to shoot dry matter ratio (RSDWR) was maximum in Flame Seedless and was least in Tas-a-ganesh at 50 per cent moisture stress. Under similar set of experiments with five grape rootstocks namely Dogridge, 1613 C, Salt Creek, St. George and VC clone, Dogridge and salt Creek maintained higher water potential and osmotic potential (less negative values) under increased moisture stress indicating their better osmotic adjustment to maintain higher heater content. Higher WUE was observed in Dogridge and Salt Creek at 50 per cent moisture stress as a result of marginal reduction in photosynthesis and greater reduction in transpiration rate. Higher ABA content accumulated in Dogridge with increased moisture stress resulted in reduced stomatal conductance thus minimizing transpiration rate. A threefold increase in RSLR was observed in Dogridge from non-stress to stress conditions while there was no much increase in other rootstocks. Reduced shoot length in Dogridge was attributed to less cytokinin content under soil moisture stress. Dogridge rootstock influenced the photosynthesis behaviour of Flame seedless and Sharad Seedless when they were budded on it. Sharad Seedless recorded maximum WUE when budded on Dogridge followed by Flame Seedless on Dogridge at 50 per cent moisture stress. Flame seedless and Sharad Seedless budded on Dogridge recorded highest ABA accumulation at 50 per cent moisture stress than when they are budded on Salt Creek and VC clone. Rootstocks strongly influenced the carbon isotope discrimination of scion varieties budded on them.

Saralamma and Singh (1986) reported that the effect of suspended solids added to clarified musts on fermentation rates and qualities of wines from grape cultivars Thompson Seedless and Arka Kanchan was studies with two fermentation temperature (13 and 18°C) and compared with cold settled and centrifuged juice. The suspended solids such as grape solids and bentonite were added at 15 and 2 per cent and 400 mg and 800 mg/1, respectively. In case of Thompson Seedless musts fermented at 13°C both the maximum and overall fermentation rates were high when either grape solids or bentonite was added. When fermented at 18 °C, only bentonite addition increased the fermentation rates. However, the cold settled juice fermented faster than the juice with suspended solids at 18 °C in case of Arka Kanchan fermented at 13 °C the maximum fermentation rates increased with higher solids contents, but overall fermentation rates didn't show much difference. At 18 °C both maximum and overall fermentation rates were higher in all the treatments except in juice with 1 per cent solids. Varietal comparison showed that the musts of Thompson Seedless fermented faster at 13 °C while Arka Kanchan at 18 °C. The alcohol content, total acidity, pH and tannin were not affected by addition of suspended solids. Volatile acidity was in all the treatments and was within the range. Residual sugar was low in all the treatments indicating that the fermentation was complete in all the samples. Sensory evaluation of wines indicated that in case of Thompson Seedless must the quality of wine was better when fermented at low temperature with bentonite. However, at high temperature, wine from centrifuged must was better. Better quality wine obtained from Arka Kanchan, when centrifuged juice was fermented at both the temperatures.

Karabulut *et al*. (2005) reported that the germination of *Botrytis cinerea* spores on potato dextrose agar after a 30 s immersion in 10 or 20 per cent ethanol was 87 and 56 per cent, respectively, compared to 99 per cent among untreated controls. After similar immersion in 0.5 or 1.0 per cent potassium sorbate, 84 and 68 per cent of the spores germinated, respectively. Addition of 0.5 and 1.0 per cent potassium sorbate to 10 and 20 per cent ethanol solution significantly increased the inhibition of spore germination. The germination of spores after 30 s immersion in 20 per cent ethanol plus 0.5 per cent potassium sorbate was 9.7 per cent. The incidence of gray mold, caused by *B. cinerea*, on detached berries of 'Flame Seedless' grapes immersed for 30 s in water, 10 and 20 per cent ethanol, and 0.5 or 1.0 per cent potassium sorbate was 55.2, 42.1, 31.0, 37.7, or 24.4 per cent, respectively. Addition of 0.5 and 1.0 per cent potassium sorbate to 10 and 20 per cent ethanol reduced decay to 10 per cent or less and was more effective than either alone. After 30 days of storage at 1°C, the combination of 20 per cent ethanol either with 0.5 or 1.0 per cent potassium sorbate was equal in efficacy to commercial SO_2 generator pads in reducing the incidence of gray mold on 'Thompson Seedless' grapes. None of the combinations of ethanol and potassium sorbate injured the berries.

13
Wine Making

Wine is fermented juice of fresh fruits. It is considered as a health drink composed of water, alcohol, pigments, esters, vitamins, carbohydrates, minerals, acids, tannins with medicinal and therapeutic values. Generally it is found that a quart of vine with 10 per cent alcohol gives about 850 calories of energy, that is, approximately equivalent to a pound of meat or 20 ounce of milk. Inspite of its quality, the wine market in India is still considered recent with hard liquor claiming a huge chunk of the market. In western countries wine has been a part of the diet and represented as a safe and healthy beverage. In India Govt. regulations make wine making and wine marketing an obstacle. Like Maharashtra state other state Govt. should also modify and simplify the modes and regulations of wine industry.

In Indian Rigveda there is a mention of vine as 'Somarasa' Ayurvedic medicines are prepared in the media of fermented fruit juices.

In recent days Indian have started appreciating wine. The upper class Indian is convinced that good food has to be accompanied by wine. Wine consumption has doubled in the last 2-3 years; the wine-pubs like beer-pubs have started appearing in Indian metropolitan cities. Indian wine lovers are realising that a good wine can very well match with deshi Indian food.

As climatic conditions in India are suitable for growing a number of different varieties of wine grapes, best Indian wines can be made from a blend of grape varieties. By addition of cardamom, cloves, and other spices medicated wines or vermouths can be prepared to popularize wine culture in India so that they can be better accepted. A wine cocktail can be prepared by mixing a bottle of wine with sugar and fruits like orange, apple and maccrated for two hours before serving. Such kind of experiments can bring wine culture in India. So that India starts moving away from hard liquor and discover the pleasure in savouring delicate wines with Indian deshi foods. Similarly wines can be prepared from various tropical fruits like pomegranate, pineapple, sapota etc.

Scope for Indian Wines

India produces over 6.5 million tons of fruits every year and nearly 30-40 per cent fruits may lose their marketability due to undersized fruits, spotted fruits, cracked fruits, immature fruits and post-harvest losses accounts for around 2.5 million tons. Such losses can be validated by preparing wine out of such fruits. By considering 50 per cent juice from such fruits, it can be converted into wine as value added product for local markets. A good quality wine can be prepared from pomegranate, apple, pineapple etc.,

Export of Wines

The European market is the main consumer of wine with its own quality standards. European Union had prohibited import of wine from Third World countries. The European union in its regulation no. 1470/1990 of July 5th 1999 has amended regulation number 3201/90 giving detailed rules for the description and presentation of wine and grape. The European Union has accepted the Indian request and India has appeared in the list of countries that can export wines to European Union. The European Commission has notified "Central Food Technological Institute" (CFTRI) Mysore, as the Indian agency for pre-shipment inspection, as per EI regulation of February 1999. The wine to be exported to European Union will have to be certified by CFTRI. However, the approved varieties of wine grape are enlisted for wine making. The Indian wine exporters can get their names enlisted in the list of wine exporters from Ministry of Commerce, Government of India.

Wineries of India

The Growers winery located in Ragunathpura in Doddaballapur taluk in Bangalore district of Karnataka is one of the main wineries in Karnataka Shaw Wallace winery. In Hyderabad in Andhra Pradesh there has been an important winery producing 'Golconda' wine. The area under grape is maximum in Maharashtra state. In recent times the prices for table grapes and resins has come down and Govt. of Maharashtra has promoted the wine Industries. The M/s Indage India (M/s Champagne India) Narayanagaon, Baramati winery in Baramati, SULA winery in Nasik, Indogrape winery in Pandarapur are some of the main wineries in Maharashtra. The new wineries are coming up in Maharashtra.

Viti-culture for Wine Makers

It is often said that great wines are made in vineyards. The industry of the world is built upon one species *i.e. Vitis vinifera*. L. While selecting a wine variety the important trails to be considered are its tractability, distinctive flavour, special characters and economics of sales, etc. It is found that highly calcareous soils produce best quality wines. However, the new planting of wine grape varieties should be taken up on root stocks. Among various root stocks are Dogridge-A, Dogridge B, Ramsay (Salt Creek), Saint George, etc. Dogridge-B is found to be more suitable. The rootstock improves the vigour, fruiting canes, drought tolerance and will develop ability to leach out harmful ions. To get better quality of wine grape nutrients are to be replenished organically rather than inorganically alone. Yield levels should not exceed 6-7 tons/acre.

Wine Grape Varieties

Red Wine Varieties	*White Wine Varieties*
1. Zinfandel	1. Chardonnay
2. Cabernet-sauvignon	2. Chennin blanc
3. Grenache	3. Sauvignon-blanc
4. Pinot Noir	4. Riesling-white
5. Ruby red	5. Pinot-blanc
6. Marlot	6. Muscat-blanc
7. Ruby cabernet	7. Colombard
8. Carignon	8. Ugni-blanc

Characteristics of Grape Required for Making Different Types of Wines

Type of Wine	*Degree Bric*	*Titrable Acidity g, tartaric/lit*	*pH*
Sparkling wine	18-20.0	7.0-9.0	2.8-3.2
White table wine	19.5-23.0	7.0-8.0	3.0-3.3
Red label wine	20.5-23.5	6.5-7.5	3.2-3.4
Sweet table wine	22.0-25.0	6.5-8.0	3.2-3.4
Dessert wine	23.0-26.0	5.0-7.5	3.3-3.7

Different Types of Wine

Wine generally contains alcohol in the range of 8-16 per cent. Depending upon its sugar content, alcohol content are of various types of wines as dry wine, fortified, wine, spiraling wine etc.,

- ☆ **Dry Wine:** A wine in which all the sugar of juices are practically fermented.
- ☆ **Sweet Wine**: A wine in which some of the sugar is left unfermented or additional sugar is added after fermentation.
- ☆ **Fortified wine:** A wine to which brandy or some other alcoholic spirit is added after the fermentation. *E.g.* Sherry, Port Wine contains 17-23 per cent alcohol.
- ☆ **Sparkling wine:** A wine in which considerable amount of CO_2 is present arising fermentation directly in the bottle. It is also called as carbonated wine *E.g.* champagne contain 11-12 per cent.

- ☆ **Red wine:** A wine made from the red grapes wherein the pomace (skins, seeds and pieces of stems) is left during fermentation, strong flavour is because of tannins from grape skin *e.g.* Burgundy.
- ☆ **White wine:** A wine that is prepared from white grapes or from the juice of red grapes from which the pomace is removed.

Home Wine-making

Wine making at home is more of an art rather than science. Any good Indian cook can become easily an oncologist. Many times homemade wines are sold in hotels at costly price than wines made from wineries. The whole procedure of wine making is given step-by-step for the easy adoption of methodology.

Preparation of Starter Culture

A wine yeast culture of *Saccharomyces ellipsoideus* maintained on Yeast Extract peptone Dextrose Agar (YEPD). A loopful inoculum of wine yeast culture is transferred into a test tube containing 5ml of YEPD broth and incubated overnight at 25°C. (The YEPD broth can be prepared by adding 10g of yeast extract 20g of peptone and 20g of dextrose to one litre of water and adjusting the pH to 5.4 before autoclaving.) The 5ml yeast culture is further added to 100 ml sterilized grape juice (5 per cent inoculum) in 250 ml flask and kept for incubation at 25°C for 22 hours on a shaking incubator at 120 rpm. This can be used to inoculate the fermentor after activating the yeast cells at 40°C for 1.5 hours.

Preparation of Juice/Must

The fruits are either crushed by hand or by using pulpier. The extracted juice or pulp is called "must". If the fruit pulp contains more pectin then pectin's enzyme is added at 0.5 per cent level and kept for incubation at 40°C for 3-4 hours for clarification of the juice.

Amelloration

The must/juice should be tested for its sugar content and total acidity. The pH should be adjusted between 3.2-3.3; if the pH of the juice is still lower the acidity can be reduced with calcium carbonate or sodium bicarbonate (baking soda). If the pH value is on the higher side, tartaric acid or citric acid can be added. If juice/must is diluted the total soluble solids (TSS) has to be raised to 22-24 brix by adding sugar.

Addition of SO_4

Addition of potassium metabisulphite (KMS) is done at the rate of 200 PPM (100 PPM SO_2). However, after addition of KMS, allow a gap of about 3-4 hours before inoculating the must with wine yeast.

Fermentation

The most is filled in the laboratory fermentor to its 3/4th capacity and inoculated with desired activated yeast culture @ 5 per cent. The lab fermentor or flasks or glass bottles must be equipped with air locking facility. In industries or wineries it is often accomplished for filling a fermentory with a special one-way value. The fermentation is allowed to carry out at 21 to 24°C. Temperature regulation is very important as finest quality wines are produced at lower temperatures. The air lock during fermentation is provided to prevent the acetification *i.e.*, conversion of alcohol into vinegar.

Siphoning/Racking

After allowing fermentation for 7 to 8 days–(2 to 3) days in case of red grape to reduce tannin content. The clean liquid is siphoned or racked from the sediment called "less" which contains the skin, seeds, yeast cells, organic precipitate etc. This can also be achieved by filtering through muslin cloth. The fermentation is allowed to continue for another two weeks. It can be racked once again to remove the sediments and filled in the bottles. The bottles can be stored in the refrigerator for maturation.

Maturation

The newly–made wine is harsh in taste and has yeasty flavour. The process of maturation makes the wine tasty and fruity in flavour besides clarification. After storing in refrigerator for one week, racking is repeated and stored again in the refrigerator. Racking is repeated for 2 to 3 times at an interval of 15-20 days. Racking can be continued further at monthly intervals. During interacting period no headspace should be kept in the bottle and should be closed tightly. The period of maturation may extend to over a year in case of pigmented wines. The matured wines can be pasteurized and stored for long time at 15 to 16°C for again. The aging is a complex process in which precipitation of the suspended solids take place besides the formation of steps that impart the fruit flavour to the wine. Usage of oak wood chips during maturation is used for grape wine making. The wood chips are also used again for other fruit wines.

Sl.No.	*Items*	*Pomegranate*	*Grape*
1.	Cost of 1 kg fruits (Rs)	10.00	20.00
2.	Yield of juice in ml/kg fruit	500.00	700.00
3.	Cost of added chemical (Rs)	2.00	1.00
4.	Processing cost (1/3 of raw material cost in Rs)	4.00	7.00
5.	Wine yield from 1 kg fruit in ml	400.00	600.00
6.	Cost of one litre of wine (Rs)		

Wine and Health

1. An occasional wine drinking for a glass of wine a day is said to stave off neuro-degenerative diseases such as Alzheimer's and Parkinsons.
2. Alcohol concentration in wine may not be anti-microbial but phenolic compounds present in wine which are anti-microbial and helps in preventing many diseases like dysentery, diarrhoea and have protective effects against cardiovascular diseases.

3. There is positive correlation found between moderate wine consumption and improved health is possible due to decrease in the prevalence of age-related muscular degeneration.
4. The consumption of moderate amount of wine can reduce the risk of certain type of cancer diseases and increased consumption may provide some cancer diseases.
5. Moderate wine consumers may live on an average 2.5 years longer than teetotalers and considerably longer than heavy drinkers.

Tips to Wine Drinkers

1. Most homemade wines are good to have when they are atleast a year old. They normally stay well for 3-4 years.
2. It is not necessary to finish your bottle of wine the day you open it. It keeps well for 2-3 weeks with airtight stoppers.
3. Never keep your wine bottle at room temperature. The ideal temperature is between 12 to 15 degree centigrade. The bottle can simply be kept in refrigerator.
4. Occasional drinking or a glass of wine a day is found to be good for health.

(*Courtesy*: Dr. A. B. Patil, DE, UHS, Bagalkot).

14

Transport and Marketing

The main means of marketing of table grape by road, for local and short distance market or long distance market; only 5 per cent if the produce is transported by rail and through air is negligible. The produce for international market is transported by refrigerated contain by road.

Cooperative gape marketing societies are in different states with a primary objective is the avoid commission to agents. There are two types of selling grape in international market one through producer's cooperative societies and traders. The grapes for export should be packet, pre-cool, store and ship in refrigerated containers.

Ladaniya *et al.* (2005) reported that the price spread and losses in grapes and marketing economics of raisins in Maharashtra were studied. Major marketing channels were: 1) Grower - commission agent - retailer - consumer for local and distant markets and 2) Grower - C.A. - trader - retailer - consumer for distant upcountry markets. Marketing efficiency was 1.27 when produce was sold by producer directly to retailer. When produce was sold to trader in vineyard, efficiency was 0.50 with share of producer 33.65 per cent in consumer's rupee. In longer channels, effiency was 0.48, with producer's share 32.50 per cent in consumer' rupee. Raisin making was profitable and earned Rs. 50,500/additional than marketing grapes. Marketing of raisins was through commission agents. Losses in grapes were 1–1.25 per cent at farm, 5.5–8.65 per cent at wholesale and 12.25–16 per cent at retail level. Negative correlation co-efficient was recorded in monthly prices and arrivals at national level and major markets in Maharashtra.

Murat Kocturk (2006) stated that in recent years, the importance of marketing of the raisins has become important because of yield increase and export incentives. The grape growers that have been organized for raisins with the rate of 50 per cent have not been able to organize effectively in the table grape marketing. Eighty seven per cent of the growers have been selling their product to traders and brokers without any guarantee of payment. The growers of the table grapes have established an agricultural development cooperative (shortly named ALA-KOOP) with the aim of marketing their products in the center of AlaOehir county and surrounding villages in 2000. This cooperative could not be activated because of the lack of interest of their members. Meanwhile, with the increase of the grape export in the last decade in AlaOehir county, 42 companies have been founded and these companies take an important role in rural industry and development. It is observed that the cooperatives which have been founded with the aim of organizing the growers are in need of professional management and financial support to make them successful.

Koshteh and Abbas (2004) reported that the grapes are one of the most important horticulture products. Due to different varieties production is possible in different regions and climates. In Sistan and Baluchestan province of Iran, grape production constitutes one of the most important horticulture products in this province. Products are distributed in the province itself, adjacent provinces and Tehran. Market of grapes in the province itself can be divided it into two sub markets: Sistan region and Baluchestan region. The results about marketing efficiency examined by the Method of Shepherd and Futrell demonstrate that the efficiency in Sistan market is 31 per cent higher than that of Baluchestan market. In comparison with other regions of the country, the marketing process of the grapes produced in Sistan and Baluchestan province faces more obstacles and difficulties. The negative role played by mediators is one of these problems. Other factors impeding marketing efficiency are: wholesale pricing system that does not take into account real production costs and an undesirable retail pricing system. Furthermore, the inefficiency in packing and transportation, lack of proper processing industries and limited facilities for storage lead to the sharp decrease in price of grapes. This paper identifies limitations that are impediments at grapes marketing process in the study area and concludes some recommendations to improve market share in the concerned markets.

Suresh and Murthy (2011) opined that many Indian policy makers and trade analysts today realize that the horticulture development has increased employment and foreign exchange earnings for the country. In terms of horticulture crop production, India is considered the most important country of the world. Grapes are widely cultivated in India because of the prevalence of congenial agro climatic conditions. In India, Tamil Nadu is one of the most important states in terms of grape production after the states of Maharashtra and Karnataka. In Tamil Nadu, grape is cultivated nearly in 2,600 ha (FAO Statistics, 2009) with an output of 91,000 tons, and has achieved the highest grape productivity among other grape producing states of India. The review of various studies shows that earning potential in grapes cultivation is high. But growers are not interested in cultivating grapes as they face a large number of marketing problems. In order to understand and analyze these problems, this study examines the marketing practices of grapes growers, marketing cost, margin, price spread and marketing efficiency of grapes in Tamil Nadu.

Ladaniya *et al.* (2005) reported that the price spread and losses in grapes and marketing economics of raisins in Maharashtra were studied. Major marketing channels were: 1) Grower - commission agent - retailer – consumer for local and distant markets and 2) Grower - C.A. - trader - retailer - consumer for distant upcountry markets. Marketing efficiency was 1.27 when produce was sold by producer directly to retailer. When produce was sold to trader in vineyard, efficiency was 0.50 with share of producer 33.65 per cent in consumer's rupee. In longer channels, efficiency was 0.48, with producer's share 32.50 per cent in consumer' rupee. Raisin making was profitable and earned Rs. 50,500/- additional than marketing grapes. Marketing of raisins was through commission agents. Losses in grapes were 1-1.25 per cent at farm, 5.5-8.65 per cent at wholesale and 12.25-16 per cent at retail level. Negative correlation co-efficient was recorded in monthly prices and arrivals at national level and major markets in Maharashtra.

Kocturk (2006) reported that in recent years, the importance of marketing of the raisins has become important because of yield increase and export incentives. The grape growers that have been organized for raisins with the rate of 50 per cent have not been able to organize effectively in the table grape marketing. Eighty seven per cent of the growers have been selling their product to traders and brokers without any guarantee of payment. The growers of the table grapes have established an agricultural development cooperative (shortly named ALA-KOOP) with the aim of marketing their products in the center of AlaOehir county and surrounding villages in 2000. This cooperative could not be activated because of the lack of interest of their members. Meanwhile, with the increase of the grape export in the last decade in AlaOehir County, 42 companies have been founded and these companies take an important role in rural industry and development. It is observed that the cooperatives which have been founded with the aim of organizing the growers are in need of professional management and financial support to make them successful.

Stephen (2007) stated that a needs assessment of the Umpqua Valley wine grape industry in 2001 revealed the greatest impediment to our sustainability was a lack of coordinated marketing. By working together creatively, board members developed successful marketing strategies and effective ways to fund their strategies. After 3 years, survey data revealed that the number of visitors at participating wineries increased by 35 per cent and wine sales increased 40 per cent on average. Fundraising events contributed $90,000 for marketing, and seven new wineries and 11 new vineyards were established. This proves that sustainable programs can be home grown in a cooperative environment.

Kamble *et al.* (2007) reported that the constraints in production and marketing of grapes have been identified by selecting 15 villages (8 from Tasgaon and 7 from Miraj tehsil of Sangli district in Maharashtra). Out of the total grape growers, 6 grape growers (2 for each variety of Sonoka, Thompson Seedless and Tas-A-Ganesh) were selected randomly from each village, making the total sample of 90 grape growers. The data collected for the year 2004–2005 have been analyzed by tabular method of analysis. The study has identified non-availability of skilled labour at proper time, lack of technical knowledge, non-availability of bank loans, non-availability of pure variety grafts at nearest places, and high cost of various inputs, *viz.* insecticides, pesticides, micronutrients as the major constraints in production of grapes in the Sangli district. The marketing problems have been revealed as high commission charges by intermediaries (48 per cent), high cost of transportation (44 per cent), high cost of packing material (10 per cent), low net returns, lack of pre-cooling and cold storage facilities, and longer chain of intermediaries in the marketing. The study has suggested that the number of market intermediaries should be reduced through establishment of co-operative marketing. Also, the state government should enact a policy on grape marketing similar to that adopted by Government of Andhra Pradesh, *viz.* provision of stalls to the growers in the urban markets.

Koshteh and Abbas (2004) stated that the grapes are one of the most important horticulture products. Due to different varieties production is possible in different regions and climates. In Sistan and Baluchestan province of Iran, grape production constitutes one of the most important horticulture products in this province. Products are distributed in the province itself, adjacent provinces and Tehran. Market of grapes in the province itself can be divided it into two sub markets: Sistan region and Baluchestan region. The results about marketing efficiency examined by the Method of Shepherd and Futrell demonstrate that the efficiency in Sistan market is 31 per cent higher than that of Baluchestan market. In comparison with other regions of the country, the marketing process of the grapes produced in Sistan and Baluchestan province faces more obstacles and difficulties. The negative role played by mediators is one of these problems. Other factors impeding marketing efficiency are: wholesale pricing system that does not take into account real production costs and an undesirable retail pricing system. Furthermore, the inefficiency in packing and transportation, lack of proper processing industries and limited facilities for storage lead to the sharp decrease in price of grapes. This paper identifies limitations that are impediments at grapes marketing process in the study area and concludes some recommendations to improve market share in the concerned markets.

Jianying Feng *et al.* (2012) reported that the wholesale market is one of the most important stakeholders in the supply chain of table grapes in China. An evaluation conceptual framework of grape distribution efficiency was established in this paper, which contains logistic efficiency, capital efficiency and information efficiency. Face-to- face interviews with wholesalers were conducted in seven grape wholesale markets of Beijing to acquire basic data. Factor analysis method was employed to evaluate grape distribution efficiency of different wholesale markets. The results indicate Xinfadi shows higher distribution efficiency than other wholesale markets in Beijing. The study also proves that grape distribution efficiency is highly influenced by wholesalers' business experience and integration condition of wholesale markets.

15

Scope for Export of Grape

Present Status of Grape Growing in the Country

As per the data and information available, it is estimated that the total area under grape cultivation in the country is about 50,000 hectare with annual production of about 10-12 lakh metric tonnes of fresh grape.

Distribution of Area

1. Karnataka

Karnataka is the second largest state in the country having about 10,000 hectares under grape. Bangalore, Bagalkot, Kolar and Bijapur are the main grape growing districts.

Varieties

Bangalore purple, Bangalore blue and Thompson seedless are the main varieties under cultivation. Other varieties like Arakavati, Neelmani, Flame seedless, Sharad Seedless, Sonaka, Clone-II, Red globe are now becoming popular among the growers.

2. Maharashtra

Nasik, Sangli, Solapur, Pune, Osmanabad are the major grape growing districts of Maharashtra having about 16,000 hectares with annual production of about 3 lakh metric tonnes per year.

Major Varieties under Cultivation

Thompson seedless, Tas-e-Ganesh, Sonaka are the main varieties. Manik Chaman, Kishmish Charani, Sharad Seedless, Flame Seedless are also getting popular. All these varieties are seedless and their total share is more than 80 per cent in the acrege and production.

3. Andhra Pradesh

Hyderabad is well known for grape cultivation since a long time and the cultivation has remained concentrated in about 75 km. radius of the twin city of Secundrabad and Hyderabad. The total area is about 6,000 hectare and the main variety grown is Anab-e-Shahi. In recent years, Dilkush, Flame seedless, Thompson Seedless and Sonaka are getting increasing demand among the growers and consumers.

4. Grape growing in North

Punjab, Haryana, and Himachal Pradesh are known for grape cultivation. However the expansion was not boosted mainly due to climatic conditious. Anab-e-Shahi and Perlette etc. are the varieties grown here. Total area in these states together does not exceed beyond 10,000 hectares.

Importance and Scope for Export

Starting from 1985 the country is exporting in an increasing and highly dignified manner especially after 1990. The export is done mainly to the Middle Eastern countries and European market. Also new markets *viz.* Hong Kong, Singapore and Japan are showing much interest.

Pomegranate, mango, are the major fruit crops having economic importance. But as compared to others, grape is enjoying a special privilege in exports. Apart from fresh table grape other products of grapes like raisin, juice and wine products also have export potential.

Policies of Central and State Governments' Foreign collaboration, growers union, cooperative sector, state agricultural universities, innovative growers and scientist, extension workers together are joining hands with research and development organizations. The National Research Centre, the APEDA, the Karnataka State Marketing Board have a very potential role to play in export.

Putting all efforts together we have a scope to export one lakh tonnes of fresh table grapes every year in the near future.

Planning and Production

Instead of collecting selected crops from diversified vine yards for export we should select vine yards for export of grapes.

This can be done by planning in advance. There should be a set of growers in different agro-climatic situations with each having 40 ha vine yard from where 800-1000 tonnes of grape of export quality could be produced. These areas should be managed with similar package of practices for one variety.

Converting the present plantation on this line and bringing new area under this programme a planning is required. It should be remembered that this cultivation is exclusively for export purposes.

After achieving the target the post-harvest treatment and other infrastructure should also be developed simultaneously. The planning should be framed with proper execution and supervision. The production and procurement should be frequently monitored by a committee on a scientific level and commercial with a strong administrative back-up.

Work should be simultaneously done in Karnataka, Maharashtra, Andhra Pradesh, Tamil Nadu with each having a regional sub-committee with financial support. It should have its headquarter at Bangalore with the support of National Grape Federation of India.

Grape Export

1. Grape production in Karnataka is quality-wise good.
2. Good production and quality leads to more demand for grape in the country and also outside for exports.
3. However as long as we send the grape, its rate increases.
4. By Exporting grape the farmers get good returns specially foreign exchange or dollars.
5. European and Middle East countries' demand is increasing day-by-day.
6. The countries which are nearer to us are Japan, Hong Kong, and The Philippines which are importing grape from us.

Export of Grape in a Wide Network

The grape which is produced in India has trade links with Bangladesh, Pakistan, Nepal, Myanmar and Sri Lanka.

Among these countries Sri Lanka requires sea route whereas others by road/ railways. Apart from these neighbouring countries, other European countries are our customers, besides Saudi Arabia, Iran, Iraq, Kuwait, Syria, and European countries *viz.*, England, France, Italy, Spain, Germany etc.

The countries which have less demand are Bangladesh, Nepal and Sri Lanka.

Medium Demand: Saudi Arabia, Iran, Iraq and Kuwait

More Demand: England, France, Italy and Germany.

While exporting grape there are several points one has to keep in mind and they are:

1. *Monetarily Strong Ones*: The countries to which we are exporting their financial condition should be good which means demand will also be good. In Middle Eastern countries, there are petroleum products which mean good whereas in European countries, there is good purchasing power due to their sound economy.

If the population of a country is enormous then the purchasing power will be less and that is to be borne in mind.

2. **Necessity** means food, clothing and shelter. If that is there then the luxury items like grape they can afford to get.
3. The demand for our grape in some of the countries has be to studied first, then only go for export and find how the demand is existing and list down the countries which require our grape.

European Countries Grape Demand

Among the European countries demand from England for our grape is good because from April-June onwards nowhere in the world grape is available except India. Since last 10-15 years in England the market study showed that our market was having encouraging trend from last 5-7 years. The grape consumption rose from 70 to one lakh tonnes and every year the consumption trends increased. About 88 per cent of demand is fulfilled, from countries *viz.*, Spain (25 per cent) Italy (25 per cent) South Africa (13 per cent), Greece (11 per cent), Cyprus (7 per cent), Chili (7 per cent), which are supplying to the super market during December to June. France also imports grape at the same time when England imports. From Spain, South Africa and Chile the grape is exported to Netherlands and Belgium which are main exporters to Italy, South Africa and Chile. So it appears that European market is a big market for grape. European people consume lot of fresh grape because of their good purchasing power.

Sl.No.	*Exporting Countries*	*Yearly Export*	*Where they Export*
1	South Africa	45.5 thousand tones	Germany, England, (75 per cent -80 per cent)
2	Chile	3000 tonnes	America, Netherlands, England (70-5 per cent)
3	Australia	12-1500 tonnes	England (50-60 per cent)

It appears that from South Africa, Chile, Australia grape is being exported to England. So one has to understand the Government rules and regulation and what precautionary measures one has to take and which variety they prefer. One has to consider all these information not only for today but also for the future. Among the European countries, England's demand is fulfilled by Chile, South Africa and Australia; grape is available from January and from those countries England has captured big departmental stores. In India we are producing export quality grape at par with other departmental stores in England.

Today in India the availability of grape starts from January and continues upto May with peak availability in the month of March. So we can export grape from January to March in large quantity. From April to June the demand for grape will be reduced. In India we are growing several varieties. To establish a variety in a particular region it will take many years to test its suitability. For example from Australia there is a variety called, Mermen seedless, from Afghanistan, Kishmish long beli and from California centenial seedless. For all these a separate package of practice is required from IIHR, Bangalore. Arkavati seedless and few year back Kishmish Beli and

Kishmish Charanai were brought to India from Russia and there testing is going on. Kishmish beli as compared to Thompson seedless comes to harvest early in January and it is available in the market and it will suit our climatic situation in India. Early availability of grape is possible by modification of pruning in time. At present we are taking up pruning in October and early pruning has to be planned in August so that by December the grape will be ready for harvest.

If October pruning is done in August then we are going to face some problems like incidence of disease but we can control it. However, the advantageous thing is that we can harvest them by January to March which facilitates exports into the European markets. In Karnataka temperature, rainfall and humidity are the three important factor for grape cultivation which are most suitable and here seedless grape is being cultivated very well and is heading time to market in January for consumption.

Quality Consideration

It is the quality and not the quantity that is important for exports. Everything about quality should be taught to the growers who are cultivating grape for export purposes. The requirement from various markets will have to be surveyed and considering the local situations, including climate and soil, pocketwise recommendation of qualitative analysis ought to be laid out on the following points:

1) Variety 2) season 3) productivity 4) optimum yield (5) bunch size/shape (6) appearance and weight (7) berry size, shape and colour (8) TSS, acidity, and specific gravity (9) taste, flavour and appearance (10) adherence and chemical residue. These can be further elaborated considering the present knowledge and experience:

1. Variety: Thompson seedless and Tas-e-Ganesh some more introductions should also be considered.
2. Bunch shape: oblong/conical
3. Bunch weight: 400-600 gm.
4. Berry; spherical: greenish, diameter above 18 mm weight 3 to 5gm.
5. Firmness: Firm, strong, and pulpy.
6. TSS: 18° to 22° brix, acidity-0.5 to 0.65.
7. No spot-bruises, no chemical residue.
8. No taste and other oddities.
9. There should be standard relations in (1) yield/ha (2) quality and (3) cost of production.

Production and Quality

In Karnataka seedless grape, on an average the farmers are producing around 20 tonnes to 25 tonnes per hectare. There is lot of controversy regarding quality as less production coincides with good quality. Similarly more production tallies with poor quality grape. But for export if a farmer is producing 25-30 tonnes per hectare, it will not be possible to produce quality fruits. For producing quality fruits for export one has to keep the following points in mind:

1. **Grape Variety:** We are growing about 4-5 seedless grape varieties, out of which Tas-e Ganesh and Thompson seedless, being the varieties that are having very good shape and good shelf life specially the berries that are longish. The leaves of these varieties are thick. They stay preserved for a long time, hence helps in getting good crop.
2. **Yield:** In our place the yield starts from the second year onwards and continues till 15 years, thereafter when it becomes old the yield declines. After 6-7 years the setting will change depending upon the nutrition practice. The garden, in which vines are of 3 to 7 years, will produce export quality grapes.

   To build such vigour the vines have to be rejuvenated.
3. **Requirement of soil and water:** for good quality grape production, the soil and water quality should be good. In our place many of our grape gardens are having lime nodules in the bottom and high pH and EC. Under such situation utmost care is required. If the soil contains 10 per cent lime avoids cultivation of grape. The water which is free from chlorine and sodium should be utilized. Coincidence of good soil with bad water and bad soil makes the situation difficult for high quality grape cultivation leading to higher cost of cultivation with less return. In such gardens where export quality production is possible the farmers are requested to go for testing of is possible the farmers are requested to go for testing of water and soil to have proper balanced application of fertilizer for availability of NPK and micronutrients. These three points are fundamental and only based on these the improvement is possible.

   The details of how to produce good quality grape will be discussed in the next chapters. Second important points to be considered is, if your garden age is more than 7-8 years then rejuvenate such gardens by going for thinning. If the main stem is affected by stem borer or dieback, such vines are also cut at the bottom to rejuvenate in a new plant or if you want to convert the variety into another, *e.g.*, Thompson seedless, then it can be grafted.

   Third point is that when the soil is not problematic then usually water as a rule also will not be problematic. The soil and water character will not change immediately but it takes many years to see the difference. But in case of get change then the problem starts. If the pH of the soil is high then to get a good grape crop is very difficult and it requires high expenditure under such circumstances, requiring application of green manure and bio-fertilizers.

The above three points are very important that one has to keep in mind.

References

Abang, M. M., Winter, S., Green, T.K.R., Hoffmann, P., Mignpuna, H. D. and Wolf, G.A., 2002. Molecular identification of *Colletotrichum gloeosporioides* causing yam anthracnose in Nigeria. *Plant Pathology*, **51:** 63-71.

Adolfo Francisco Muñoz-Rodríguez, Rafael Tormo and María Inmaculada Silva. 2011. Pollination Dynamics in *Vitis vinifera* L. *Am. J. Enol. Vitic*. 62:113-117.

Adsule, P. G., Karibasappa, G. S, Banerjee K. and Mundankar, K., 2006. Status and Prospects of raisin industry in India. Paper presented at the *International Symposium on Grape Production and Processing*, at Baramati, Maharashtra (India) held on Feb 6-11, 2006.

Adsule, P. G., Karibassapa, G. S., Banerjee, K. and Mundankar. Status and prospectus of raisin industry in India. National Research Centre for grapes, Pune.

Adsule, P.G. and Anuradha Upadhaya. Grape Cultivation on root stock, National Research Centre, Pune.

Adsule, P.G., Karibasappa, G.S., Banerjee, K. and Mundankar, K., 2006. Status and prospects of raisin industry in India. ISHS Acta Horticulturae 785: *International Symposium on Grape Production and Processing, held at Baramati* (Pune), Maharashtra, India During 6–11th February 2006.

Agarwal, R. and Tilak, K. V. B. R., 1988. Establishment of *Azospirillum brasilense* in roots of finger millet *Eleusine coracana. Zentralblatt fur Microbiologie,* **143** : 539-544.

Agnew, R. H., Mundy, D. C. and Balasubramaniam, R., 2004. Effects of spraying strategies based on monitoring disease risk on grape disease control and fungicide usage in Marlborough. *New Zealand Plant Protection*. 57: 30-36.

Ahlawat, V. P. and Sindhu, S. S., 1990. Nutritional survey of grape orchards of Haryana. *Haryana J. Hort. Sci.*, **19**(1-2) : 96-100.

Ahlawat, V. P. and Yamadagni, T., 1988. Effect of various levels of N and K application on growth, yield and petiole composition of grapes cv. Perlette. *Prog. Hort.*, **20**(3-4) : 190-196.

Ahlawat, V. P., Sharma, S. and Yamadagni, 1982. Assessment of some grape cultivars for their mineral content. *Prog. Hort.*, **14**(2-3) : 153-154.

Ahlawat, V. P., Sindhu, S. S., Gupta, O. P and Tharej, R. K., 1990. Nutrient element status of soils and leaves of ber (*Zizyphus marutitina*) trees in Haryana. *Haryana J. Hort.Sci.*, **19**(1-2) : 106-111.

Ahmad I. Athamneh, Bruce W. Zoecklein and Kumar Mallikarjunan. 2008. Electronic Nose Evaluation of Cabernet Sauvignon Fruit Maturity. Journal of Wine Research. **19**(1): 69-80(12)

Ainsworth, G. C., 1971. *Ainsworth and Bisby's dictionary of the fungi*. VIth Edn. Commonwealth Mycological Institute Kew, Surrey, England. pp.663.

Alagawadi, A. R. and Gaur, A. C., 1992. Inoculation of *Azospirillum brasilense* and phosphate solubilizing bateria on yield of sorghum (*Sorghum biocolor* L. Moench) in dry land. *Tropical Agriculture*, **69:** 347-350.

Alagawadi, A. R. and Gaur, A. G., 1988. Interaction between *Azospirillum brasilense* and phosphate solubilizing bacteria and their influence on yield and nutrient uptake of sorghum. *Zentrablatt fur Mikrobiologie*, **143** : 637-643.

Alagawadi, A. R. and Krishnaraj, P. U., 1998. Field performance of two local rhizobacteria isolates in sorghum. *37th Ann. Conference of Association of Microbiolohists on India*, Mangalore, p. 278.

Al-Dujaili, J. A., Al-Khafaji, M. A., Al-Juboory, K. H., 1987. The effect of localized GA_3 application on the yield and quality of Thompson Seedless grape (*Vitis vinifera*). *Iraqi J. Agric. Sci., Zanco*, **5:** 51-57.

Aleweldt, A, Ibac, E. and E. Roul 1982. Investigation on gas exchange in grapevine. II Influence of temperature, leaf age and daytime on net photosynthesis and transpiration. *Vitis*, **21**:93-100.

Alexopoulos, C. J. and Mims, C. W., 1979. *Introductory Mycology*, Wiley Eastern Limited, New Delhi. pp.367.

Ananthanarayana, 1978. Dynamics of Mg in relation to crops in soils of Karnataka. *Ph. D.Thesis*, Uni. Agric. Sci. Bangalore (India).

Anderson. H. W., 1956. *Diseases of fruit crops*. McGraw Hill Book Company. Inc. pp387-392.

Anirudh Thakur, N.K. Arora and Som Pal Singh, 2006. Evaluation of some grape varieties in the arid irrigated region of northwest India. ISHS Acta Horticulturae 785: *International Symposium on Grape Production and Processing, held at Baramati* (Pune), Maharashtra, India During 6–11th February 2006.

Anita Zamboni, Mariasole Di Carli, Flavia Guzzo, Matteo Stocchero, Sara Zenoni, Alberto Ferrarini, Paola Tononi, Ketti Toffali, Angiola Desiderio, Kathryn S. Lilley, M. Enrico Pè, Eugenio Benvenuto, Massimo Delledonne and Mario Pezzotti. 2010. Identification of Putative Stage-Specific Grapevine Berry Biomarkers and Omics Data Integration into Networks. *Plant Physiol.* 154(3): 1439-1459

Anonymous, 1970. *Official Method of Analysis*, Association of Official Analytical Chemistry, Washington, DC, 10th Ed.

Anonymous, 1988. Agricultural Status of District Caner and Spti Valley-oniterim of Status Report (NARP). Dr. Y.S.Parmar University of Horticulture and Forestry, Horticultural Regional Research Station, Sharbo, Caner H.P. 172 107. P90

Anonymous, 1989. *FAO Production Yearbook*, **43:** 203-204.

Anonymous, 1990. Statistical Abstract of Caner District. The Directorate of Economics and Statistics, Govt. of Himachal Pradesh, Shimla. P 124.

Anonymous, 2004. Fertilizer trial in grape, Indian Institute of Horticulture Research, Bangalore. Research reports, All India Co-ordinated Fruit Improvement Project (Cell-II). *Fruit Res. Workshop*, Nagpur, pp. 268-277.

Anonymous, 2004a, *Drakshavrutta Sakmaranika*, Maharashtra Rajya Draksha Bagaitdar Sangh. pp.140.

Anonymous, 2004b, Grape cultivation-Present and Future-an overview. *Pestology*, **28:** 13-20.

Anonymous, 2005. Food and Agriculture Organization, *Production Yearbook*, **54** : 166.

Anonymous, 2006. Proceedings of *International Symposium on Grape Production and Processing*, at Baramati, Maharashtra (India) held on Feb 6-11, 2006.

Anthony L. Robinson, Douglas O. Adams, Paul K. Boss, Hildegarde Heymann, Peter S. Solomon and Robert D. Trengove. 2012. Influence of Geographic Origin on the Sensory Characteristics and Wine Composition of *Vitis vinifera* cv. Cabernet Sauvignon Wines from Australia. *Am. J. Enol. Vitic.* 63 (4): 467-476.

Apte, R. and Iswaran, V., 1971. Culture of *Rhizobium* inoculants with those of Beijerinckia and *Azotobacter*. *Proc. Ind. Nation. Sci. Acad. Bull.*, **40** : 482-485.

Arjunan, G., Karthikeyan, G., Dinakaran, D. and Raguchander, T., 1999. *Diseases of horticultural crops*, TNAU, Coimbatore, pp.391.

Arora, C. L. and Sekhon, C. S., 1981. Influence of soil characteristics on DTPA extractable micronutrient cations in some soil series of Punjab. *J. Ind. Soc. Soil Sci.*, **29**(4) : 453-461.

Arora, C. L., Brar, M. S. and Dhatt, A. S., 1992. Secondary and micronutrient status of pear orchards in Punjab. *Ind. J. Hort.*, **49**(2): 150-154.

Asghar, H., Zahir, Z., Arshad, M. and Khalid, A., 2002. Relationship between in vitro production of auxins by rhizobacteria and their growth promoting activities in *Brassica juncea* L. *Biology and Fertility of Soil*, **35** : 231-237.

Aulakh, P. S., Mehrotra, N. K. and Vij, V. K., 2003. Comparative Performance of some Early Ripening Grape Cultivars Grown under Arid Irrigated Region of Punjab, *Indian Journal of Horticulture.* **60**(4): 343-345 .

Awasthi, R. P., Gurudev Singh and Sharma, R. P., 1985. A survey of the mineral nutrients status of Mandarin orchards in Naurpur area of Himachal Pradesh. *Punjab Hort. J.*, **51**(4) : 27-36.

Bagnoli, E. B., 2011. Survey on *Scaphoideus titanus* egg distribution on grapevine Gargani IOBC/wprs Bulletin, 67: 233-238.

Bakshi, J.C., Singh. I. S. and Nangia, R. P., 1970. Relative susceptibility of grape varieties to anthracnose. *Indian Journal of Horticulture*, **27:** 58–60.

Balachander, D. and Nagarajan, P., 1999. Dual inoculation of vesicular-arbuscular mycorrhiza and *Rhizobium* in greengram. *Legume Res.*, **22** : 177-180.

Balamurugan, 1999. Effect of *Azospirillum* and nitrogen on growth and yield of bhendi cv. Pusa Sawani. *M. Sc.* (*Agri.*) *Thesis*, Tamil Nadu Agricultural University, Madurai.

Balamurugan, S. and Gunasekaran, S., 1996. Effect of combined inoculation of *Rhizobium* sp. and phosphobacteria at different levels of phosphorus in groundnut. *Madras Agric. J.*, **83** : 503-505.

Balapande, S. S., Deshpande, S. B. and Pal, K., 1996. Factors and process of soil degradation in Vertisols of the Purna valley. Maharashtra, India. *Land Degradation and Development*, **7** : 313-324.

Balesdent, M. H., Jedryczka, M., Jain, L., Mendes Pereira, E., Bertrandy, J. and Rouxel, T., 1998. Conidia as a substrate for internal transcribed spacer based PCR identification of members of the *Leptosphaeria maculans* species complex. *Phytopathology*, **88:** 1210-1217.

Banerjee Kaushik, Oulkar, D. P. and Adsule P.G. 2008. Persistence and residue dynamics of mancozeb and its toxic metabolite ethylene thiourea (ETU) in table grapes. Acta. 785, ISHS, Belgium.

Barea, J. M., Navamo. E. and Monotoya, E., 1976. Production of plant growth regulators by rhizosphere phosphate solubilizing bacteria. *Journal of Applied Bacteriology*, **40**: 129-134.

Barett, E. L., Solanes, R. E., Tang, J. S. and Palleroni, N. J., 1986. *Pseudomanas fluorescens*, its resolution into distinct component groups and relationship of these to other *P. fluorescens* biovars to *P. putida* and to *Psychrophilic pseudomanads* associated with food spoilage. *J. Gen. Microbiol.*, **130**: 2709-2721.

Barreto, M. S., Nookaraju, A., Joglekar, A. M., Karibasappa, G. S., Agrawal, D.C., 2006. Variability among *Vitis vinifera* cultivars to in vitro propagation. ISHS Acta Horticulturae 785: *International Symposium on Grape Production and Processing, held at Baramati* (Pune), Maharashtra, India During 6–11th February 2006.

Barrs, H. D. and weatherley, P. E., 1962. A re-examination of the relative turgidity technique for extimating water deficit in leaves. *Aust. J. Biol. Sci.* 15:413-428.

Bashan, Y. and Levanony, H., 1995. Migration, colonization and absorption of *Azospirillum brasilense* to wheat roots. In : *Lectins-Biology, Biochemistry, Clinical Biochemistry* (Ed. By T. C., Bog – Hansen and D. L. J., Freed, Vol. 6 Sigma, Chemical Cp., St. Louis, Mo, pp. 69-84.

Bates, L.S.Waldren R.P and Teare I.D.1973. Rapid determination of free proline in water stress studies. *Plant and soil*. 38: 205

Baus, O., Reineke, A. and Berkelmann-Löhnertz, B. 2011. Control of grey mould by application of gibberellin acid inhibitor (Regalis®) and management practices in *Vitis vinifera* (cv. Riesling). *IOBC/wprs, Bulletin*. 67:49-53

Beattie, J. M. and Forshey, C. G., 1954. A survey of the nutrient element status of concord grape. *Proc. Am. Soc. Hort. Sci.*, **64** : 21-28.

Becking, J. H., 1974. Family 2. Azotobacteriaceae. In: *Bergey's Manual of Determinative Bacteriology*, 8th Ed. (Eds.) R. E. Buchanan and N. E. Gibbons. The Williams and Wilkins Co., Baltimore, pp. 253-261.

Becking, J. H., 1981. Studies on the nitrogen fixing bacteria of the genus Beijerinckia geographical and ecological distribution in soils. *Pl. and Soil*, **14** : 49-81.

Bedi, P. S., Singh, G. and Suryanarayana, D., 1969a, Field testing of Aureofungin and other chemicals to control anthracnose disease of grape in the Punjab. *Hindustan Antibiotics Bulletin*, **11**:251-253.

Bedi, P.S., Singh, G. and Suryanarayana, D., 1969b, Studies on the chemical control anthracnose in Punjab. *Indian Phytopathological Society Bulletin*, **5**: 19–22.

Begum, S., Ibrahim M., Samad, A., Khan, M. R. Khaleda, S. Alam, AKMS and Chowdhury, A. N., 2007. Effect of Pre Harvest Application of Some Growth Regulators on the Yield and Quality Characteristics of Grape (*Vitis vinifera* L.) *Bangladesh J. Sci. Ind. Res.* **42**(1), 59-66.

Belimov, A. A., Kojemiakov, A. P. and Churaliyera, C. V., 1995. Interaction between barley and mixed culture of nitrogen fixing and phosphate solubilizing bacteria. *Plant and Soil*, **173** : 29-37.

Beniwal, B. S., Gupta, O. P. and Ahlawat, V. P., 1992. Effect of foliar application of urea and potassium sulphate on physico-chemical attributes of grapes (*Vitis vinifera* L.) cv. Perlette. *Haryana J. Hort. Sci.*, **21**(3-4 : 161-165.

Beresford, R. M., Evans, K. J., Wood, P.N. and Mundy, D.C., 2006. Disease assessment and epidemic monitoring methodology for bunch rot (*Botrytis cinerea*) in grapevines. *New Zealand Plant Protection*. 59: 355-360

Bernizzoni Fabio, Matteo Gatti, Silvia Civardi and Stefano Poni, 2009. Long-term Performance of Barbera Grown under Different Training Systems and Within-Row Vine Spacings. *Am. J. Enol. Vitic*. **60** (3): 339-348

Bhandari, A. R. and Randhawa, N. S., 1978. Micronutrient status of apple orchards of Simla hills. *Ind. J. Hort.*, **75** : 321-327.

Bhandari, A. R. and Sharma, V. S., 1978. Micronutrient status of apple orchards of Simla district. *Ind. J. Hort.*, 75 : 16-21.

Bhargava, B. S. and Chadha, K. L., 1988. Developing leaf nutrient guide in fruit crops. *Fert. News*, **33** : 21-29.

Bhargava, B. S. and Raghupathi, H. B., 1993. Phosphorus status of soils and plants in vineyards of peninsular India. *J. Ind. Soc. Soil Sci.*, **41**(4) : 689-694.

Bhargava, B. S. and Raghupathi, H. B., 1995. Current status and new norms of nitrogen nutrition for grape vine (*Vitis vinifera*). *Ind. J. Agric. Sci.*, **65**(3) : 165-169.

Bhargava, B. S. and Raghupathi, H. B., 2001. Soil and plant diagnostic norms of perlette grape. *Haryana J. Hort. Sci.*, **30**(3-4) : 165-167.

Bhargava, B. S. and Sumner, M. E., 1987. Proposal for sampling grape (*Vitis vinifera* L.) petiole for nutritional *Oliagnosis communs. Soil Sci. Plant Anal.*, **18** : 581-591.

Bhatia, S. K., Yadav, S., Ahlawat, V. P. and Dahiya, S. S., 2001. Effect of foliar application of nutrients on the yield and fruit quality of winter season Guava cv. L-49. *Haryana J. Hort. Sci.*, **30** (1&2) : 6-7.

Bhobia, S. K., Godara, R. K., Sulkhbir Singh, Beniwal, L. S. and Surender Kumar, 2005. Effect of organic and inorganic nitrogen on growth, yield and NPK content of guava cv. Hisar surkha during winter season. *Haryana J. Hort. Sci.*, **34**(3-4) : 232-233.

Bhujabal, B.G., Draksha Virut, Contnental publications, Pune.

Bhujbal, B. C., 1977. Studies on nutrient status of Thompson Seedless variety of grape (*Vitis vinifera* L.) in Western Maharashtra. *Ph. D. Thesis,* Mahatma Phule Krishi Vidyapeeth, Rahuri.

Bhushan, B. and Singh B. 2005. Effect of three species of arbuscular mycorrhiza on bud sprout and ripening in grapevine (*Vitis vinifera* L.) cv. Perlette. Find all citations by this author (default).

Biological Agriculture and Horticulture, 23(1):73-83

Black, C. S., 1965. *Methods of Soil Analysis* (Part 2); Agronomy Monograph 9, American Society of Agronomy, Medison, Wisconsin, USA, pp. 149-157.

Bleyer, K. and Kast, W. K., 2011. Experiences about the effect of growth regulators on cluster structure and fruit rot diseases of the vine varieties Pinot Meunier, Pinot Blanc, Pinot Noir and Rhine-Riesling. *IOBC/wprs, Bulletin.* 67:41-47

Blum, A. 1974. Genetic responses in sorghum to drought stress. I. Responses to soil moisture stress. *Crop Sci.* 14:361-364.

Böll S., Lange T., Hofmann, H. and Schwappach P. 2011. Variety-dependent responses of seeded vine cultivars to different growth regulators. IOBC/WPRS Bulletin 67: 55-61.

Bollamma, K. T. and Yadav, I. S., 1997. Micropropagation of salt tolerant grape rootstock variety 'Ramsey'. *M. Sc. Thesis, IIHR, Bangalore*

Bollen, G. J. And Van Zaayen, A., 1975. Resistance to benzimidazole fungicides in pathogenic strains of *Verticillium fungicola. Netherlands Journal of Plant Pathology*, **81:** 157-167.

Boora, R. S., Devi Singh, Siddiqui, S. and Verma, S. L., 2002. Response of sapota to NPK fertigation. *Haryana J. Hort. Sci.*, **31**(1-2) : 15-17.

Boraiah, B. and Srinivas, K., 2001. Response of grapes to micro-irrigation under different evaporation replenishment rates. *IIHR, Bangalore.*

Bottomley, A.M., 1937. Grape anthracnose. *Farming South Africa*, **137:** 338 – 339.

Brar, S. S., Bindra, A. S. and Rehalia, A. S., 1979. Mineral nutrient status of some grape cultivars. *Prog. Hort.*, **11**(3) : 17-20.

Bravdo, B. A. and Hepner, Y., 1987. Irrigation management and fertigation to optimize grape composition and vine performance. *Acta Horticulturae*, **206** : 49-67.

Breksa Andrew P., Find all citations by this **author** (default).

Bremner, J.M., 1960. Determination of nitrogen in soil by the Kjeldahl method. *J. of Agril. Sci.*, **55:** 11-13.

Briand, F., Vergely, C., Charmillot, P.-J. and Kehrli, P., 2011. Small insect enclosure field cages: A simple method to assess mating disruption. IOBC/wprs Bulletin, 67: 289-293

Bric, T. V., Sridhar, R. and Mahadevan, A., 1991. Growth hormones. In : *Methods in Physiological Plant Pathology*, 2nd Edition Sivakami Publishers, Madras, pp. 24-44.

Brook. P. J., 1973. Epidemiology of grapevine anthracnose caused by *Elsinoe ampelina. New Zealand Journal of Agricultural Research*, **16:** 333-342.

Brown, J.G., 1922. *32nd Annual Report. Arizona Agricultural Experimental Statistics*, pp. 606 – 615.

Brown, M. E. and Burlingham, S. K., 1962. Production of plant growth substances by *Azotobacter chrococcum. J. Gen. Microbiol.*, **53** : 135-144.

Brown, M. E. and Burlingham, S. K., 1968. Production of plant growth substances by *Azotobacter chrococcum. J. Gen. Microbiol.*, **53** : 135-144.

Brown, M. E., 1974. Seed and root bacterization. *Annu. Rev. Phytopathol.*, **12** : 181-198.

Bruce W. Zoecklein, Yamuna S. Devarajan, Kumar Mallikarjunan and Denise M. Gardner, 2011. Monitoring Effects of Ethanol Spray on Cabernet franc and Merlot Grapes and Wine Volatiles Using Electronic Nose Systems *Am. J. Enol. Vitic.* **62** (3): 351-358

Burrill, T.J., 1886. Grape rots. *American Pomology Society Processing*, **20:** 47– 49.

Butler, E. J. 1905. Pilzkrankheiten in Indien im Jahre 1903. *Zeitsch fur Pflanzenkr*, **15:** 44-48.

Caffi, T., Rossi, V. and Lusitani, M. (2011). Long-term survival of *Plasmopara viticola* oospores. *IOBC/wprs, Bulletin.*67: 111-114.

Calonnec, A., Schnee, S., Cartolaro, P. and Langlais, M., 2011. On the dynamics of a powdery mildew epidemic. *IOBC/wprs, Bulletin.*67: 123-130.

Camargo, U. A., Protas, L. M. R. and Mello, 2006. Grape growing and processing in Brazil. ISHS Acta J.F.S. Horticulturae 785: *International Symposium on Grape Production and Processing, held at Baramati* (Pune), Maharashtra, India During 6–11th February 2006

Camargo, U.A., Ritschel, P.S., 2006. New table and wine grape cultivars: World scenario with emphasis on Brazil, ISHS Acta Horticulturae 785: *International Symposium on Grape Production and Processing, held at Baramati* (Pune), Maharashtra, India During 6–11th February 2006.

Carbonneau, A. 1985. The early selection of grape vine rootstocks for resistance to drought conditions. Am. J. Enol. And Vitic. 36(3): 195-198.

Chadha, K. L. and Pareek, O. P., 1993. *Adv. Hort.*, Malhotra Publishing House, New Delhi, 1:1-3.

Chadha, K. L. and Shikhamany, S. D., 1999. *The Grape, Improvement, Production and Post-harvest management*. Malhotra Publishing House, New Delhi, pp.579.

Chadha, K. L., 2006. Grape Research-Now and Then Paper presented at the *International Symposium on Grape Production and Processing*, at Baramati, Maharashtra (India) held on Feb 6-11, 2006.

Chadha, K. L., 2006. Indian viticulture scenario. ISHS Acta Horticulturae 785: *International Symposium on Grape Production and Processing, held at Baramati* (Pune), Maharashtra, India During 6–11th February 2006.

Chadha, K.L. and Shikhamani, S.D. The Grape, Malhotra publication. New Delhi.

Chakraborthy, S. K., Sinha, H. C. and Prasad, R. N., 1979. Copper status of soils of Assam.*J. Ind. Soc. Soil Sci.*, **27** :23-25.

Chanana, Y.R. and Gill, M. I. S., 2006. High quality grapes can be produced in Punjab. ISHS Acta Horticulturae 785: *International Symposium on Grape Production and Processing, held at Baramati* (Pune), Maharashtra, India During 6–11th February 2006

Chandha, K.L. 1984. Grape research in India-Priorities and suggestions, approach for future. *Indian J. Hort.* 41(3&4): 145-159.

Chandra Sekhar, D., Jagdishwar Reddy, D., Rahman, S. J., Ranga Reddy, A. and Narendranath, V.V., 2006. Ecology and management of red spider mite, *tetranychus urticae* koch on grape. ISHS Acta Horticulturae 785: *International Symposium on Grape Production and Processing, held at Baramati* (Pune), Maharashtra, India During 6–11th February 2006.

Chandrasekhara Rao, K., 1992. Epidemiology of some common diseases of grape around Hyderabad. *Proceedings of International Symposium on: Recent Advances in Viticulture and Oenology*, A.P. Grape Growers' Association, Hyderabad. pp. 323-329.

Chandrasekhara Reddy, B. M. and Chadha, K.L., 1998. Investigations on the effect of rootstocks in Anab-e-Shahi grape (*Vitis vinifera* L.). *M. Sc. Thesis, IIHR, Bangalore*

Cheema, S. S., Kapur, S.P., Chohan, J.S. and Jeyarajan, R., 1978. Studies on the cultural and pathogenic variation of *Sphaceloma ampelinum*, the causal organism of the anthracnose disease of grape. *Indian Phytopathology*, **3:** 163- 166.

Cheema, S. S., Singh, P. and Dhillon, W. S., 2003. Effect of crop regulation and canopy management on fruit quality and disease incidence in grape. *Ind. J. Hort.*, **60**(3): 208-213.

Cheema, S.S., Kapur, S.P. and Chohan, J.S., 1981. Effect of different media on the sporulation of *Sphaceloma ampelium*, the causal organism of grape anthracnose. *Journal of Research*, **18:** 95–96

Chinnathambi Srinivasan and Michael G. Mullins, 1981. Physiology of Flowering in the Grapevine – a Review. *American Journal of Enology and Viticulture*. **32** (1) 47-63.

Chitkara, S. D., Singh, J. P. And Bakshi, 1972. Influence of different levels of nitrogen on vigour, shoot composition relationship with fruit but differentiation, yield and quality of fruit in Thompson Seedless grape (*Vitis vinifera* L.). *Haryana J. Hort.Sci.*, **1**(1/4) : 1-11.

Chohan, J.S., 1965. Diseases of grape in the Punjab and their control. *Punjab Horticulture Journal*, **5:** 181-184.

Chohan, J.S., 1974. Diseases of grape in Punjab and their control. *Journal of Research PAU*. **11:** 179 – 180.

Chokha Singh, Sharma, V. P., Usha, K. and Sagar, V. R., 2002. Effect of macro and micronutrients on physico-chemical characters of grape cv. Perlette. *Ind. J. Hort.*, **59**(3): 258-260.

Chougule, R.A., Tambe, T.B. and Kshirsagar, D.B., 2006. Effect of canopy management on yield and quality attributes of Thompson seedless grapes. ISHS Acta Horticulturae 785: *International Symposium on Grape Production and Processing, held at Baramati* (Pune), Maharashtra, India During 6–11th February 2006.

Christine Böttcher, Robert A. Keyzers, Paul K. Boss and Christopher Davies. 2010. Sequestration of auxin by the indole-3-acetic acid-amido synthetase GH3-1 in grape berry (*Vitis vinifera* L.) and the proposed role of auxin conjugation during ripening. *J Exp Bot.* 61(13): 3615-3625

Chuche, J., Boursault, A. and Thiéry, D., 2011. Preliminary study of the aggregative behaviour of *Scaphoideus titanus* larvae. IOBC/wprs Bulletin, 67: 239-244

Chuche, J., Sauvion, N. and Thiéry, D., 2011. Electropenetrography, a tool to investigate the feeding behaviour of sucking insects: development of this technique to *Scaphoideus titanus*. IOBC/wprs Bulletin, 67: 299-303

Claude Dubois, Claire Manginot, Jean-Louis Roustan, Jean-Marie Sablayrolles and Pierre Barre, 1996. Effect of Variety, Year, and Grape Maturity on the Kinetics of Alcoholic Fermentation. *Am. J. Enol. Vitic.* 47 (4):363-368.

Cocco, A., Coinu, M., Lentini, A., Serra, G. and Delrio, G., 2011. Mating disruption field trials to control the vine mealybug *Planococcus ficus*. IOBC/wprs Bulletin, 67: 215-221

Conradie, W. J. and Saayman, D., 1989. Effect of long-term nitrogen, phosphorus and potassium fertilization on Chenin Blanc vines. *Am. J. Enol. Vitic.*, **40**(2) : 85-90.

Costantini, E. A. C., Pellegrini, S., Bucelli, P., Storchi, P., Vignozzi, N., Barbetti, R. and S. Campagnolo, 2009. Relevance of the Lin's and Host hydropedological models to predict grape yield and wine quality. *Hydrol. Earth Syst. Sci.*, 13. 1635-1648.

Craig N. Austin and Wayne F. Wilcox, 2011. Effects of Fruit-Zone Leaf Removal, Training Systems, and Irrigation on the Development of Grapevine Powdery Mildew. *Am. J. Enol. Vitic.*, 62:193-198

Cristina González-Flor, Lydia Serrano Porta and Gil Gorchs Altarriba, 2013. Predicting Berry Quality Attributes in cv. Xarel·lo Rain-Fed Vineyards Using Narrow-Band Reflectance-Based Indices. *Am. J. Enol. Vitic.* 64 (1): 88-97

Daane, K. M., Bentley, W.J., Millar, J.G., Walton, V.M., Cooper, M.L., Biscay, P. and Yokota, G.Y., 2006. Integrated management of mealybugs in california vineyards. ISHS Acta Horticulturae 785: *International Symposium on Grape Production and Processing, held at Baramati* (Pune), Maharashtra, India During 6–11th February 2006

Dabas, A. S. and Jindal, P. C., 1985. Effect of B and Mg sprays on fruit bud formation, berry set, berry drop and quality of Thompson Seedless grapes. *Ind. J. Agric. Res.*, **19**(1) : 40-44.

Daniel Molitor, Marc Behr, Lucien Hoffmann and Danièle Evers, 2012. Impact of Grape Cluster Division on Cluster Morphology and Bunch Rot Epidemic. *Am. J. Enol. Vitic.*, 63:508-514.

Datar, V.V. And Ashtaputre, J.V., 1985. Managing anthracnose of grapes (*Vitis vinifera* L.) caused by *Elsinoe ampelina* (de Bary) Shear. *Maharashtra Journal of Horticulture*, **2:** 24 – 26.

De Bary, A., 1874. Ueber den sogenannten Brenner (pech) der Reben. *Oenologie*, **4:** 165-167.

De Boer, M., Vander Sluis, I., Van Loon, L. C., Bakker Pashm, 2004. Combining flueroscence pseudomonas. Strains to enhance the suppression of fusarium wilt of raddish. *European J. Plant Path.*, **15** : 201-210.

De Freitas, J. R., Banerjee, M. R. and Germida, J. J., 1997. Phosphate solubilizing rhizobacteria enhance the growth and yield but not phosphorus uptake of canola (*Brassica napus* L.). *Biology and Fertility of Soils*, **24** : 358-364.

Dekker, J. and Georgopoulos, S.G., 1982. *Fungicide resistance in crop protection* (Eds.), *PUDOC*, Wageningen.pp.273.

Delp, C. J., 1980. Coping with resistance to plant disease control agents. *Plant disease*, **64:** 652-657.

Delp, C. J., 1989. *Fungicide resistance in North America* (Ed), *APS Press*. Minnesota. pp. 133.

Deokate, A.S., Khilare, V.C. and Gangawane, L.V., 2002. Management of anthracnose of grape by carbendazim in combination with other agrochemicals. *Indian Phytopathology*, **55** (1): 107-108.

Desale, A. G., 1980. Effect of bacterization with *Azotobacter* and *Azospirillum* cultures under various levels of nitrogen on growth and yield of sorghum and maize. *M. Sc.(Agri.) Thesis*, MPVK, Rahuri (MS), India.

Deshmukh, R. B., 2006. *Grape research in Maharashtra*. Paper presented in the *International Symposium on Grape Production and Processing*, at Baramati, Maharashtra (India) held on Feb 6-11, 2006.

Deshpande, S. D. and Shukla, B.D., 2006. Modified atmosphere packaging of table grapes in polymeric films under ambient conditions for increased storage life. ISHS Acta Horticulturae 785: *International Symposium on Grape Production and Processing, held at Baramati* (Pune), Maharashtra, India During 6–11th February 2006.

Devananda, B. J., 2000. Role of plant growth promoting rhizobacteria on growth and yield of pigeonpea (*Cajanus cajan* L.) cultivars. *M. Sc. (Agri.) Thesis*, Univ. Agric. Sci., Dharwad.

Devarajan, Yamuna S., Zoecklein, Bruce W. and Mallikarjunan Kumar, 2011. Electronic Nose Evaluation of the Effects of Canopy Side on Cabernet franc (*Vitis vinifera* L.) Grape and Wine Volatiles. *Am. J. Enol. Vitic*. 62:1.

Dhanumjayarao J.K., Jindal, P. C., Room Singh, Srivastava, G. C. and R. C. Sharma. 2006. Biochemical variability studies for disease resistance in grape germplasm against powdery mildew (*Uncinula negator*) (Schw) bur along with some varietal characters. *Indian J. Agric. Res.*, **40** (3): 212 - 215,

Díez-Navajas, A.-M. and Ortiz-Barredo, A., 2011. Downy, powdery mildew and botrytis risk assessment by climate parameters, pest and spores monitoring in Rioja Alavesa vineyard (Northwest of Spain). IOBC/wprs Bulletin, 67: 21-29.

Dimitrov, I., 1975. Studies of drought resistance in some vine rootstocks. *Gradinarska I Lozarika Nauka*, **12**(6): 117-125.

Dobereiner, J. and Depolli, H., 1980. *Diazotrophic rhizocoenosis*. In : *Nitrogen Fixation* Eds. W. D. P., Steward and Gallon, J. R., Acad. Press, pp. 301-333.

Dobereiner, J., Day, J. M. and Dart, P. J., 2001. Nitrogenase activity in the rhizosphere of sugarcane and other grasses. Plant Soil, **37** : 191-196.

Doroftei, V.I., Nevryanskaya, A.D. and Shishkanu, G.V., 1993. Photosynthetic pigments in the leaves of grape under different soil moisture conditions. Buletinul Academiede stibnite and Repulicii Moldova – Stiinte Biologicesi Chimice, No. 3: 8-11.

Doshi, P. J. and Adsule, P.G., 2006. Effect of storage on physicochemical parameters, phenolic compounds and antioxidant activity in grapes. ISHS Acta Horticulturae 785: *International Symposium on Grape Production and Processing, held at Baramati* (Pune), Maharashtra, India During 6–11th February 2006.

Dudi, O. P., Sukhbir Singh, Satpal Baloda and Bal Singh, 2005. Effect of nitrogen and FYM on fruit quality and yield of kinnow mandarin. *Haryana J. Hort. Sci.*, **34**(3-4) : 224-226.

Duplessis, S.J., 1940. Anthracnose of the vine. *Farming South Africa*, **168:** 99–100.

During H. 1980. Stomatal frequency of leaves of *Vitis* sps. and cultivars. Vitis, 19(2): 79-98.

During H. 1985. Aspects of drought tolerance in grape breeding. IVth International symposium on Grapevine Breeding. pp. 113-116.

During H. 1986. Testing for drought tolerance in grapevine scions. Angewandte Botanik, 60:103-111.

Duso, C., Pozzebon, A., Baldessari, M. and Angeli, G., 2011. Current status of grapevine leafminers in North-Eastern Italy. IOBC/wprs Bulletin, 67: 203-206

Dwight Wolfe and Gerald R. Brown. 1992. Training system affects performance of table grape cultivars. *Hort Science.* 27 (11): 1178.

Dwivedi, D. and Jobri, B.N., 2003. Anti-fungal from florescent pseudomonads. Biosynthesis and Regulation. *Curr. Sci.*, **85** : 1693-1703.

E.A.C. Costantini, P.G. Arcara, P. Cherubini, F. Campostrini, P. Storchi, M. Pierucci, Soil and climate functional characters for grape ripening and wine quality of "Vino nobile di montepulciano". *ISHS Acta Horticulturae* 427: Strategies to Optimize Wine Grape Quality

Earnapalli, V. N., Jagadeesh, K. S., Krishnaraj, P. U. and Moon, S. S., 2005. Mechanisms in the biocontrol of early blight of tomato by native isolates of tomato. *Proc. of Asian Conference on Emerging Trends in Plant-Microbe Interactions*, December 8-10, 2005, University of Madras, Chennai.

Ecleford, M. O., 1927. Thermophillic bacteria in milk. *Am. J. Hygiene*, **7:** 201-202.

Eibach, R. and Alleweldt, G. 1985. Influence of water supply on growth, gas exchange and substance production in bearing grapevines. Vitis 24(3):183-198.

Ellis, M. A. and Erincik, O., 2005. *Anthracnose of Grape. http://ohioline.usu.edu.*

Elmer, P.A.G., Parry, F.J. and Reglinski, T. 2009. Development of biologically-based products for control of *Botrytis* in wine grapes. *New Zealand Plant Protection.* 62: 397

El-Mokadem, M. T., Helemish, F. A., Abou-bakr, Z. Y. M. And Sheteawi, S. A., 1990. Associative effect of *Azospirillum lipoferum* and *Azotobacter chrococcum* with *Rhizobium* spp. on mineral composition and growth of chickpea (*Cicer arientinum*) on sandy soils.

Erfani Moghaddam, J., Ebadi, A. and Fatahi Moghaddam, M., 2008.An Investigation into New Seedless Grapevine Genotype Production via Using Controlled Crosses. *Journal of Science and Technology of Agriculture and Natural Resources, Water and Soil Science.* **12(45)**: 591-601.

Fermaud, M., Deytieux-Belleau, C., Roudet, J., Darrieutort, G.and Geny, L., 2011. Modelling the effect of the grapevine growth and susceptibility Water activity at the fruit surface: a potential indicator of grape berry susceptibility to *Botrytis cinerea*. *IOBC/wprs, Bulletin*.**67:** 155-161

Ferree, D., Steiner, T.,Gallander, J., Scurlock, D., Johns, G. and Riesen, R., 2002. Performance of 'Seyval Blanc' Grape in four training systems over Five years, *HortScience*. **37** (7): 1023-1027

Fleisch, A., Pérez, L., Kreiter, P., Odor, A. V. and Sentenac, G., 2011. Biological characteristics of *Heliococcus bohemicus* and *Ericydnus sipylus* in controlled conditions. IOBC/wprs Bulletin, **67:** 223-232

Fornasiero, D., Pozzebon, A., Pavan, F. and Duso, C., 2011. Susceptibility and sensibility of grape cultivars to the leafhopper *Empoasca vitis*. IOBC/wprs Bulletin, 67: 267-270

Fourie, J. F. 2006. Harvesting, handling and storage of table grapes (with focus on pre- and post-harvest pathological aspects). ISHS Acta Horticulturae 785: *International Symposium on Grape Production and Processing, held at Baramati* (Pune), Maharashtra, India During 6–11th February 2006.

Freeman, S., Katan, T. and Shabi, E., 1998. Characterization of *Colletotrichum* species responsible for anthracnose diseases of various fruits. *Plant Disease*, **82:** 596-605.

Gadagi, R., 2000. Studies on *Azospirillum* isolates of ornamental plants and their effect on *Gaillardia pulchella* var. *Picta fouger*. *Ph.D. Thesis*, Univ. of Agril. Sci., Dharwad.

Gaikwad, M. B., Kalbhour, V. P., Khanolagale, M. T. and Patil, A. V., 1981. Studies on nutritional requirements of Thompson Seedless grape (*Vitis vinifera* L.). *Abst. of Papers, Nation. Symp. on Trop. and Sub-trop Fruit Crops*, January, 1981, p. 53.

Gaikwad, R. M., 1999. Effect of phosphate solubilizing microorganisms on growth nutrient uptake (N and P) and yield of brinjal (*Solanum melongena* L.). *M. Sc. (Agri.)Thesis*, M. P. K. V., Rahuri.

Gaind, S. and Gaur, A. C., 1991. Thermotolerant phosphate solubilizing microorganisms and their interaction with mung bean. *Plant and Soil*, **133** : 141-149.

Galal, Y. G. M., 1997. Dual inoculation with strain of *BradyRhizobium japonicum* and *Azospirillum brasilense* to improve growth and biological nitrogen fixation of soybean (*Glycine max* (L.)). *Bio. and Fertility of Soils*, **24** : 317-322.

Ganapathy, T. and Narayana Swamy, P., 1990. Effect of plant products on the incidence of major diseases of groundnut. *International Arachis News Letter*, **77:** 20 – 21.

Gandhi, S. R., 1956. Grape culture in India. *Farm Bulletin, ICAR No.***12.** pp56.

Garande V.K., Gawade M.H., Sapkal K.T., Gurav S.B., 2002. Effect of IBA and number of internodes on rooting of stem cuttings of grape rootstocks. *Agricultural Science Digest*. 22(3): 176-178.

Gardner, Denise M., Zoecklein Bruce W. and Mallikarjunan Kumar, 2011. Electronic Nose Analysis of Cabernet Sauvignon (*Vitis vinifera* L.) Grape and Wine Volatile Differences during Cold Soak and Post-Fermentation. *Am. J. Enol. Vitic.* **62**:1.

Gathala, M. K., Yadav, B. L. and Singh, S. D., 2004. Mineral nutrient status of pomegranate orchard in Jaipur district of Rajasthan. *J. Ind. Soc. Soil Sci.*, **52**(2) : 206-208.

Gatta, B. La., Arace, E., Gambacorta, G., Notte, E. La. and Notte, P. La., 2007. Phenolic characterization of some native varieties of grapes and wines from southern Italy. ISHS, Acta Horticulturae 754: International Workshop on Advances in Grapevine and Wine Research.

Gaur, A.C., 1990. Phosphate solubilizing microorganisms as Biofertilizers. Omega Scientific Publishers, New Delhi, p. 176.

Geeta, S. K., 2001. Analysis of PGPs mutants of MPS bacteria and their effect on plant growth and nutrient uptake. *M. Sc. (Agri.) Thesis*, Univ. Agric. Sci., Dharwda.

Georgopoulos, S. G., 1982. Detection and measurement of fungicide resistance: In "Fungicide resistance in crop protection" Dekker, J. and GEORGOPOULOS, S. G. (Eds)., PUDOC, Wageningen.pp.24-31.

Gessler C., and Jermini, M. 2011. Blackrot - downy mildew control in small vineyards in Southern Switzerland. *IOBC/wprs, Bulletin*.67: 139-142.

Ghewande, M. P., 1989. Management of foliar diseases of Groundnut (*Arachis hypogaea* L.) using plant extracts. *Indian Journal of Agricultural Sciences*, **59:** 133-134.

Gill, M. I. S. and Arora, N. K. 2009. Performance of different grape varieties under north Indian conditions. *Back !Abstract !Descriptors !Top !*

Gill, M. I. S., Sharma, J. K., Poonam Sachdev and Mehan, V. K., 2006. H-516: A promising grape hybrid under N-Indian conditions ISHS Acta Horticulturae 785: *International Symposium on Grape Production and Processing, held at Baramati* (Pune), Maharashtra, India During 6–11th February 2006.

Gill, S. S., Barr, S. S. and Singh, S. N., 1987. Effect of gibberellic acid application on fruit quality of Thompson Seedless grapes. *Punjab Hort. J.*, **27** : 37-41.

Glick, B. R. 1995. The enhancement of plant growth by free-living bacteria. *Can. J. Microbiol.*, **41**:109-117.

Glick, B. R., Jacobson, C. B., Schwarze, M. K. and Pasternak, J. J., 1997a and 1994b, Does the enzyme 1-aminocyclopropane-1 carboxylate deaminase play a role in plant growth promotion by Pseudomonas putida GR 12-2. In : *Improving Productivity with Rhizosphere Bacteria* (Ed.) M. H. Ryder, P. M., Stephens and G. D. Bowen, Commnwealth Scientific and Industrial Research Organization, Adelaide, Australia, pp. 150-152.

Gopalaswamy, N. and Rao, V. N. M., 1972. Effect of graded dose of K on yield and quality of grapes cv. Anab-e-Shahi. *South Ind. Hort.*, **20:** 41-49.

Govindan, M. and Bagyaraj, D. J., 1995. Field response of wetland rice to *Azospirillum* inoculation. *J. Soil Biol. and Ecol.*, **15** : 17-22.

Govindarajan, K., Venkataramana, C. R., Kandaswamy, P., Raniperumal and Krishnamurthy,P., 1979. Classification of soils of Madurai Agricultural College Farm. *Madras Agric. J.*, **66**(8): 756.

Govindarao, P. and Dakshina Murthy, V., 1964. A note on the influence of certain environmental factors on the incidence of diseases of grapevine. *Andhra Agricultural Journal*, **11**: 245 – 249.

Govindwar, S. P. 2009. Occupational pesticides exposure of sprayers of grape gardens in western Maharashtra (India): effects on liver and kidney function. *Find all citations in this journal (default).*

Goyal, J.P., Pathak, V.N. and Sharma, 1971. Incidence of anthracnose on different varieties of grapes in Jobner, Rajasthan. *Tropical Agriculture*, **127**: 105 – 106.

Gump, B. H., Zoecklein, B. W., Fugelsang, K. C. and Whiton, R. S., 2002. Comparison of Analytical Methods for Prediction of Prefermentation Nutritional Status of Grape Juice. *American Journal of Enology and Viticuture.* 53(4): 325-329

Gupta, A., Meyer, J. M. and Goel, R., 2002. Development of heavy metal resistant mutants of phosphate solubilizing Pseudomonas sp. NBR14014 and their characterization. *Curr. Microbio.*, **45**: 323-327.

Gupta, J. S., Agarwal, M. B., Dixit, R. B. and Agrwal, M., 1981. Effect of metabolites from different host plants on conidial germination of *Colletotrichum graminicola* and *C. capsici. Geobios*, **8**: 226 – 228.

Gurme, P.N. and Kore, S.S., 1977. Incidence of anthracnose and varietal resistance in grape in Marathawada, Maharashtra State. *Journal of Maharashtra Agricultural Universities*, **2**: 177.

Halder, A.K., Mishra, A.K., Bhattacharyya, P. and Chakrabarthi, B.K., 1990. Solubilization of rock phosphate by Rhizobium and Bradyrhizobium. *J. of Gen. And Appl. Microbiol.*, **36** : 81-82.

Harris, A. R. and Adkins, P. G., 1999. Versatility of fungal and bacterial isolates for biological control of damping off diseases caused by *Rhizoctonia solani, Phythium* spp. *Biol. Control*, **15**: 10-18.

Hartmann, P. J., McNair, H. M. and Zoecklein, B. W., 2002. Measurement of 3-Alkyl-2-Methoxypyrazine by Headspace Solid-Phase Microextraction in Spiked Model Wines. *American Journal of Enology and Viticulture.* **53**(4): *285-288.*

Hein, D. F., Breuer, M., Hummel, H. E., Greiner, A., Wendorff, J. H., Hellmann, C., Vilcinskas, A., Kratt, A., Kleeberg, H. and Leithold, G., 2011. *Electrospun nanofibers* as novel carriers of insect pheromones: communication disruption strategy against *Lobesia botrana*. IOBC/wprs Bulletin, **67**: 183-187.

Hemantha Kumar, J., Sarwar, H. A. K., Chandrasekhara Rao, K. and Ramesh Babu, T., 1998. Acquired Resistance in *Gloeosporium ampelophagum* and strategies for its management. *Indian Journal of Plant Protection*, **26**: 170-173.

Hemanthakumar, J., Sarwar, H. A. K., Chandrasekhara Rao, K. and Ramesh Babu, T., 2001a, Cross resistance of carbendazim resistant isolates of *Gloeosporium ampelophagum* (Pass.) Sacc., to other fungicides. *Indian Journal of Plant Protection*, **29:** 136-139.

Hemanthkumar, J., Sarwar, H. A. K., Chandrasekhara Rao, K., and Ramesh Babu, T., 2001b, Management of carbendazim resistance in *Gloeosporium ampelophagum* (Pass.) Sacc., the causal agent of grapevine anthracnose. *Indian Journal of Plant Protection*, **29:** 154-156.

Hemawathi, M., 1997. Effect of organic manures and biofertilizer on growth and productivity of chrysanthemum (*Chrysanthemum morifolium* Ramat) cv. Local yellow. *M. Sc. (Agri.) Thesis*, Uni. Agric. Sci. Bangalore (India).

Henshall, W.R., Beresford, R.M.,. Chynoweth, R.W. and Ramankutty, P., 2005. Comparing surface wetness inside and outside grape canopies for region-wide assessment of plant disease risk. *New Zealand Plant Protection* 58: 80-83

Hidalgo Marlene B., Find all citations by this author (default).

Higa, T. and Wididana, G. N., 1991. Changes in the soil microflora induced by effective microorganisms. *Proc. First Int. Conf. Kyusei Nature Farming*, Washington, USA (Ed.) J. F. Parr, S. B. Hornick and C. E. Whitman, pp. 153-162.

Hill, G. N., Beresford, R. M. and Evans, K. J., 2010. Tools for accurate assessment of botrytis bunch rot (*Botrytis cinerea*) on wine grapes. *New Zealand Plant Protection* 63: 174-181.

Himanshu Kumar Singh, Singh, A. K. P. and Sinha, A. K., 2005. Effect of fertigation on fruit growth and yield of papaya with drip irrigation. *Haryana J. Hort. Sci.*, **34**(1-2): 7-8.

Hipparagi Kulapati and Singh Rajendra, 1991. Studies on shoot growth and berry development in different genotypes of grape. IIHR, Bangalore

Hommay, G., Kienlen, J. C., Gertz, C., Bihry, C. and Pizzo, J., 2011. Factors affecting the post-release dispersal of *Trichogramma cacoeciae* Marchal in the vineyard. IOBC/wprs Bulletin, 67: 207-214.

Hooda, P. S., Ahlawat, V. P. and Sindhu, S. S., 1990. Growth and mineral composition of three ultivars as influenced by salinity. *Haryana J. Hort. Sci.*, **19**(1-2) : 55-61.

Hossein Karbalaei Khiavi, Haji Shikhlinskiy, Asadollah Babaei Ahari and Mohammad Akrami, 2012.Evaluation of different grape varieties for resistance to powdery mildew caused by *Uncinula necator. African Journal of Agricultural Research.* (29):4182-4186.

Husameddin Al-Said and Prakash, G. S., 2008. Propagation of some grape rootstocks in enriched organic media and their subsequent graft success. *M. Sc. Thesis, IIHR, Bangalore*

Huth, C., Schirra, K.-J., Seitz, A. and Louis. F. (2001). A beneficial species becomes a pest - the common earwig *Forficula auricularia* (Linnaeus 1758). IOBC/wprs Bulletin, 67: 249-256.

Iacano, F, Buccella A and E. Peterlunger 1998. Water stress and rootstock influence on leaf gas exchange of grafted and ungrafted grapevines. Scientia Horticulturae, 75:27-39

Illmer, P. and Schinner, F., 1992. Solubilization of inorganic phosphates by microorganisms isolated from forest soils. *Soil Biol. Biochem.*, **24:** 389-395.

Iswaran. V, 1980. Determination of soil moisture by oven drying method. In : A Laboratory Handbook for Agricultural Analysis. P.P.85-86.

Ivanova, E. G., Doronina, N. V. And Trotsenko, Y. A., 2001. Aerobic methylobacteria are capable of synthesizing auxins. *Microbiologiya*, **70** : 345-347.

J. Tonietto, 2006. Geographical indicators for grapes. ISHS Acta Horticulturae 785: *International Symposium on Grape Production and Processing, held at Baramati* (Pune), Maharashtra, India During 6–11th February 2006.

Jackson, D. I. and. Lombard, P. B., 1993. Environmental and Management Practices Affecting Grape Composition and Wine Quality - A Review. *Am. J. Enol. Vitic.* 44 (4):409-430.

Jackson, D. I., 1986. factors affecting soluble solids acid, pH and colour in grapes. *Am. J.Entol. Vitic.*, **37**(3) : 179-183

Jackson, M. L., 1967. *Soil Chemical Analysis*, Prentice Hall of India Pvt. Ltd., New Delhi, p.205.

Jackson, M. L., 1973. *Soil Chemical Analysis*, Prentice Hall of India Pvt. Ltd., New Delhi.

Jahagirdar Shamarao, Jamadar, M. M. and Lakshmana, D., 2004a, Performance of Shield SC against anthracnose of grape in North Karnataka. *Pestology*, **28:** 14.

Jahagirdar Shamarao, Ravikumar, M. R., Srikant Kulkarni, Sharma, N. N. And Chandramouli, B., 2004b, Bioefficacy and phytotoxicicityof RIL 004, a new molecule in the management of anthracnose of grape in northern Karnataka caused by *Colletotrichum gloeosporioides*. *Pestology*, **28:** 38 – 40.

Jameel Akhtar and Singh, M. K., 2007. Studies on the variability in *Colletotrichum capsici* causing chilli anthracnose. *Indian Phytopathology*, **60:** 63-67.

James, M. and Beattile, 1954. A survey of the nutrient element status of concord grapes in Ohio. *Am. Soc. Hort. Sci.*, **63:** 21-28.

Janaki, D., Velu, V. and Savithri, P., 2002. Yield and quality of grapes cv. Mascat as affected by boron application. *Madras Agric. J.*, **89**(7-9) : 386-389.

Jaskowska, H., 1995. Occurrence and characters of *Azospirillum* spp. in the rhizosphere of cereals. *Acta Microbiologica, Polonica*, **44:** 69-77.

Jayashankar, S., Li, Z. and Gray, D.J., 2000. *In-vitro* selection of *Vitis vinifera* 'Chardonnay' with *Elsinoe ampelina* filtrate is accompanied by fungal resistance and enhanced secretion of chitinase. *Planta*, **211:** 200-208.

Jeetram, Dal Singh, Jain, P. K. and Ahlawat, V. P., 1993. Growth and nutrient composition of perlette grapes (*Vitis vinifera* L.) as affected by nitrogen levels and pruning intensities. *Haryana J. Hort. Sci.*, **22**(4): 280-284.

Jenkins, A.E. and Bittancourt, A.A., 1943. Gourmand and burgeons treatment of *Sphaceloma ampelinum*. *Mycologia*, **35:** 272 – 276.

Jermini, M., Angst, A., Raynal, M., Gessler C. and Broggini, G.m 2011. First study on the population genetic structure of *Guignardia bidwellii*. *IOBC/wprs, Bulletin*.67: 143-148.

Jeyarajan, R., Singh, R. and Singh, R., 1970. Incidence of anthracnose and *Cercospora* leaf spot diseases of grape varieties. *Madras Agricultural Journal*, **57:** 160 – 164.

Jianying Feng, Jintao Song, Weihua Jiao and Weisong Mu., 2012. Grape distribution efficiency in wholesale market, Beijing, China. *Journal of Food, Agriculture and Environment*. 10 (1): 67-70.

Jindal, P.C. and Bhavani Shankar, 2002. Screening of grape germplasms against anthracnose (*Sphaceloma ampelinum* de Bary). *Indian Journal of Agricultural Research*, **36:** 145-148.

Jisha, 2002. Efficiency of *Pseudomonas fluorescence* and *Trichoderma* in mixtures for plant growth promotion and disease. *Ind. J. Microbiol.*, **35** : 86-89.

Johri, B. N., Rao, C.V.S. and Goel, R., 1997. Florescent pseudomonads in plant disease management. In : *Biotechnology Approaches Soil Microorganisms for sustainable crop production*, Dadarwal K.R. ed). Scientific Publishers, Jodhpur, India, pp. 193-221.

Jones, P.S., Ley, S.V., Morfan. E. D. and Santafianos, D., 1989. The chemistry of neem tree. In: 1988 Focus on phyto-chemical pesticides, *Vol. I. The neem tree*, CRD Press Florida. pp. 19 – 45.

Joseph, P. J., Vrinda, T.S., Sivaprasad, P. and Heera, G., 2003. Abstracts and short papers. *Sixth Intl. PGPR workshop*, 5-10 October, 2003, Indian Institute of Spices Research, Calicut, Kerala, India.

Justin J. Scheiner, Justine E. Vanden Heuvel, Bruce Pan and Gavin L. Sacks, 2012. Modeling Impacts of Viticultural and Environmental Factors on 3-Isobutyl-2-Methoxypyrazine in Cabernet franc Grapes. *Am. J. Enol. Vitic*. **63**:94-105

Justin J. Scheiner, Justine E. Vanden Heuvel, Bruce Pan and Gavin L. Sacks, 2012. Modeling Impacts of Viticultural and Environmental Factors on 3-Isobutyl-2-Methoxypyrazine in Cabernet franc Grapes. *Am. J. Enol. Vitic.* 63 (1): 94-105.

Kamble, B. H., Jadhav, M. S., Yadav, D. B., 2007. Constraints in production and marketing of grapes in Sangli district of Maharashtra. *Agricultural Economics Research Review*. 20:595.

Kanamadi Vijayakumar and Singh Rajendra, 1987. Studies on comparison of grape hybrids and their parents for qualitative characters. *M. Sc. Thesis*, IIHR, Bangalore

Kapoor, B. S., Singh, H. B. and Goswami, S. C., 1981. Academic in sodic profile. *J. Ind. Soc. Soil Sci.*, **28** : 513-515.

Karabulut Ozgur Akgun, Gianfranco Romanazzi, Joseph L. Smilanic, Amnon Lichter. 2005. Postharvest ethanol and potassium sorbate treatments of table grapes to control gray mold. *Postharvest Biology and Technology*. **37** : 129–134.

Karibasappa G. S., Find all citations by this author (default).

Karibasappa, G. S. and Adsule, P. G., 2006. Evaluation of wine grape genotypes by national research centre for grapes at their farm at Pune, Maharashtra, India

Kast, W. K. and Bleyer, K., 2011. The expert system OiDiag-2.2 – a useful tool for the precise scheduling of sprays against powdery mildew of vine (*Erysiphe necator* Schwein.). *IOBC/wprs, Bulletin.* 67:79-84.

Kehrli, P., Kessler, S., Schaerer, S. and Delabay, N., 2011. Distribution and host plant preferences of *Hyalesthes obsoletus*, the vector of bois noir disease, in Switzerland. IOBC/wprs Bulletin, 67: 3-8.

Kerem Baki and Bruce W. Zoecklein. 2012. Effect of Ethanol Sprays on Cabernet Sauvignon (*Vitis vinifera* L.) Grape Maturity and Wine Quality. *M.Sc. (Thesis).Enology-Grape Chemistry at Virginia Tech.*

Khammas, K. M., Ageron, E., Grimont, P. A. D. and Kaiser, P., 1989. *Azospirillum irakense* spp. nov. a nitrogen fixing bacterium associated with rice roots and rhizosphere soil. *Res. Micribiol.*, **140** : 679-693.

Khan, Z. H., Mazumder, A. R., Hussain, M. S. and Mohiuddin, A. S. M., 1997. Total and DTPA-extractable Fe, Mn, Cu and Zn contents in some soils of Bangladesh. *J. Ind. Soc. Soil Sci.*, **45**(3) : 485-489.

Khandagale, M. T., 1977. Effetc of various levels of nitrogen, phosphorus and quality of Thompson Seedless grape (*Vitis vinifera* L.). *M. Sc. (Agri.) Thesis*, Mahatma Phule Krishi Vidyapeeth, Rahuri.

Khilari, J. M., 2006. Grape disorders – Some observations. ISHS Acta Horticulturae 785: *International Symposium on Grape Production and Processing, held at Baramati* (Pune), Maharashtra, India During 6–11th February 2006.

Kim, K. S., Hill, G. N. and Beresford, R. M., 2010. Remote sensing and interpolation methods can obtain weather data for disease prediction. *New Zealand Plant Protection* 63: 182-186.

Kim, K. Y., Jordan, D. and McDonald, G. A., 1998a, Effect of phosphate solubilizing bacteria and vesicular arbuscular mycorrhizae on tomato growth and soil microbial activity. *Biol. and Fertility of Soil*, **26**(2) : 79-87.

King Petra, D., Daniel J. McClellan and Richard E. Smart, 2012. Effect of Severity of Leaf and Crop Removal on Grape and Wine Composition of Merlot Vines in Hawke's Bay Vineyards. 63 (4):500-507.

Klein, I., 2006. Autonomous threshold tension irrigation (ATTI). ISHS Acta Horticulturae 785: International Symposium on Grape Production and Processing. held at Baramati (Pune), Maharashtra, India During 6–11th February 2006

Klien, L. and Spieler, W. S., 1987. Effect of fertilization on fruit yield, quality and nutrient status of grape cv. Perlette. *Ind. J. Hort.*, **59**(3) : 270-274.

Kocturk Murat, O., 2006. The Marketing of Table Grapes and Farmers Organization in Turkey (AlaOehir Case Study). *Journal of Biological Sciences*, 6: 961-967.

Kore, S. S. and Gurme, P.N., 1971. A strain of *Gloeosporium ampelophagum* causing grapevine anthracnose. *Indian Journal of Mycology and Plant Pathology*, **1:** 132-134

Kore, S. S. and Gurme, P.N., 1978. Factors influencing growth and sporulation of *Gloeosporium ampelophagum*. *Indian Journal of Mycology and Plant Pathology*, **8:**222-224.

Kortekamp A. and Köckerling, J., 2011. Investigation of contamination pathways of Esca-associated fungal pathogens and of fungicide-based control strategies. *IOBC/wprs, Bulletin*.67: 163-169.

Koshteh K. and Abbas, A., 2004. Strategy and limitations to improve grapes marketing efficiency (A case study of S and B, Iran) ISHS Acta Horticulturae 655: XV International Symposium on Horticultural Economics and Management, Berlin, Germany.

Koshteh, K. and Abbas A., 2004. Strategy and limitations to improve grapes marketing efficiency (A case study of S and B, Iran). *ISHS Acta Horticulturae 655: XV International Symposium on Horticultural Economics and Management.*

Krishnaraj, P. U. and Gowda, T. K. S., 1990. Occurrence of phosphate solubilization in the endorhizosphere of crop plants. *Curr. Sci.*, **59** : 933-934.

Kulkarni, N. S., Sawant, I. S., Sawant, S.D. and Adsule, P.G., 2006. Bio-efficacy of neem formulations (azadirachtin 1 per cent and 5 per cent) on important insect pests of grapes and their effect on shelf life. ISHS Acta Horticulturae 785: *International Symposium on Grape Production and Processing, held at Baramati* (Pune), Maharashtra, India During 6–11th February 2006.

Kulkarni, N.S., Sawant, S. D., Adsule, P.G., 2006. Seasonal incidence of insect pests on grapevine and its correlation with weather parameters. ISHS Acta Horticulturae 785: *International Symposium on Grape Production and Processing, held at Baramati* (Pune), Maharashtra, India During 6–11th February 2006.

Kulkarni, S., Sarangmath, P. A., Salakinakoppa, S. R. and Gaddi, A. V., 2000. Response of chickpea (*Cicer arientinm* L.) to rock phosphate and phosphate solubilizers in typic chromustert. Legume Res. **23**(1) : 21-24.

Kumar S, Thind, T.S. and Mohan Chander., 1995. Morphogenic and pathogenic variations in *Gloeosporium ampelophagum*. *Indian Phytopathology*, **48:** 331-334.

Kumar, R. and Rajan, S. 2006.Grape cultivars flame seedless and pusa navrang can be commercially grown in north Indian conditions. ISHS Acta Horticulturae 785: *International Symposium on Grape Production and Processing, held at Baramati* (Pune), Maharashtra, India During 6–11th February 2006

Kumar, S. and Thind, T. S., 1992. Reduced sensitivity in *Gloeosporium ampelophagum* to carbendazim in Punjab. *Plant Disease Research*, **7:** 103-105.

Kumar, S. Thind, T.S. and Mohan Chander, 1994. Occurrence of *Gloeosporium ampelophagum* and *Colletotrichum gloeosporioides* incitants of grape anthracnose during different months of Punjab. *Plant Disease Research*, **9:** 222-224.

Kundu, B. S. and Gaur, 1980. Establishment of nitrogen fixing and phosphate solubilizing bacteria in rhizosphere and their effect on yield and nutrient uptake of wheat crop. *Pl. and Soil*, **57** : 223-230.

Kunwar, R. and Sinha, M. M., 1987. Macronutrient status and fruit quality of some grape cultivars grown in Srinagar valley of Garhwal hills (India). *Haryana J. Hort. Sci.*, **16**(1-2) : 59-64.

Ladaniya M.S., Wanjari Vinod, Mahalle Bipinchandra, 2005. Marketing of grapes and raisins and post-harvest losses of fresh grapes in Maharashtra. *Indian Journal of Agricultural Research.* 39 (3)167-176.

Ladaniya, M. S., Vinod Wanjari and Bipinchandra Mahalle, 2005. Marketing of grapes and raisins and post-harvest losses of fresh grapes in Maharashtra. *Indian J. Agric. Res.*, **39** (3): 167 – 176.

Lal, B. M., Sahu, D. and Das, N. B., 1959. Studies on available copper in some Indian Soils. *Curr. Sci.*, **28** : 494-495.

Lavania, M. L. and Singh, R., 1977. Effect of different organo-mineral fertilizers on growth, yield and quality of perlette grape - II. Effect on yield and quality. *Vitis*, **16** : 69-78.

Lavee, S., 1987. Usefullness of growth regulators for controlling vine growth and improving grape quality in intensive vineyards. *ISHS Acta Horticulturae 206: Symposium on Grapevine Canopy and Vigor Management, XXII IHC*

Leandro Camargo Neves, Vanuza Xavier Da Silva, Ronaldo Moreno Benedette, Marcos André De Souza Prill, Rogério Lopes Vieites and Sérgio Ruffo Roberto, 2008. Conservation of grapes "Crimson Seedless" and "Italia", submitted to different types of packings and sulfur dioxide (SO_2). *Rev. Bras. Frutic.* 30(1): 65-73.

Lecomte, P., Darrieutort, G., Laveau, C., Blancard, D., Louvet, G., Goutouly, J.-P., Rey P. and Guérin-Dubrana. L., 2011. Impact of biotic and abiotic factors on the development of Esca decline disease. *IOBC/wprs, Bulletin.*67: 171-180

Legler, S. E., Caffi, T., Kiss, L., Pintye, A. and Rossi, V., 2011. Methods for screening new *Ampelomyces* strains to be used as biocontrol agents against grapevine powdery mildew. *IOBC/wprs, Bulletin.*67: 149-154

Leinhos, V. and Vacek, O., 1994. Biosynthesis of auxins by phosphate solubilizing rhizobacteria from wheat and rye. Microbiological Research, **149:** 31-35.

Linder, C., Lorenzini, F. and Kehrli. P., 2011. Can *Harmonia axyridis* affect the taste of European wines. IOBC/wprs Bulletin, 67: 271-274.

Linder, C., Schaub, L. and Klötzli-Estermann, F., 2011. Effectiveness of hot water treatments against the eggs of *Scaphoideus titanus* Ball. IOBC/wprs Bulletin, 67: 17-20.

Linderman, R. G., 1986. Rhzobacteria from field grown mungbean isolation and identification. *Hortic. Sci.*, **21** : 1299-1302.

Lindsay, W. L. and Norvell, W. A., 1978. Development of DTPA soil test for Zn, Fe, Mn and Cu. *J. Am. Soc. Soil Sci.*, **42** : 421-428.

Lopez-Tamames E, Carro-Marino N, Gunata YZ, Sapis C, Baumes R, Bayonove C. 1997. Potential aroma in several varieties of Spanish grapes. *Journal of Agricultural and Food Chemistry*. 45(5):1729-1735.

Lorenzon, M., Pozzebon, A. and Duso, C.,2011. Development and reproduction of the predatory mites *Kampimodromus aberrans, Typhlodromus pyri* and *Amblyseius andersoni* on different food sources. IOBC/wprs Bulletin, 67: 261-265

Lucchi, A. and Santini, L. 2011. Life history of *Lobesia botrana* on *Daphne gnidium* in a Natural Park of Tuscany. IOBC/wprs Bulletin, 67: 197-202

Lucchi, A., Pozzolini, E., Anfora, G., Mazzoni, V., Tasin, M., Leonardelli, E. and Ioriatti, C., 2011. Feeding behaviour of *Lobesia botrana* on leaves and shoots of grapevine. IOBC/wprs Bulletin, 67: 275-280.

M. Carmo Vasconcelos, Marc Greven, Chris S. Winefield, Mike C.T. Trought and Victoria Raw. 2009. The Flowering Process of *Vitis vinifera*: A Review. *American Journal of Enology and Viticulture*. 60:411-434

M. V. McKenry and Safdar A. Anwar, 2006. Nematode and Grape Rootstock Interactions Including an Improved Understanding of Tolerance. *J. Nematol*. **38(3):** 312–318.

Madhava Rao, V.N., Sivanappan, R.K., Palaniswamy, K.P. and Natarajan, P., 1975. A preliminary note on water requirements of grapes. South Indian Hort. 23(1-2): 54-55.

Magarey, R. D., Emmett, R.W., Magarey, P.A. And Franz, P. R., 1993. Evaluation of control of grapevine anthracnose caused by *Elsinoe ampelina* by pre-infection fungicides. *Australian Journal of Plant Pathology*, **22:** 48

Magdalena Kaplan, 2011. The effect of the method of application of growth regulators on fruit quality of 'Einset seedless' grape (*Vitis* sp. L.) *Polish Botanical Society Journals Acta Agrobotanica*. 64 (4).

Mahadevan, A. and Sridhar, R., 1982. Hydrolytic enzymes. In : *Methods in Physiological Plant Pathology*, 2nd Edition Sivakami Publishers, Madras, pp. 24-44.

Maharkar, V. K. and Patil, V. K., 1984. Nutritional status of Thompson Seedless grape orchard surveyed in Osmanabad district of Maharashtra. *Prog. Hort.*, **16**(3-4) : 238-242.

Mahdi Alizadeh and Sanjay Kumar Singh, 2009. Molecular assessment of clonal fidelity in micropropagated grape (*Vitis* spp.) rootstock genotypes using RAPD and ISSR markers. *Iranian Journal of Biotechnology*, 7(1).

Mahesh Kumar, K. S., 1997. Studies on microbial diversity and their activity in soil under bamboo plantation. *M. Sc (Agri). Thesis*, Univ. Agric. Sci., Dharwad.

Malik, R. P., Ahlawat, V. P. and Nain, A. S., 2000. Effect of foliar spray of urea and zinc sulphate on yield and fruit quality of Kinnow – A Mandar in Hybrid. *Indian Journal of Ecology* **36** (1):15-17

Manjunatha G.C.and Shikhamany, S.D.,1991. Studies on shoot growth and bunch development as influenced by the time of pruning and hydrogen cyanamide application in Thompson Seedless grape (*Vitis vinifers* L.). IIHR, Bangalore

Manjunatha, B. and Prakash, G. S., 2003. Studies on pre and post pruning growth regulator treatments on yield and quality of seedless grapes (*Vitis vinifera* L. cv. Sonaka). *IIHR*, Bangalore

Mansfield Anna, K. and Zoecklein Bruce, W., 2003. Effect of Fermentation, Post fermentation, and Post bottling Heat Treatment on Cabernet Sauvignon Glycoconjugates. *American Journal of Enology and Viticulture. 54(2):99-104.*

Mansfield, A. K., Zoecklein B. W. and Whiton, R. S., 2003. Quantification of glycosidase activity in selected strains of *Brettanomyces bruxellensis* and *Oenococcus oeni. American Journal of Enology and Viticulture.* 53(4): 303-307.

María José Carmona, Jamila Chaïb, José Miguel Martínez-Zapater and Mark R. Thomas, 2008.A molecular genetics perspective of reproductive development in grapevine. *J Exp Bot.* July 1. 59:2579-2596

Mariagiovanna Fragasso, Donato Antonacci, Sandra Pati, Maria Tufariello, Antonietta Baiano, Lucia R. Forleo, Angelo R. Caputo and Ennio La Notte. 2012. Influence of Training System on Volatile and Sensory Profiles of Primitivo Grapes and Wines. *Am. J. Enol. Vitic.* 63 (4): 477-486.

Martin Amanda, Mallikarjuan Kumar and Zoecklein Bruce, 2008. Discrimination of Wines Produced from Cabernet Sauvignon Grapes Treated with Aqueous Ethanol Post-Bloom Using an Electronic Nose. *International Journal of Food Engineering.* **4**(2).

Mathew R and Find all citations by this **author** (default).

Mathur, R.L., Mathur, B. N., Singh, R.R., Sharma, B.S. and Godara, P.S., 1973. Varietal reaction of grapes to *Gloeosporium ampelophagum* causing anthracnose. *Indian Journal of Mycology and Plant Pathology*, **3:** 179–181.

Matthews, M.A. and Anderson, M.M., 1989. Reproductive development in grapes (*Vitis vinifera* L): Response to seasonal water deficits. American Journal of Enology and Viticulture, 40 (1): 52-60.

Mazzoni, V., Anfora, G., Ioriatti, C., Virant-Doberlet, M. and Lucchi, A., 2011. Vibrational signals associated to the mating behaviour of *Hyalesthes obsoletus* Signoret (Hemiptera Fulgoromorpha). IOBC/wprs Bulletin, 67: 245-248

Mazzoni, V., Trona, F., Ioriatti, C., Lucchi, A., Eriksson, A. and Anfora, G., 2011. Attractiveness of different colours to *Scaphoideus titanus* Ball (Hemiptera: Cicadellidae) adults. IOBC/wprs Bulletin, 67: 281-284.

Mc Carthy, M.g., 1997. The effect of transient water deficit on berry development of cv.Shiraz (*Vitis vinifera* L.). Australian Journal of Grape and Wine Research, 3(3): 102-108.

Mello, 2001. Compatible strains of *Pseudomonas fluorescens* suppresses fusarium wilt of radish and strain combination. *Plant Pathol.*, **18** : 25-29.

Meriaux S.Rutten. P.Rollin H., and Weber, M., 1976. Effect of drought on grapevine behavior. Comptes – Rendus-des-Seances de.1. Academic 'd'Agriculture –de France, 62(4): 280-289.

Michaelgrainge, and Saleem Ahmed, 1988. Poisonous plants and plants controlling human and animal pest. *Hand book of plants with pest control properties, Section III,* pp. 309.

Miller, G. L., 1972. Use of dinitro salicylic acid reagent for determination of reducing sugar. *Annu. Chem.*, **31:** 426-428.

Mohan Chander and Thind, T.S., 1995. Development of carbendazim resistance in *Gloeosporium ampelophagum* and strategies for its management. *Indian Journal of Mycology and Plant Pathology*, **25:** 25-33.

Mohanraj, D., 1970. Studies on anthracnose of grapevine. *M.Sc. (Agri.) Thesis, submitted to Tamil Nadu Agricultural University Coimbatore*.pp.234.

Mohapatra Soudamini, Ashok K Ahuja, Deepa, M., Jagadish, G. K., Prakash, G. S. and Sampath Kumar, 2010. Behaviour of trifloxystrobin and tebuconazole on grapes under semi-arid tropical climatic conditions. *Pest Management Science*: 66(8): 910–915.

Monga, P. K., Josan, J. S., Paramjit Sing and Harish Kumar, 1990. Effect of N, P and K doses on fruit yield and quality in grapes cv. Perlette under arid irrigated region of Punjab. *Ind. J. Hort.*, **47**(4): 401-404.

Mori, N., Reggiani, N., Pozzebon, A., Duso, C. and Pavan, F., 2011. Influence of nettle control along a ditch on spatial distribution of *Hyalesthes obsoletus* Signoret in a neighbouring vineyard. IOBC/wprs Bulletin, 67: 295-298

Morris, T. R., Sims, C. A. and Striegler, R. K., 1987. Effect of cultivar, Maturity, cluster thinning and excessive potassic fertilization on yield and quality of Arkansas wine grapes. *Am. J. Vitic.*, **38**(4): 260-264.

Mortensen, E. and U. A. Randoff, 1940. Grape production inTexas Agric.Exp.station. Circ.No. 89.

Motsara, M. R., 2002. Available nitrogen, phosphorus and potassium status of Indian soils as depicted by soil fertility maps. *Fert. News*, **47** : 15-21.

Mundankar K.Y. and Karibasappa, G. S., 2006. Information System for Grape Germplasm in India. ISHS Acta Horticulturae 785: *International Symposium on Grape Production and Processing, held at Baramati* (Pune), Maharashtra, India During 6–11th February 2006.

Mundankar, K.Y. and Karibasappa, G.S. 2006. Information system for grape germplasm in India. ISHS Acta Horticulturae 785: *International Symposium on Grape Production and Processing, held at Baramati* (Pune), Maharashtra, India During 6–11th February 2006.

Mundankar, K.Y. Sawant, S.D. Sawant, I.S. and Sharma, J. 2006. In expert system for the management of powdery mildew disease of grapes in India Acta. 785, ISHS, Belgium. 297-300.

Mundy, D. C. and Agnew, R. H., 2002. Effects of mulching with vineyard and winery waste on soil fungi and botrytis bunch rot in Marlborough vineyards. *New Zealand Plant Protection.* 55: 135-138.

Mundy, D. C., Haycock, S. R., McLachlan, A. R. G., Wood, P. N. and Raw, V., 2012. Tendrils as a source of seasonal carryover of *Botrytis cinerea* in vineyards. *New Zealand Plant Protection* 65: 236-240

Mundy, D. C., 2008. A review of the direct and indirect effects of nitrogen on botrytis bunch rot in wine grapes *New Zealand Plant Protection.* 61: 306-310

Mundy, D. C., Agnew, R. H. and Wood. P.N., 2012. Grape tendrils as an inoculum source of *Botrytis cinerea* in vineyards - a review. *New Zealand Plant Protection.* 65: 218-227

Mundy, D.C. and Beresford, R.M., 2007. Susceptibility of grapes to *Botrytis cinerea* in relation to berry nitrogen and sugar concentration. *New Zealand Plant Protection.* 60: 123-127

Muneshwar Singh and Sekhon, 1991. DTPA extractable micronutrients cations in twenty soil series of India. *J. Ind. Soc. Soil Sci.*, **39** : 129-133.

Murat Kocturk, O., 2006. The Marketing of Table Grapes and Farmers Organization in Turkey (AlaOehir Case Study). *Journal of Biological Sciences*, 6: 961-967.

Murthy, B. N. S., Smita, R., 2006. Genotypic influence on regeneration in grape. ISHS Acta Horticulturae 785: *International Symposium on Grape Production and Processing, held at Baramati* (Pune), Maharashtra, India During 6–11th February 2006.

Muthuswamy, S. and Abdul Khader, J. B. M. M., 1960. Severity of infection of some fungus diseases in certain varieties of grape at Coimbatore. *South Indian Horticulture*, **8:** 24 – 25.

Myburgh, P.A., Van Zyfand, J.L. and Conradie, W.J., 1996. Effect of Soil Depth on Growth and Water Consumption of young. *Vitis vinifera* L. cv. Pinot noir. *S. Afr. J. Enol. Vitic.*, 17(2). 53-62

Nache Gowda, V., Keshava, S.A. and Shyamalamma, S., 2006. Growth, yield and quality of Bangalore blue grapes as influenced by foliar applied polyfeed and multi-k. ISHS Acta Horticulturae 785: *International Symposium on Grape Production and Processing*, held at Baramati (Pune), Maharashtra, India During 6–11th February 2006

Nahas E., 1996. Factors determining rockphosphate solubilization by microorganisms isolated from soil. *World J. of Microbio. and Biotech.*, **12** : 567-572.

Nanjundaswamy, M. J., 1996. Status and distribution of secondary and micronutrients in research farms of agro-climatic zone 8 of Karnataka. *M. Sc.* (*Agri.*) *Thesis*, Univ. Agric. Sci. Dharwad (India).

Narayana Reddy, Y., Bhavani Shankar, V. and Padmalatha, V., 2006. Studies on the possible involvement of polyamines in the shelf life of grapes (*Vitis vinifera* l.). ISHS Acta Horticulturae 785: *International Symposium on Grape Production and Processing, held at Baramati* (Pune), Maharashtra, India During 6–11th February 2006.

Narayana Reddy, Y., Ramprasad, V. and Reddy, M.G.D.M., 2006. Studies on extension of shelf life of grape through alternative oxidase inhibitors alone and in combination with polyamines. ISHS Acta Horticulturae 785: *International Symposium on Grape Production and Processing, held at Baramati* (Pune), Maharashtra, India During 6–11th February 2006.

Narendra Babu H. K. and Prakash, G. S. 1996. Studies on the response of grape genotypes in relation to different levels of moisture stress. *M. Sc. Thesis, IIHR, Bangalore*

Naresh Babu, Anamika Sharma and Shamsher Singh, 2004. Effect of different nitrogen doses and their split applications on growth, yield and quality of Jahajee banana (Musa paradisiacal). *South Ind. Hort.*, **52**(1-6): 35-40.

Narsian, V., Thakkar, J. and Patel, H. H., 1994. Isolation and screening of phosphate solubilizing fungi. *Ind. J. Microbiol.*, **34:** 113-118.

Naud, O., Cartolaro, P., Delière, L. and Léger, B., 2011. Modelling and experimenting crop protection decision workflows: some lessons from GrapeMilDeWS research. IOBC/wprs Bulletin, 67: 93-100

Neelam, T. and Meenu, S., 2003. Phosphate solubilization exopolysacharide production and indole acetic acid secretion by rhizobactera isolated from *Triogella foenumgraceum. Ind. J. Microbiol.*, **43**: 37-40.

Negi, J. P., 1999. *Indian Horticulture Database*, National Horticulture Board, Gurgaum, p. 37.

Nei, M. and Li, W.H., 1979. Mathematical model of studying genetic variation in terms of restriction endonucleases. *Proceedings of National Academy of Sciences, USA*, **76:**5269-5273.

Neilsen, Stevenson, D. S. and Gehringer, A., 1987. The effect of NPK fertilization on element uptake, yield and fruit composition of Foch grapes in British Columbia. *Can. J. Plant Sci.*, **67**(2): 511-520.

Nicholls, C.I., Altieri, M. A. and Ponti, L., 2006. Enhancing plant diversity for improved insect pest management in northern California organic vineyards. ISHS Acta Horticulturae 785: *International Symposium on Grape Production and Processing, held at Baramati* (Pune), Maharashtra, India During 6–11th February 2006.

Nijjar, G. S. and Singh, N., 1979. Effect of nitrogen, phosphorus, potassium fertilization on the growth, yield and quality of Thompson Seedless grapes. *Ind. J. Hort.*, **36**(1): 114- 118.

Nitant, H. C. and Dargan, K. S., 1974. Influence of nitrogenous fertilizers on yield and nitrogen uptake of wheat in saline sodic soils. *J. Ind. Soc. Soil Sci.*, **22:** 121-124.

Novello, V. and de Palma, L., 2006. Growing grapes under cover, ISHS Acta Horticulturae 785: *International Symposium on Grape Production and Processing, held at Baramati* (Pune), Maharashtra, India During 6–11th February 2006.

O'neill, N. R., Van Berkum, P., Lin, J.J., Kuc, J., Ude, G.N. Kenworthy, W. and Saunders, J. A., 1997. Application of amplified restriction fragment length polymorphism for genetic characterisation of *Colletotrichum* pathogens of alfalfa. *Phytopathology*, **87:** 745-750.

Obanor, F. O., Williamson, K., Mundy, D.C., Wood, P. N. and Walter, M. 2004. Optimisation of PTA-ELISA detection and quantification of *Botrytis cinerea* infections in grapes. *New Zealand Plant Protection.* **57:** 130-137

Okon, Y., 1985. *Azospirillum* as a potential inoculant for agriculture. Trends in Biotechnology, **3** : 223-228.

Okon, Y., Albrecht, S.K. and Burris, R.H., 1977. Methods for growing *Spirillum lipoferum* and for counting it in pure culture and in association with plants. *Appl. and Environ. Microbiol.*, **33** : 85-87.

Olmo, H. P., 1970. F.A.O. Report to the Govt. of India on grape culture. *United Nations Development Programme, Rome. Report No. T.A.* 2825-2899.

Olmo, H.P., 1970. Grape culture, FAO Report No. TA 2825

Oulkar, D. P., Banerjee, K., Ghaste, M. S., Ramteke, S. D., Naik, D. G., Patil, S. B., Jadhav, M. R. and Adsule, P. G., 2011. Multiresidue analysis of multiclass plant growth regulators in grapes by liquid chromatography/tandem mass spectrometry. *Journal of AOAC International*, 94(3):968-977.

Pakhale S. S., Find all citations by this author (default).

Palaniappan, R. and L. R. Varalakshmi, 2005.Soil and Water Management in Grapes under Adverse Conditions, *M. Sc. Thesis, IIHR, Bangalore.*

Palanichamy V., Jindal, P. C. and Singh Room, 2004. Studies on severity of pruning in grapes (*Vitis vinifera* L.) cv. Pusa Navrang - A teinturier hybrid. *Agricultural Science Digest*.24

Palma, M., Taylor, L.T., Zoecklein, B.W. and Douglas, L. S., 2000. Supercritical Fluid Extraction of Grape Glycosides. *Journal of Agricultural and Food Chemistry.* 48(3):775-779.

Pampanagouda, B., 2000. Studies on anthracnose of grape caused by *Sphaceloma ampelinum. M.Sc. (Agri.) Thesis submitted to the University of Agricultural Sciences, Dharwad*. pp.113.

Parashar, A., Awasthi, C. P., Singh, A. B. and Gupta, D. M. K., 1999. Effect of *Rhizobium* inoculation and phosphorus application on quality of broadbean (*Vicia faba* L.). *Legume Res.*, **22** : 162-166.

Parr, J. F., Hornick, S. B. and Kaufman, D. D., 1994. Use of microbial inoculants and organic fertilizers in agricultural production. In : *Proc. of the Int. Sem. on the Use of Microbial and Organic Fertilizers in Agric. Prod.*, Food and Fertilizer Technology Centre Publicaiton, Taiwan, pp. 13-18.

Parveen, G. and Alam, M.M., 1993. Bioactivity against plant pathogens. *In: Neem Research and Development* (*Eds. Randhawa. N.S. and Parmer, B.S.*) *Publication No.3 Society of Pesticides Science, India*, pp. 144 – 153.

Patel, V. B. and Chadha, K. L., 2002. Effect of graded doses of nitrogen on tissue nutrient status of grape. *Ind. J. Hort.*, **59**(3) : 247-252.

Patel, V. B., Chadha, K. L. and Jogendra Kumar, 2003. Micronutrient content in petiole and leaf lamina as affected by applied nitrogen in grape (*Vitis vinifera* L.) cv. Perlette. *Ind. J. Hort.*, **60**(3) : 214-221.

PATHAK, V. N., 1980. *Diseases of fruit crops*. Oxford and IBH Publishing Co. Pvt. Ltd.,. New Delhi, pp.309.

Patil, A. B., Efimov, V. N. and Orobekov, 1992. Effect of combined inoculations on mineral nutrition and yield of leguminous crops. *Ph. D. Thesis*, State Agrarian University, St. Petersburg, Russia.

Patil, A. J., Find all citations by this author (default).

Patil, A.B. and Patil, S., 2006. Chemical and organoleptic analysis of homemade wine from table grape cultivar Thompson seedless. ISHS Acta Horticulturae 785: *International Symposium on Grape Production and Processing, held at Baramati* (Pune), Maharashtra, India During 6–11th February 2006.

Patil, D. R., Sulikeri, G. S. and Patil H.B. 2006. Studies on the weed management practices in Thompson seedless grape vineyard. ISHS Acta Horticulturae 785: *International Symposium on Grape Production and Processing, held at Baramati* (Pune), Maharashtra, India During 6–11th February 2006.

Patil, D. R., Sulikeri, G. S., Patil, H. B. and Balikai, R.A., 2006. Studies on the integrated nutrient management in Thompson seedless grapes. ISHS Acta Horticulturae 785: *International Symposium on Grape Production and Processing*, held at Baramati (Pune), Maharashtra, India During 6–11th February 2006.

Patil, J. A,. Find all citations by this author (default).

Patil, R. B., Konde, B. K., Pawar, H. K. and Khade, K. K., 1989. Field studies on the effect of *Azotobacter chroococcum* and / or *Azospirillum brasilense* inoculations under fertilizer N levels on growth, yield and nitrogen fixation in ginger (*Zingiber officinales* Rose). *Paper Presented at State Level Seminar on Ginger Development*, held at Nahan (Dist. Sirmur), H. P., March, pp. 2-3.

Patil, S.G. Karkamkar, S.P. and Rao, V.S., 2006. Selection of new promising grape varieties for their commercial cultivation. ISHS Acta Horticulturae 785: *International Symposium on Grape Production and Processing, held at Baramati* (Pune), Maharashtra, India During 6–11th February 2006.

Patiram, R. C., Upadhyaya, C. S., Singh and Munnaram, 2000. Micronutrient cation status of Mandarin orchards of Sikkim. *J. Ind. Soc. Soil Sci.*, **48**(2) : 246-249.

Pauline Aliya, Sanyal, D. and Akali Sema, K., 2004. Influence of micronutrient application on quality of papaya cv. *Ranchi. Haryana J. Hort. Sci.*, **33**(1&2) : 25-26.

Pearson, B. C. and Goheen, A. C., 1998. *Compendium of grape diseases*. (Eds.) APS Press, USA, pp.93.

Peres, N. A., Timmer, L. W., Adaskaveg, J. E. and Correll, J. C., 2005. Life styles of *Colletotrichum acutatum*. *Plant Disaese*, **89:** 784-796.

Perry, RL, Lyda, S.O. and H.H.Bowen 1983. Root distribution in four Vitis cultivars. Plant & soil, 71:63-74.

Pikovskaya, R. I., 1948. Mobilization of phosphorus in soil in connection with vital activity of some microbial species. Mikrobiologiya, **17** : 362-370.

Piper, C. S., 1950. *Soil and Plant Analysis*, International Society Publications, New York. Piper, C. S., 1966. *Soil and Plant Analysis*, Hans Publishers, Bombay, p. 368.

Planzinski, J. and Rolfe, B. G., 1985. Interaction of *Azospirillum* and *Rhizobium* strains leading to inhibition of nodulation. *Appl. Env. Microbiol.*, **49** : 990-993.

Poi, S. C., Ghosh, G. And Kabi, M. C., 1989. Response of chickpea (*Cicer arietinum* L.) to combined inoculation with *Rhizobium*, phosphobacteria and mycorrhizal organisms. *Zentralblatt fur Mikrobidogie*, **144:** 249-253.

Porwal, S. and Sharma, M. P., 1974. Chemical control of grape anthracnose in Rajsthan. *Pesticides*, **8: 57**-58.

Possingham, J. V., 2006. Developments in the production of table grapes, wine and raisins in tropical regions of the world. ISHS Acta Horticulturae 785: *International Symposium on Grape Production and Processing*, held at Baramati (Pune), Maharashtra, India During 6–11th February 2006.

Possingham, J.V., 2006. Developments in the production of table grapes, wine and raisins in tropical regions of the world. ISHS Acta Horticulturae 785: *International Symposium on Grape Production and Processing, held at Baramati* (Pune), Maharashtra, India During 6–11th February 2006

Pozzebon, A., Pederiva, M., Moret, R. and Duso, C., 2011. The impact of a number of insecticides on *Empoasca vitis* populations in North-Eastern Italy. IOBC/wprs Bulletin, 67: 285-288

Prabhu, P. C. and Singaram, P., 2001. Effect of micronutrients on growth and yield of grapes cv. Muscat. *Madras Agric. J.*, **88**(1-3): 45-49.

Prakash, G. S. and Chadha, K. L., 1998. Root distribution and activity in grape cultivars as influenced by systems of Training, IIHR, Bangalore.

Prakash, G.S. 1998. Dogridge – a rootstock with multipurpose uses. Drakshavritta Souvenir, pp. 111-115.

Prasad, A. and Nirvan, R. S., 1965. Grape anthracnose and its control. *The Punjab Horticultural Journal*, **15:** 181 – 190.

Prasad, G. And Govindarajan, K., 2001. *Azospirillum* technology in Indian Agriculture: Potentials, current usage and limiting factors. *Biofert. News lett.*, December, 2001, pp. 13-16.

Prasad, R., 1991. Distribution of micronutrient cations in old alluvial rice soils of Bihar. *Annu.Agric. Res.*, **12** : 427-430.

Prasad. S. V. R., Rawal, R.D., Singh, R. and Murthy, B.N.S., 1992. Reaction of grape hybrids and their parents to anthracnose disease. *Proceedings of the International Symposium on Recent Advances in viticulture and Oenology*, held from 17 – 19th February, 129 – 132.

Prathibha, C. E., 1993. Interactions of *Azospirillum* sp. and *Pseudomonas striata* or *Gomus fasciculatum* on growth, yield and nutrient uptake of hybrid cotton. *M.Sc. (Agri.) Thesis*, Univ. of Agril. Sci., Dharwad (India).

Prathibha, C. K., Alagawadi, A. R. and Sreenivas, M. N., 1994. Establishment of inoculated organisms in rhizosphere and their influence on nutrient uptake and yield of cotton. *Karnataka J. Agric. Sci.*, **8** : 22-27.

Prathibha, C. K., Alagawadi, A. R. and Sreenivasa, M. N., 1995. Establishment of inoculated organisms in rhizosphere and their influence on nutrient uptake and yield of cotton. *Karnataka J. Agric. Sci.*, **8** : 22-27.

R.G. Somkuwar, J. Satisha and S.D. Ramteke, 2006. Effect of Different Rootstocks on Fruitfulness in Thompson Seedless (*Vitis vinifera* L.) Grapes. *Asian Journal of Plant Sciences*, **5**: 150-152.

Raghavendra K.P.and Sahijram Leela, 2005. Development of seedless grape hybrids through embryo rescue and their characterization using Molecular markers. IIHR, Bangalore

Rajan, 2003. Foot rot of black pepper (*Piper nigrum* L.) caused by *Phytophthora capsici*. Rao, C. V. S., and Sinha, B. N., 1963. Influence of fluorescent pseudomonads on growth and nodulation of lentil (*Lens esculentus*) in *Fusarium* infested soil. *Indian J. Microbiol.*, **39** : 23-29.

Rajpal Singh, Godara, N. R., Ahlawat, V. P. And Dahiya, S. S., 2001. Mineral composition of ber leaves as affected by foliar application of growth regulators and nutrients. *Haryana J. Hort. Sci.*, **30**(1&2): 10-11.

Ram Kumar, Shailendra Rajan and Negi, S. S., 2002. Evaluation of early-ripening grape genotypes under subtropical North Indian conditions. *Jour. Appl. Hort., 4(1)60-62*

Ram, R. A. and Bose, T. K., 2000. Effect of foliar application of magnesium and micronutrients on growth, yield and fruit quality of mandarin orange (*Citrus reticulate* Blamo). *Ind. J. Hort.*, **57**(3): 215-220.

Ramakrishnan, T. S. and Sundaram, N. V., 1955. Grapevine diseases and their control. *Madras Agricultural Journal*, **42:** 108-116.

Ramaswamy, G. R. and Anilkumar, T.B., 1995. Spore production and germination in carbendazim sensitive and resistant strains of *Gloeosporium ampelophagum*, the grapevine anthracnose fungus. *Mysore Journal of Agricultural Sciences*, **29:**54-60.

Ramchandani, A.G., Find all citations by this author (default).

Ramhari G. Somkuwar, Devanand D. Bondage, Manisha S. Surange, Sahadeo D. Ramteke. 2011. Rooting behaviour, polyphenol oxidase activity, and biochemical changes in grape rootstocks at diff erent growth stages. *Turk. J. Agric.* 35: 281-287

Ramming David W. 2010. Antioxidant activity and phenolic content of 16 raisin grape (*Vitis vinifera* L.) cultivars and selections. *Food Chemistry*, 121(3):740-745.

Ramteke, S. D., Somkuwar, R. G. and Adsule. P. G., 2006. Effect of CPPU on bunch and berry development in Thompson seedless grafted on Dogridge rootstock. ISHS Acta Horticulturae 785: International Symposium on Grape Production and Processing

Ramteke, S. D., Somkuwar, R.G. and Adsule, P.G., 2006. Use of bio-regulators to improve the quality of sharad seedless grapes. ISHS Acta Horticulturae 785: International Symposium on Grape Production and Processing

Rana, R. K. and Chandel, J. S., 2003. Effect of biofertilizers and nitrogen on growth, yield and fruit quality of straw berry. *Prog. Hort.*, **35**(1): 25-30.

Rana, R. S. and Sharma, H. C., 1999. Effect of $FeSO_4$ sprays on yield and quality of grapes. *Madras Agric. J.*, **67**(2): 136-137.

Rangaswamy, R., 2006. A *text book of Agricultural Statistics*. New Age International (P) Ltd. Publishers, New Delhi.p.496.

Rao, K. C. and Satyanarayana, C., 1991. Epidemiology of anthracnose of grape (*Vitis vinifera* L.) caused by *Elsinoe ampelina* around Hyderabad. *Indian Phytopathology*, **44**: 501-503.

Rao. M. M., 1981. Factors governing physiological deterioration and wastage of grapes in handling and cool storage. Drakshavritta, 1 (5): 61

Rashmi, K., 2004. Biodiversity of soil bacteria from evergreen forests of Western Ghats of Uttara Kannada district with special reference to P-solubilizers. *M. Sc.* (*Agri.*) *Thesis*, Uni. Agric. Sci. Dharwad.

Rather, A.Q. 2006. Management of phytophagous and predatory mites in vineyards of Jammu and Kashmir, India. ISHS Acta Horticulturae 785: *International Symposium on Grape Production and Processing, held at Baramati* (Pune), Maharashtra, India During 6–11th February 2006.

Rattanpal, H. S., Shoba Rani, Anil Kumar and Dhaliwal, 2005. Effect of potassium and 2. 4-D spray on physical parameters of kinnow fruits. *Haryana J. Hort. Sci.*, **34**(3-4): 222-223.

Ravi Kher, Shanoobhat and Wali, V. K., 2005. Effect of foliar application of GA3, NAA and CCC on physico-chemical characteristics of guava cv. Sardar. *Haryana J. Hort. Sci.*, **34**(1-2): 31-32.

Ravikumar, M., Vijayakumar, R. M., Suresh, J., Kumar N. and. Shahul Hameed, M., 2006. Varietal screening and optimization of quality wine production technology in grapes (*Vitis vinifera* l.). ISHS Acta Horticulturae 785: *International Symposium on Grape Production and Processing*. held at Baramati (Pune), Maharashtra, India During 6–11th February 2006

Rawal, R. D. and Muniyappa, N. C., 1981. *Proceedings of Third International Symposium on Plant Pathology*, IARI, New Delhi.

Rawal, R. D. Venugopalan, R. and Saxena, A.K. 2006. A statistical model for predicting the outbreak of powdery mildew in grapes. Acta. 785. ISHS, Belgium.

Rawal, R. D., 1987. Development of grape sprayed with systemic and non-systemic fungicides. *Plant Disease Research*, **2:** 16-19.

Rawal, R. D., 1995. *Major fungal diseases of some tropical fruits:* In *Advances in Horticulture* Vol. Ed: K. L. Chadha and O. P. Pareek, Malhotra Publishing House, New Delhi, pp 396-424.

Rawal, R. D., Venugopalan, R. and Saxena, A.K. 2006. A statistical model for describing the incidence of downy mildew in grapes. Acta. 785. ISHS, Belgium.

Reinhold, B., Hurek, T., Fendrik, I., Pot, B., Gillis, M., Kersters, K., Thielemans, S. and Deley, J., 1987. *Azospirillum nalopraeferans* sp. nov. a nitrogen fixing organism associated with roots of kallar graes (*Leptochloa fusca* (L.) Kunth). *Int. J. Syst. Bactriol*, **37:** 43-51.

Rewal, H. S. and Aulakh, P. S., 1990. Development of anthracnose epidemic on grapevine in south western regions of Punjab. *Annals of Biology*, **6:** 188-187.

Reynolds Andrew G. and Justine E. Vanden Heuvel, 2009. Influence of Grapevine Training Systems on Vine Growth and Fruit Composition: A Review. *Am. J. Enol. Vitic.* 60:251-268

Rives, M. 1971. Statistical analysis of rootstock experiments as providing a definition of the terms vigour and affinity in grapes. Vitis, 9:28.-290.

Rohlf, F.J., 1998. *NTSYS-PC Numerical Taxonomy and Multivariate Analysis Version 2.0. Appl. Biostat., Inc. New York.*

Rossi, V., Caffi, T., Legler, S. E., Bugiani, R. and Frisullo, P., 2011. Dispersal of the sexual stage of Erysiphe necator in northern Italy. *IOBC/wprs, Bulletin.*67: 115-121.

Sabourry, A. B., 1978. Studies on anthracnose of grapevine (*Elsione ampelina* (de Bary) Shear) *Ph. D. thesis.* Punjab Agric. Univ., Ludhiana. pp.99.

Saha, M. N., Mandal, A. K. and Mandal, L. N., 1982. Distribution of Fe, Mn, Cu and Zn in soils of jute growing areas of Assm and West Bengal. *J. Ind. Soc. Soil Sci.*, **30:** 140-145.

Sakthivel, N. and Gnanamanickam, S. S., 1986. Evaluation of *Pseudomonas fluorescens* for suppression of sheath rot disease and enhancement of grain yields in rice (*Oryza sativa* L.). *Appl. Environ. Microbiol.*, pp. 2056-2059.

Salakinkop, S.R., Shivaprasad, P., Raghuramulu, Y. and Paneer Selvam, P., 2003. Bio inoculation for improving the vigour of coffee seedlings. *6th Intl. PGPR Workshop* Calicut.

Salisbury, F. B., 1994. The role of plant hormones in plant environment interactions. Ed. RE, Wilkinson, pp. 39-81. Marcel Dekker, New York, USA.

Salunkhe, N. M., Tambe, T.B. and Kadam, J.H., 2006. Influence of pruning times and various chemicals on yield and quality of Thompson seedless grapes. ISHS Acta Horticulturae 785: *International Symposium on Grape Production and Processing, held at Baramati* (Pune), Maharashtra, India During 6–11th February 2006.

Sandhu, A.S., Jawanda, J. S. and Cheema, S.S., 1981. Evaluation of F1 grape hybrids against anthracnose and *Cercospora* leaf spot diseases. *Punjab Horticultural Journal*, **21:** 192 – 194.

Sanghavi, K. V. and Nijjar, G. S., 1978. Effect of factorial combination of nitrogen, phosphorus and potassium on the yield and quality of Himrod grape. *Punjab Hort. J.*, **18**(1-2) : 48-51.

Sanjay Kumar and Pathak, R. A., 1992. Effect of foliar application of nutrient on the yield and quality of grapes cv. Perlette. *Prog. Hort.*, **24**(1-2) : 13-16.

Sapatnekar, H. G., Rasal, H. and Patil, P. L., 2001a, Effect of superphosphate and phosphate solubilizers on yield of greengram. *J. Maharashtra Agric. Univ.*, **26**(1): 120-121.

Sapatnekar, H. G., Rasal, P. H. and Patil, P. L., 2001b. Efficient utilization of P-sources through P-solubilizers in wheat. *J. Maharashtra Agric. Univ.*, **26**(1) : 121-123.

Saralamma P., and Singh Rajendra, 1986. Effect of temperature and insoluble solids added to clarified musts on fermentation rate and wine quality. *M. Sc. Thesis, IIHR, Bangalore.*

Saraswathy, S., Balakrishnan, K., Anbu, S., Azakia, Manavalan, R. S. and Thangaraj, T., 2004. Effect of zinc and boron on growth, yield and quality of sapota cv. PKM-1. *South Ind. Hort.*, **52**(1-6) : 41-44.

Sastry, M.N.L., Lingaraju, S. and Naik, S. T., 1989. Efficiency of some fungicides in the control of anthracnose of grape vine, *Pesticides*, **23:** 29 – 30.

Satisha, J. and Adsule, P. G., 2006. Rooting behavior of grape rootstocks in relation to IBA concentration and biochemical constituents of mother vines. ISHS Acta Horticulturae 785: *International Symposium on Grape Production and Processing, held at Baramati* (Pune), Maharashtra, India During 6–11th February 2006.

Satisha, J. and Prakash, G. S., 2005. Studies on genetic variability in water use efficiency of grape varieties, rootstocks and their stionic combinations. *M. Sc. Thesis, IIHR, Bangalore.*

Sattar, M. A. and Gaur, A. C., 1987. Production of auzins and gibberellins by phosphate dissolving microorganisms. *Zentralblatt fur Mikrobiologie*, **142:** 393-395.

Savalgi, V. V., Savalgi, V. and Pakale, N., 1994. Stimulation of peanut – *Rhizobium* symbiosis and plant productivity by *Azospirillum* strains in red loam soil. *Legume Res.*, **17:** 75-78.

Sawant, I. S., Sawant, S. D. and Adsule, P. G., 2006. Studies on fungi associated with post-harvest decay in table grapes from Maharashtra. ISHS Acta Horticulturae 785: *International Symposium on Grape Production and Processing, held at Baramati* (Pune), Maharashtra, India During 6–11th February 2006.

Sawant, Indu, S., and Sawant, S.D., 2006. Potential of *Trichoderma* spp. for management of anthracnose disease of grapevines: In the book of Abstracts on *International*

Symposium on Grape Production and Processing, at Baramati, Maharashtra (India) held on Feb 6-11, 2006, pp. 116.

Sawant, S. D. and Sawant, I. S. 2008. Potential of *trichoderma* spp. and hot water treatment for control of grapevine anthracnose. Acta. 785, ISHS, Belgium. 301-304.

Sawant, S. D. and Sawant, I. S. 2008. Use of potassium bi-carbonates for the control of powdery mildew in table grapes. Acta. 785, ISHS, Belgium

Schaefer, W.W., 2006. Hot climate/tropical wine making. ISHS Acta Horticulturae 785: *International Symposium on Grape Production and Processing, held at Baramati* (Pune), Maharashtra, India During 6–11th February 2006.

Schnee, S., Jolivet, J. and Calonnec, A., 2011. Consideration of dynamical plant-pathogen interactions for an improved management of powdery mildew epidemics in grapevine. *IOBC/wprs, Bulletin.*67: 131-138.

Seeley, H. W. and Vandemark, P. J., 1981. *Microbes in Action – A Laboratory Manual of Microbiology*, Freeman and Company, San Francisco, USA, p. 388.

Seeley, H. W. and Vandemmark, P. J., 1970. *Microbes in Action – A Laboratory Manual of Microbiology*, D. B. Taraporevala Sons and Company Pvt. Ltd., Bombay (India), pp. 86-95.

Seeliger, M. T. and French, R. J., 1971. Changes in soil chemical properties in a long-term fertilizer trial in a non-irrigated vineyard. *Aust. J. Agric. Res.*, **22** : 931-940.

Sentenac, G., Kuntzmann, P., Perez, L., Gili, A. and Kreiter, P., 2011. Biological control of Heliococcus bohemicus Sulc. (Hemiptera: Pseudococcidae) with the natural enemy Ericydnus sipylus (Walker) (Hymenoptera: Encyrtidae). IOBC/wprs Bulletin, 67: 31-40.

Shanmngavelu, K.G., Viticulture in India. Agro Botanical Publishers, Bikaner.

Shanmugavelu, K.G., 1989. *Viticulture in India*, Saraswathi Agrobotanicals Publishers, New Delhi, pp.380

Shanmugavelu, K.G., 2003. *Grape cultivation and processing*. Agobios (India). pp. 289 Sharvelle, E.G., 1961. *The nature and use of modern fungicides*. Burgess Publishing Co., Minnesota, USA. pp. 308.

Sharma Debi and Awasthi Mahesh D., 2003. Behaviour of forchlorfenuron residues in grape, soil and water. *Chemosphere.* **50**: 589–594

Sharma Jagdev and Upadhyay, A. K. 2006. Rootstock effect on Tas-a-ganesh (V *itis vinifera* l.) for sodium and chloride uptake. ISHS Acta Horticulturae 785: *International Symposium on Grape Production and Processing, held at Baramati* (Pune), Maharashtra, India During 6–11th February 2006.

Sharma Jagdev and Shikhamany, S.D., 2006. Petiole nutrient standards for Thompson seedless vines on dogridge rootstock. ISHS Acta Horticulturae 785: *International Symposium on Grape Production and Processing*, held at Baramati (Pune), Maharashtra, India During 6–11th February 2006.

Sharma Jagdev, Shikhamany, S.D., Singh, R. K. and Upadhyay, A. K., 2006. Irrigation scheduling for improving water use efficiency in drip irrigated Thompson seedless grape grown on dog ridge rootstock. ISHS Acta Horticulturae 785: *International Symposium on Grape Production and Processing*, held at Baramati (Pune), Maharashtra, India During 6–11th February 2006.

Sharma Jagdev, Upadhyay, A.K., Shikhamany, S.D. and Singh, R.K., 2006. Effect of fertilizer application through irrigation water on Thompson seedless grape yield and fertilizer use efficiency. ISHS Acta Horticulturae 785: *International Symposium on Grape Production and Processing*, held at Baramati (Pune), Maharashtra, India During 6–11th February 2006.

Sharma, A., 2007. Scavenging effect of Indian grape polyphenols on 2,2'-diphenyl-1-picrylhydrazyl(DPPH) radical by electron spin resonance spectrometry. Find all citations by this author (default).

Sharma, J. K. and Sindhu, D. S., 2005. Effect of potassium on quality improvement of grape cv. perlette. *Haryana J. Hort. Sci.*, **34**(1-2) : 3-4.

Sharma, J. R., Gupta, R. B., Surender Singh and Bhatia, S. K., 2005. Growth parameters of ber cv. Umran as influenced by soil application of phosphorus and zinc. *Haryana J. Hort. Sci.*, **34**(3-4): 227-228.

Shear, C. L., 1929. *Grape anthracnose in America Official Representation of International Congress on Viticulture*, pp. 111– 117.

Shelke, R.B., Ingle, S. T., Patil, S. R. And Sonalkar, U.V., 1997. Susceptibility to grape varieties to major disease in Vidarbha region. *Journal of Soil and Crops*, **7:** 150 – 152.

Shikhamani, S.D. and Satish. Grape on root stock, NRC, grape, Pune.

Shikhamany, S. D. and Sharma Jagdev, 2006. Interaction of sodium and potassium and potassium use efficiency in Thompson seedless grape. ISHS Acta Horticulturae 785: *International Symposium on Grape Production and Processing*, held at Baramati (Pune), Maharashtra, India During 6–11th February 2006.

Shikhamany, S. D., Chadhal, K. L. and Chittiraichelavan, R., 1985. Determination of critical levels of petiole nutrient contents for Thompson Seedless grape (*Vitis vinifera* L.). *Ind. J. Hort.*, **42** : 8-11.

Shikhamany, S. D., Chittiraichelavan, R. and Chadha, K. L., 1989. Effect of varying level of nitrogen and potash on petiole nutrient contents in Thompson Seedless grape (*Vitis vinifera* L.). *Ind. J. Hort.*, **45**(3&4) : 180-188.

Shikhamany, S. D., Chittiraichelvan, R. and Chadha, K. L., 1982. Nutritional studies on grapes. *Annu. Rep.*, Indian Institute of Hort Res. (ICAR), 1982.

Shikhamany, S.D., Somkuwar, R.G. and Venugopalan. R., 2006. Evaluation of canopy efficiency using leaf area index in Thompson seedless vines. ISHS Acta Horticulturae 785: *International Symposium on Grape Production and Processing, held at Baramati* (Pune), Maharashtra, India During 6–11th February 2006.

Shitole, D. M., Chandra Mouli, B., Sharma, N. N. and Mithyantha, M.S., 2003. Evaluation of new fungicide combinations against anthracnose (*Gloeopsorium ampelophagum*) disease of grapevine. *Pestology*, **27:** 18-19.

Shitole, D.M., Chandra Mouli, B., Sharma, N. N. and Mithyantha, M.S. 2005. Evaluation of Samarth and RIL-006/C1 against anthracnose, downy mildew and powdery mildew diseases of grape. *Pestology*, **29:** 39-40.

Shivakumar, A. P., 2006. Management of post harvest diseases of chilli. *M.Sc. (Agri.) Thesis submitted to the University of Agricultural Sciences, Dharwad*. pp.202.

Shivakumaraswamy, C.R. 2001. *M.Sc. (Agri.) Thesis*, Univ. of Agril. Sci., Dharwad (India).

Shivaprasad, C. R., Reddy, R. S., Seghal and Velayutham, M. M., 1998. Interpretation for soil and land resource management. *Nation. Bureau of Soil Survey and Land Use Planning*, Nagpur, **47:** 65-68.

Shoef, T. W. and Lium, B. M., 1976. Improved extraction of chlorophyll A and B from algae using dimethyl sulfoxide, *Lomnol. Oceunogr*, **21:** 926-928.

Sicard, D., Michalakis, Y., Dron, M. and Neema, C., 1997. Genetic diversity and pathogenic variation of *Colletotrichum lindemuthianum* in the three centres of diversity of its host *Phaseolus vulgaris*. *Phytopathology*, **87:** 807-813.

Sindhu, A. S., Tomer, N. S, Chahil, B. S. and Brar, J. S., 2002. Effect of N, P and K on physico-chemical characteristics of grapes (*Vitis vinifera* L.) during development. *Haryana J. Hort. Sci.*, **34**(1-2): 19-22.

Sindhu, P. C., Ahlawat, V. P. and Nain, A. S., 1999. Effect on yield and fruit quality of grapes (*Vitis vinifera* L.) cv. Perlette. *Haryana J. Hort. Sci.*, **28**(1&2): 19-21.

Singh Virendra, Mohammed Shafaat and Verma Plyush. 2003. Effect of Nitrogen on Growth, Yield and fruit quality of Grape cv. Thompson seedless. *International Journal of Tropical Agriculture.* 21 (1-4): 59-67

Singh, A. K., Bhatiya, S. P., Awasthi, R. P. and Bhutani, V. P., 2005. Influence of plant ioregulators and nutrients on fruit quality of apple. *Haryana J. Hort. Sci.*, **34**(3-4): 211-213.

Singh, B.P. and Joshi, L.K., 1971. Studies on the diseases of pan (*Piper betle* L.) in Jabalapur (MP) assessment of losses due to *Colletotrichum capsici*. *Science and Culture*, **37:** 581- 582.

Singh, G. C. P. and Gupta, S. P., 2000. Maintenance of guava orchards through micro and secondary nutrient management. *Haryana J. Hort. Sci.*, **29**(1&2) : 32-34.

Singh, R. S., 2000. *Diseases of fruit crops*. Oxford and IBH Publishing Co. Pvt. Ltd., New Delhi. pp. 311.

Singh, R.S. and Pande, K.R., 1966. Effect of green and mature plant residues and compost on population of *Pythium aphanidermatum* in soil. *Indian Phytopathology*, **19:** 367 – 371.

Singh, S. and Kapoor, K. K., 1994. Solubilization of insoluble phosphates by bacteria isolated from different sources. *Environment Ecology*, **12** : 51-55.

Singhrot, R. S., Singh, S. P. and Suhag, L. S., 1982. Effects of anthracnose disease on productiveness of Thompson Seedless cultivar of grape. *Indian Journal of Mycology and Plant Pathology*, **12:** 13-15.

Singram, P. and Prabhu, P. C., 2001. Effect of zinc and boron on growth and quality of grapes cv. Muscat. *Madras Agric. J.*, **88**(4-6) : 233-236.

Sinha, A. K., Singh, C. and Jain, B. P., 1999. Effect of plant growth substances and micronutrients in fruit set, fruit drop, fruit retention and craiking of litch cv. Purbi. *Ind. J. Hort.*, **56**(4): 309-311.

Sivaprasad, P., Sasikumar, S., Joseph, P. J., Menakumari, K. S. and Shahul Hameed, 2003. Characterization of efficiency testing of *Azospirillum* isolates from Acid sulphate soils – *6th International PGPR Workshop*, 5-10 October 2003, Calicut, India.

Snedecor, G.W. and Cochran, W. G., 1967. *Statistical Methods*. Oxford and IBH Publshing Co. New Delhi.pp.593.

Sohan Singh, A. S., Bindra and Brar, S. S., 1985. Nutrients removal by grape vines (*Vitis vinifera*) cv. Perlette. *J. Res.* Punjab Agric. Univ., **22**(4): 667-700.

Sokhi, S. S. and Clerjeau, M., 2005. Improvement in the control of anthracnose and powdery mildew diseases of grapevine by rational and need based use of fungicides. *http://www.cefipra.org./cefipra/project*.

Somkuwar, R. G., Satisha, J., Ramteke, S. D. and Sharma, J., 2006. Effect of rootstocks and pre-harvest treatments on storage life of Thompson seedless grapes. *ISHS Acta Horticulturae* 785: *International Symposium on Grape Production and Processing,* held at Baramati (Pune), Maharashtra, India During 6–11th February 2006.

Somogyi, M., 1952. Note on sugar determination. *J. Biol. Chem.*, **200** : 145-154.

Sonawane, R. B. and Konde, B. K., 2000. Response of grape vine to dual inocualtion of VAM fungi and diazotrophos. *J. Maharashtra Agric. Univ.*, **22**(3) : 356-357.

Song, Gi-Cheol, 2006. Grape breeding, cultivation and processing in South Korea. ISHS Acta Horticulturae 785: *International Symposium on Grape Production and Processing, held at Baramati* (Pune), Maharashtra, India During 6–11th February 2006.

Sontakke, A. V., and Norisawa, 2004. Effect of fungus *Heberoconium chaebospira* suppression of clubroot disease and verticillium yellowish in Chinese cabbage. *Hort. Sci.*, **22**(3) : 179-182.

Sreenivasaprasad, S., Mills, P.R., Meehan, B.M. and Brown, A. E., 1996. Phylogeny and systematic of 18 *Colletotrichum* species based on ribosomal DNA spacer sequences. *Genome*, **38:** 499-512.

Sridhar, T. S. and Sohi, H. S., 1970. Relative resistance and susceptibility of different grape varieties against *Gloeosporium ampelophgum*. *Indian Journal of Horticulture*. **27:** 161 – 163.

Sridhar, T.S. and Sohi, H.S., 1976. Control of diseases. *Proceedings of All India Coordinated Fruit Improvement Project Workshop*. Hyderabad. pp.51.

Srikant Kulkarni, Yashoda R. Hegde and Ramesh V. Kota., 2007. Molecular variability in *Plasmopara halstedii* causing downy mildew of sunflower. *Indian Phytopathology*, **60:** 48-51.

Srinivasarao, C. H., Subbarao, A. and Ganeshmurthy, A. N., 1995. Status and desorption kinetics of potassium in some swell-shrink soils. *J. Ind. Soc. Soil Sci.*, **43** : 356- 360.

Srivastava, K. K. and Soni, S. L., 1989. Effect of N, P and K on growth, yield and some physical characteristics of perlette grapes (*Vitis vinifera* L.). *Haryana J. Hort. Sci.*, **18**(3-4) : 192-196.

Steel, C. C. and Greer, D. H., 2006. Effect of Climate on Vine and Bunch Characteristics: Bunch Rot Disease Susceptibility, Acta. 785, ISHS, Belgium.

Stephen B. Renquist, 2007. Creative Marketing for a Small Wine Grape Region. *Journal of Extension*. 45(6).

Stevens R.M., and Harvey, G. 1994. The relative effects of waterlogging and salinity on the growth of potted vines. In the Proceeding of the 4th International Symposium on Grapevine Physiology. *Giovanni Dalmaso Foundation,* Torino, pp 607-610.

Stevens R.M., Harvey, G., Partington, D.L. and Coombe, B.G. 1999. Irrigation of grapevines with saline water at different growth stages. I. Effects on soil salinity and sodicity, vegetative growth and yield. *Aust. J. Agric. Res.* 50.

Stevens, R. M. and Harvey, G., 1996. Soil water depletion rates under large grapevines. *Australian Journal of Grape and Wine Research.* 2:155–162.

Stevens, R. M. and Cole, P. 1987. Grape must composition depends on irrigation management. *In Proceedings of the Sixth Australian Wine Industry Technical Conference* (Lee, T. ed). Australian Industrial Publishers, Adelaide, pp 159–64.

Stevens, R. M. and Harvey, G. C. 1990. Grapevine response to transient soil salinization. In Management of Soil Salinity in South East Australia (Humphreys, E., Muirhead, W.A., and Van Der Lelij, A. eds). *Australian Society of Soil Science,* Riverina Branch, pp 211-19

Stevens, R.M. and Prior, L.D., 1994. The effect of transient waterlogging on the growth, leaf gas exchange, and mineral composition of potted Sultana grapevines. *American Journal of Enology and Viticulture* 45:285–290.

Stevens, R.M., and Douglas, T., 1994. Distribution of grapevine roots and soil salt under drip and full-cover microjet irrigation systems. *Irrigation Science.* 15:147–52.

Stevens, R.M., Harvey, G. and Davies, G., 1996 Separating the effects of foliar and root salt uptake on growth and mineral composition of four grapevine cultivars on their own-roots and on Ramsey Rootstock. *Journal of the American Society of Horticultural Science.* 121:569-575.

Stewart, 1988. Distribution, proportion and management of Vertisols of India. *Adv. Soil Sci.*, **8**: 151-200.

Subbaiah, B. V. and Asija, G. L., 1956. A rapid procedure for the estimation of available nitrogen in soils. *Current Sci.*, **25** : 259-260.

Subbarao, N. S., 1995. Soil microorganisms and plant growth, Oxford and IBH Publishing Company Private Ltd., New Delhi, p. 174.

Suhag, L. S., 1971. Studies on anthracnose and its control in north India. *Ph.D. Thesis submitted to Haryana Agricultural University, Hissar*. Pp.254.

Suhag, L. S. and Grover, R. K., 1972a, Distribution of grape anthracnose in Haryana and estimation of losses in term of yield. *Haryana Journal of Horticultural Science*, **1:** 83 – 86.

Suhag, L. S. and Grover, R. K., 1972b, *Third intern. Symp. Subtrop. Trop. Hort.*, Bangalore, pp. 207-208.

Suhag, L. S. and Grover, R. K., 1973. Anthracnose of grapevine can be controlled. *Indian Farming*, **23:** 22– 24.

Suhag, L. S., and Kaushik, J. C., 1982. Studies on four pathogenic fungi on grapevine in North India. *Indian Phytopathology*, **35:** 344-346.

Suhag, L. S., Grover, R. K. and Kaushik, J.C., 1982. Variability in *Sphaceloma ampelinum* and the performance of grapevine cultivars against the pathogen. *Indian Phytopathology*, **35:** 526-528.

Suhag, L.S. and Grover, R. K., 1977a, Epidemiology of grape vine anthracnose caused by *Sphaceloma ampelinum* in North India. *Indian Phytopathology*, **30:** 460– 465.

Suhag, L.S. and Grover, R. K., 1977b, Over-wintering and control of *Elsinoe ampelina*, the cause of grape anthracnose. *Viticulture in Tropics*, Chadha, K. L., Eds. *Horticulture Society.India*, pp. 294-299.

Sukhatme, P.V. and Amble, A.N., 1985. *Statistical methods for Agricultural workers*. Publication and Information Div., ICAR, New Delhi. pp. 553.

Suleman Mohammed, Dal Singh and Ahlawat, V. P., 1993. Growth, yield and quality of grapes as affected by pruning and basal application of potassium. *Haryana J.*

Suleman Mohammed, Dal Singh and Ahlawat, V. P., 1996. Growth, yield and quality of grapes as affected by pruning and foliar spray of potassium sulphate. *Haryana J. Hort. Sci.*, **25**(3) : 80-85.

Sunaina, V. and Ajay, S., 2005. Effect of plant growth promoting rhizobacteria on black scurf disease of potato and ability to promote their growth. *J. Biol. Control*, **19:** 47-50.

Sundararaj, N., Nagaraju, S., Venkataramu, M.N and Jaganath M.K., 1972. Design and Analysis of field experiments. Pub : Univ. of Agril. Sciences, Bangalore.

Suneesh, K., 2004. Biodiversity of fluorescent pseudomonads in soils of moist-deciduous forests of Western Ghats of Uttar Kannada district. *M. Sc (Agri). Thesis*, Univ. Agric. Sci., Dharwad (India).

Surange and Kumar, 1993. Solubilization of insoluble phosphates by fungi and bacterial isolates from different sources. *Environ. Ecol.*, **12** : 51-55.

Surender Kumar, Chharia, A. S. and Godara, R. K., 2004. Effect of self-life of grapes cv. Delight as affected by various growth regulators applied at berry set stage. *Haryana J. Hort. Sci.*, **33**(1&2) : 65-66.

Suresh, E. R. and Ethiraj, S., 1987. Effect of grape maturity on the composition and quality of wines made in India. *Am. J. Enol. Vitic.* **38** (4) 329-331

Suresh, G. and Murthy S. Krishna, 2011. Marketing of Grapes in Tamil Nadu: A Case Study of Coimbatore District. *The IUP Journal of Marketing Management*. 10 (3) 72-82.

Suryakala, D., Umamaheshwari Devi, P. and Vijaya Lakshmi, K., 2004. Chemical characterization and *in vitro* antibiosis of siderophores of rhizosphere florescent pseudomonads. *Ind. J. of Microbiol.*, **44** (2) : 105-107.

Sy, A., Girud, E., Jourand, P. Garcia, N., WIllems, A., Lajudie, P. D. E., Prin, Y., Neyra, M., Gills, M., Catherine, B. M. and Dreyfus, B., 2001. Methylotrophic methylobacterium bacteria nodulate and fix atmospheric nitrogen in symbiosis with legumes. *J. Bacteriol.*, **183:** 214-220.

Takeoka Gary R., Find all citations by this **author** (default).

Tamhankar, S. A., Argade, N. C., More, M. N., Dhanorkar, V.M., Patil, S.G., Rao, V.S., Karibasappa, G.S. and Agrawal, D.C., 2006. DNA profiling of the grape varieties grown in India using ISSR Markers. ISHS Acta Horticulturae 785: *International Symposium on Grape Production and Processing, held at Baramati* (Pune), Maharashtra, India During 6–11th February 2006.

Tamibar and Okon, Y., 1993. Tryptophan conversion ton indole-3-acetic acid via indole-3-acetamide in *Azospirillum brasilense* sp. 7. *Can. J. Microbiol.*, **39** : 81-86.

Tamilvendan, K. and Purushothaman, D., 1996. Developing phosphorus solubilizing strains in *Azospirillum*. *Proc. of the National Seminar on Microorganisms in Sustainable Agriculture*, Madurai, pp. 79-89.

Tan, M. K., Timmer, L. W., Broadbent, P., Priest, M. and Cain, P., 1996. Differentiation by molecular analysis of *Elsinoe* spp. causing scab diseases of citrus and its epidemiological implications. *Phytopathology*, **86:** 1039-1044.

Tanaka, S., Katumotok and Yukawa, Y., 1974. Studies on the grape anthracnose *Elsinoe ampelina* cultural aspects of the fungus. *Bull. Fac. Agric. Yamaguchi Chiu No.***25:** 917-946.

Tanjeet Singh Chahal and Bal, J. S., 2005. Effect of NPK a growth and yield of cape groseberry (*Physalis peruviana* L.) *Haryana J. Hort. Sci.*, **34**(3-4) : 246-249.

Tarafdar, J. C. and Marschner, H., 1995. Dual inoculation with *Aspergillus fumigatus* and *Glomus mosseae* enhances biomass production and nutrient uptake in wheat (*Triticum aestivum* L.) supplied with organic phosphorus as Na-phytate. *Plant Soil*, **173** : 97-102.

Taupeau, V., 1986. Sensibilite de la vigne a Panhracnose (*Gloeopsorium ampelophagum*)- Essan de muse en evidence de variabilities clonale et somaclonale. *M.Sc. Thesis INRA de Montpelliter, France*. Pp.197.

Thakur Anirudh, Arora, N.K., Sidhu, A.S. and Brar, J.S., 2006. Effect of potassium sprays on the quality of perlette grapes. ISHS Acta Horticulturae 785: *International Symposium on Grape Production and Processing*, held at Baramati (Pune), Maharashtra, India During 6–11th February 2006

Thiéry Denis, 2011. Gaps in knowledge for modern integrated protection in viticulture: lessons from controlling grape berry moths. IOBC/wprs Bulletin, 67: 305-311

Thiéry, D., Delbac, L., Villemant, C. and Moreau, J., 2011. Control of grape berry moth larvae using parasitoids: should it be developed. IOBC/wprs Bulletin, 67: 189-196

Thind, S. K., Monga, P.K., Nirmaljit Kaur and Arora, J.K., 2001. Efficacy of fungicides in controlling grape anthracnose. *Journal of Mycology and Plant Pathology*, **31:** 211-212.

Thind, S.K., Monga, P.K., Nirmaljit Kaur, P.K. and Kumar, H., 1997. Evaluation of grape varieties against anthracnose and its fungicidal control. *Plant Disease Research*, **12**: 99 – 100.

Thind, T. S. and Bedi, J. S., 1988. Comparative efficacy of some fungicides against anthracnose of grapes in Punjab. *Indian Phytopathology*, **41:** 416 – 419.

Thind, T. S. and Mohan, C. 1998 Management of carbendazim resistance in *Gloeopsorium ampelophagum* in India. *Proceeding of 9th International Congress on Pesticides Chemistry,* The Food Environment Challenge, London Vol.1 pp. 4D-022.

Thind, T. S., Mohan Chander, Clerjeau, M., Sokhi, S. S. and Jailloux, F., 1998. Management of carbendazim resistance in *Gloeosporium ampelophagum* with new fungicides. *Indian Phytopathology Golden Jubilee Proceedings*, pp163-164.

Thind, T. S., Munshi, G. D., Sokhi, S.S., Mohan, Chander and Grewal, R.K., 1992. Seasonal periodicity of grapevine diseases in Punjab and their management. Proceedings of *Intentional Symposium on Recent Advances in Viticulture and Oenology* held at Hyderabad, India, from 14-17 Feb, 1992, pp 305-310.

Thool, V. R., Thool, R.C., Khandlikar, W.S. and More, A. G., 2008. Development of Microcontroller Based Grape Dryer. *IAALD AFITA WCCA2008. World Conference on Agricultural Information and IT*.

Tien, T. M., Gaskins, M. H. and Hubbell, D. H., 1979. Plant growth substances produced by *Azospirillum brasilense* and their effect on the growth of pearl millet. *App. Env. Microbiol.*, **37** : 1016-1024.

Tirello, P., Vettore, S., Pozzebon, A., Lorenzon, M. and Duso, C., 2011. Effects of some insecticides on *Kampimodromus aberrans*: laboratory and field studies. IOBC/wprs Bulletin, 67: 257-260.

Tripathi Pramila, Dubey, N. K. and. Shukla, A. K., 2008. Use of some essential oils as post-harvest botanical fungicides in the management of grey mould of grapes caused by *Botrytis cinerea. World J. Microbiol Biotechnol. 24:39–46.*

Tripathi, M. and Johri, B. N., 2002. *In vitro* antagonistic potential of fluorescent pseudomonads and control of sheath blight of maize caused by *Rhizoctonia solani. Ind. J. Microbiol.*, **42** : 207-214.

Tuite, J., 1969. *Plant pathological methods: Fungi and Bacteria*. Burgess Publishing Co. Minnepolis, USA.pp.239.

Turkben Cihat and Bahattin Cetin, 2001. Handling and Marketing Problems of Seedless Raisin Grapes in Turkey. *Small Fruits Review*. 1(4):73-79.

Upadhyay Anuradha, Deokar, K.P., Suvarna Reddy, Sawant, I.S., Kshirsagar, M.D. Saboji, M.D. and Karibasappa, G.S., 2006. Identification of micro satellite markers associated with downy mildew resistance in grape – An example of association mapping in perennial crops. ISHS Acta Horticulturae 785: *International Symposium on Grape Production and Processing, held at Baramati* (Pune), Maharashtra, India During 6–11th February 2006.

Upadhyaya, J. and Yadav, G.R., 1982. Chemical control of grape anthracnose *Sphaceloma ampelinum*. *Progressive Horticulture*, **14:** 23 – 25.

Usha K, Find all citations by this **author** (default).

Van Helden, M., Fulchin, E., Verpy, A., Gil, F. and Garcia, C., 2011. Adult monitoring improves control of the flavescence dorée leafhopper *Scaphoideus titanus* in Gironde (France) while using less pesticide. IOBC/wprs Bulletin, 67: 9-16.

Van Helden, M., Villate, L., Laveau, C., Morin, E., Darrieutort, G., Van Leeuwen, C., 2011. Monitoring nematode populations to adapt fallow periods against *Xiphinema* vectors of Grapevine fanleaf virus (GFLV). *IOBC/wprs, Bulletin.* 67:63-67.

Vasantha Kumar, G. K., 2006. *Status of viticulture in Karnataka-Past, Present and Future:* Paper presented at the *International Symposium on Grape Production and Processing*, at Baramati, Maharashtra (India) held on Feb 6-11, 2006.

Vasanthakumar, S.K., 2003. Studies on beneficial endorhizosphere bacteria on solanaceous crop plants. *M.Sc.* (*Agri.*) *Thesis*, Univ. of Agril. Sci., Dharwad (India).

Vasse Justine, Find all citations by this **author** (default).

Veena and Devanand, B. J., 2000. Role of plant growth promoting rhizobacteria on growth and yield of pigeonpea (*Cajapanus cajan* L.) cultivars. *M. Sc.* (*Agri.*) *Thesis*, Uni. Agric. Sci. Dharwad (India).

Veena, S. C., 1999. Development of inoculum consortia for enhanced growth and nutrient uptake of sorghum (*Sorghum bicolor* (L.) Moench). *M. Sc.* (*Agri.*) *Thesis*, Univ. Agric. Sci., Dharwad (India).

Veena, S. C., 2002. Development of inoculum consortic for enhanced growth and nutrient uptake of sorghum (*Sorghum bicolor* L.). *M.Sc.* (*Agri.*) *Thesis*, Univ. of Agril. Sci., Dharwad (India).

Venkataram Prasad, S. and Singh Rajendra, 1989. Phenotypic variability for qualitative and quantitative attributes of certain grape hybrids. *M. Sc. Thesis*, IIHR, Bangalore.

Venkataram, S. and Singh Rajendra, 1991. Phenotypic variability for qualitative and quantitative attributes of certain grape hybrids, IIHR

Venkatesh, H., Jamadar, M. M. and Shamarao Jahagirdar., 2001. Meteorological influence on anthracnose of grape in northern Karnataka. Paper presented in the *National Seminar of Agro meteorological Research for Sustainable Agricultural production*, held at GAU, Anand, on Sep 27-28, 2001 pp.148.

Venugopalan, R. and Rawal, R. D., 2006. A statistical model for predicting the outbreak of anthracnose in grapes. Paper presented in the *National symposium on Improving Input Use Efficiency in Horticulture*, IIHR, Bangalore, held on Aug 9- 11, 2006 pp.191-192.

Verghese Smita, R. and Murthy, B.N.S., 2004. Comparison of micropropagation efficacy among different grape hybrids with special reference to Shweta Seedless. *M. Sc. Thesis, IIHR, Bangalore*

Viala and Pacottet, 1906. Culture and development of anthracnose. *Review of Viticulture*, **22:** 117 – 121.

Vijay, N. and Venugopalan, R., 2010. Investigation and development of non-linear statistical models for disease forecasting in grapes. *IIHR, PG Education*, 317-318.

Vilches Ana, Find all citations by this **author** (default).

Vincent, J. M., 1927. Distortion of fungal hyphae in the presence of certain inhibitor. *Nature*, **159:** 800.

Viret, O., Dubuis, P.-H., Fabre, A.-L., Bloesch, B., Siegfried, W., Naef, A., Hubert, M., Bleyer, G., Kassemeyer, H.-H., Breuer, M., Krause, R., 2011. An interactive platform for a better management of grapevine diseases and pests. *IOBC/wprs, Bulletin*. 67: 85-91.

Walker, R. R. and Blackmore, D. H., Clingeleffer, P. R. and Lacono, F., 1997. Effect of salinity and ramsey rootstock on ion concentration and carbon diaonide assimilation in leaves of drip irrigated field grown grapevine (*Vitis vinifera* L.) cv. Sultana. *Aust.J. Grape and Wine Res.*, **3**(2): 66-74.

Wang Ting, Lv Xiu-Lan, Meng Duo, Tu Xun-Liang, Wu Xiao-Hua and Wu Shi- Lei, 2011. Research on Effects of Plant Growth Regulators on Quality of Manicure Finger Grape. *Plant Sciences Research*, **3**: 10-16.

Wang, Y., He, D., Lamikanara, O. and Lu, J., 1998. Resistance to Chinese *Vitis* species to *Elsinoe amopelina* (de Bary) Shear. *Horticulture Science*, **33:** 123– 126.

Wani, P. V., 1980. Studies on phosphate solubilizing microorganisms review. *J. Maharashtra Agric. Univ.*, **5** : 144-147.

Wani, P. V., More, B. B. and Patil, P. L., 1979. Effects of seed inoculation with some P solubilizing microorganisms on phosphate uptake and yield of gram. *J. Maharashtra Agric. Univ.*, **3** : 271-272.

Warusavitharana, A. J., Tambe, T. B. and Kshirsagar, D. B., 2006. Effect of cytokinins and brassinosteroid with gibberellic acid on yield and quality of Thompson seedless grapes. ISHS Acta Horticulturae 785: International Symposium on Grape Production and Processing.

Wasnik, H. M. and Bhargava, B. S., 1981. Changes in the characteristics of vineyard soil on continuous application of Fertilizers. *J. Ind. Soc. Soil Sci.*, **29**(4) : 554-556.

Weaver Robert J., Wasfy Shindy and W. Mark Kliewer, 1969. Growth Regulator Induced Movement of Photosynthetic Products into Fruits of 'Black Corinth' Grapes. *Plant Physiology*. 44 (2): 183-188.

Wheeler, B.E.J., 1969. An Introduction to plant diseases. John Wiley and Sons Limited, London, pp. 301.

Whiton, R. S. and Zoecklein, B. W., 2000. Optimization of Headspace Solid-Phase Microextraction for Analysis of Wine Aroma Compounds. *American Journal of Enology and Viticulture,* **51**(4): 379-382.

Whiton, R. S. and Zoecklein, B. W., 2002. Determination of Ethyl Carbamate in Wine by Solid-Phase Microextraction and Gas Chromatography/Mass Spectrometry. *American Journal of Enology and Viticulture.* 53: 60-63.

Whiton, R. S. and Zoecklein, B. W., 2002. Evaluation of Glycosyl-Glucose Analytical Methods for Various Glucosides. *American Journal of Enology and Viticulture.* **53**(4): 315-317.

Williams, J.G.K., Kubelik, A. R., Livak, K.J., Rafalski, J.A. and Tingry, S.V., 1990. DNA polymorphism amplified by arbitrary primers are useful as genetic markers. *Nuc. Acid. Res.*, **12:** 6521-6535.

Winkler, A. J., 1965. *General viticulture*. Cambridge University Press. London. Pp.380–381.

Wolfe Dwight and Brown Gerald R., 1992.Training system affects performance of table grape cultivars. *HortScience*. 27 (11): 1178-1183.

Wuest, P. J., Cole, H. Jr. and Sanders, P. L., 1974. Tolerance of *Verticillium malthousei* to benomyl. *Phytopathology*, **64:** 331-334.

Yarwood, C. E., 1946. Detached leaf culture. *Botanical Review*, **12:** 51–56.

Yashoda and Avasthe, R. K., 1995. Altitudinal distribution of micronutrients in the soils of Sikkim. *J. Ind. Soc. Soil Sci.*, **113**(3) : 374-377.

Yates, P., Panse, V. G. and Sukhatme, P. V., 1937. Statistical Methods for Agricultural Workers, Indian Council of Agricultural Research, New Delhi.

Yesh Pal Singh, Tiwari, J. P. and Misra, K. K., 2003. Effect of micronutrients on fruit yield and physico-chemical characters of mango cv. Dashehari. *Prog. Hort.*, **35**(1) : 34-37.

Zhan Wu Dai, Nathalie Ollat, Eric Gomès, Stéphane Decroocq, Jean-Pascal Tandonnet, Louis Bordenave, Philippe Pieri, Ghislaine Hilbert, Christian Kappel, Cornelius van Leeuwen, Philippe Vivin and Serge Delrot., 2011. Ecophysiological, Genetic, and Molecular Causes of Variation in Grape Berry Weight and Composition: A Review. *Am. J. Enol. Vitic.* **62**:413-425.

Zhang, A.W., Riccioni, L., Pedersen, W. L., Koilpara, K.P. and Hartman, G.L., 1998. Molecular identification and phylogenetic grouping of *Diaporthe phaseolorum* and *Phomopsis longicolla* isolates from soybean. *Phytopathology*, **88:**1306-1314.

Zhang, D., Zhang, T. Q. and Zhang, F.J., 1991. Biological characteristic and epidemiology of *Elsinoe ampelina* (de Bary) Shear. on grapes. *Acta Agriculture-Shanghai*, **6:** 59-63.

Zoecklein B. W. and Fugelsang, K. C., 2003. Population Dynamics and Effects of *Brettanomyces bruxellensis* Strains on Pinot noir (*Vitis vinifera* L.) Wines. *American Journal of Enology and Viticulture.* **54**(4):294-300.

Zoecklein Bruce W., Wolf Tony K., Lisa Pélanne, Miller Kay M. and Sandy S. Birkenmaier, 2008. Effect of Vertical Shoot-Positioned, Smart-Dyson, and Geneva Double Curtain training systems on Viognier grape and wine composition. *Am. J. Enol. Vitic.* 59:1.

Zoecklein Bruce, W. Tony K. Wolf, Lisa Pélanne, M. Kay Miller and Sandy S. Birkenmaier, 2008. Effect of Vertical Shoot-Positioned, Smart-Dyson, and Geneva Double-Curtain Training Systems on Viognier Grape and Wine Composition. *Am. J. Enol. Vitic.*, 59 (1): 11-21

Zoecklein, B. W., Douglas, L. S. and Jasinski, Y. W., 2000. Evaluation of the Phenol-Free Glycosyl-Glucose Determination. *American Journal of Enology and Viticulture*, 51 (4): 420-423.

Zoecklein, B. W., Williams J. M. and Duncan, S. E., 2000. Effect of Sour Rot on the Composition of White Riesling (*Vitis vinifera* L.) Grapes. *Small Fruits Review.* **1**(1): 63-77.

Zoecklein, Bruce W., Devarajan, Yamuna, S., Kumar Mallikarjunan and Denise M. Gardner. 2008. Monitoring the effect of ethanol spray on Cabernet franc and Merlot grapes and wine volatiles using electronic nose systems. *International Journal of Food Engineering.* 4 (2): 14.

Zoffoli, J. P. 2006. Postharvest handling of table grape. ISHS Acta Horticulturae 785: *International Symposium on Grape Production and Processing, held at Baramati* (Pune), Maharashtra, India During 6–11th February 2006.

Appendices

Annexure-I
Soil Analysis Report

Soil Type				*Clay,* *Clay loam,* *Silt loam,* *Silt clay loam,* *Sandy loam*
Parameter	*Unit*	*Value*	*Normal Range*	*Remark*
pH	–	–	4.00-5.00	Acidic, use of lime is necessary
			5.00-6.00	Acidic
			6.00-7.00	Optimum
			7.00-8.00	Alkaline
			8.00-8.60	Very alkaline saline
			> 8.60	
$CaCO_3$	Per cent		Less than 1.00	Optimum
			1.00-2.00	Sufficient
			3.00-8.00	High
			8.00-15.00	Reduce availability of P, Zn
			More than 15.00	Reduce availability of P, Fe, Zn, Mn
Organic carbon			Less than 0.50	Deficient
			0.50-1.00	Low
			1.0-2.00	Sufficient
			More than 2.00	High
EC	dsm^{-1}		<1.00	Optimum
			1-2.0	Harmful
			2.0-4.0	Root stock needed
			more than 4.0	Not suitable for grape
Parameter	Unit	Value	Normal Range	Remark
Nitrogen	ppm		Less than 160	Low
			160-320	Sufficient
			320-480	High
			More than 480	Very high
Phosphorus	ppm		Less than 10	Deficient
			10-50	Low
			50-100	Sufficient
			More than 100	High
Potassium	ppm		Less than 180	Low
			180-500	Sufficient
			500-800	High
			More than 100	Very high

Contd...

Annexure-I–*Contd...*

Parameter	*Unit*	*Value*	*Normal Range*	*Remark*
Calcium	ppm		Less than 500	Low
			500-1000	Sufficient
			More than 1000	High
Magnesium	ppm		Less than 250	Low
			250-500	Sufficient
			More than 500	High
Sulphur	ppm		Less than 10	Low
			10-50	Sufficient
			50-100	High
Parameter	Unit	Value	Normal Range	Remark
			More than 100	Very high
Iron	ppm		Less than 2.50	Low
			2.50-5.00	Sufficient
			More than 5.00	High
			More than 25.00	Very high
Magnesium	ppm		Less than 2.00	Low
			2.50-5.00	Sufficient
			More than 5.00	High
			More than 50.00	Very high
Zinc	ppm		Less than 1.00	Deficient
			1.00-5.00	Sufficient
			More than 5.00	High
			More than 20.00	Very high
Copper	ppm		Less than 0.20	Low
			0.20-0.50	Sufficient
			More than 0.50	More than required
			More than 10.00	Very high
Sodium	ppm			Optimum harmful

Annexure-II
Water Evaluation Report

Parameter	*Unit*	*Value*	*Normal Range*	*Evaluation/rank*
pH			6.00-7.00	Safe
			7.00-8.00	Manageable
			>8.00	Alkaline
Electrical conductivity	dSm^{-1}		<1.00	Safe
			1.00-2.00	Problematic
			>2.00	Root stock necessary
			>3.80	Tolerance limit of Bangalore Dogridge
			>6.00	Yield reduction of 10 per cent even after use of Bangalore Dogridge, yield reduction by 25 per cent even after use of Bangalore Dogridge
Chloride	mel^{-1}		3.0	Safe
			3.00	
			3.0 to 4.00	Tolerance limit of grape variety on its
			>4.00	Use of own roots
			18-20	Rootstock necessary tolerance limit
			21-24	Dogridge as rootstock, yield reduction by 10 per cent
			26-28	Even after use of Bangalore Dogridge, yield reduction by 25 per cent even after use of Bangalore Dogridge
Nitrate-N	ppm		<10	Safe
			>10	Harmful-unsafe for drinking but useful for irrigation
Carbonate	mel^{-1}			
Bi-Carbonate	mel^{-1}			
Calcium	mel^{-1}			
Magnesium	mel^{-1}			
Sodium	mel^{-1}			Low and safe
				High and unsafe
Potassium	mel^{-1}		–	–
SAR	mel^{-1}		<10	Low and safe
			>10	High and unsafe

Annexure-III
Petiole Analysis Report
(October Pruning)

Parameter	*Unit*	*Value*	*Normal Range*	*Remark*
Nitrogen	per cent		Less than 0.87	Deficiency of nitrogen
			0.87-1.31 1.32-2.21	Hidden hunger, not sufficient for growths and production
			2.22-2.66 >2.66	More than required for growth development of bunches very high for growth development of bunches, growth will be fast.
NO_3-N	ppm		less than 105	Deficiency of nitrogen
			105-400	Hidden hunger, need for more N containing fertilizers and manure.
			400-747	Sufficient/optimum
			748-1168	More than required
			1200-2400	High
			More than 2400	Very high
Phosphorus	per cent		Less than 0.19	Deficiency of phosphorus
			0.020-0.37	Low status, not enough for growth and of bunches
			0.38-0.75	Sufficient, apply recommended manure fertilizer
			0.76-0.95	High, for higher production P level is useful
			More than 1.00	Very high, apply P as per recommended.
Potassium	per cent		Less than 0.60	Deficiency of potassium
			0.60-1.13	Hidden hunger, need for more nitrogen fertilizers and manure.
			1.11-2.20	Sufficient use as per recommended
			2.21-2.73	More than required, still use of K fertilizers is necessary
			More than 3.00	Very high, use limited quantity of K
Calcium			Less than 0.53	Deficiency of calcium
			0.54-0.73	Hidden hunger, use super phosphate and foliar spray of calcium containing compounds
			0.74-1.14	Sufficient containing compound
			1.15-1.35	More than required. If Ca deficiency is noticed use foliar spray or dipping of calcium containing compound
			More than 1.50	Very high

Contd...

Annexure-III–*Contd...*

Parameter	*Unit*	*Value*	*Normal Range*	*Remark*
Magnesium	per cent		Less than 0.30	Deficiency of magnesium
			0.31-0.49	Hidden hunger, need to use Mg.
			0.50-0.80	Sufficient status, use mg as per recommendations
			0.81-1.00	More than required
			More than 1.00	Very high
Sulphur	per cent		Less than 0.07	Deficiency of sulphur
			0.07-0.13	Hidden hunger, use sulphate of potash
			0.14-0.27	Sufficient/optimum
			0.28-0.34	More than required
			More than 0.40	Very high
Iron	ppm		Less than 10	Deficiency of iron
			10-29	Hidden hunger, use of Fe compound needed
			30-80	Sufficient
			81-200	More than required
			More than 200	Very high
Fe^{++} Iron	ppm		Less than 10	Deficiency of ferrous
			10-25	Hidden hunger, need for Fe
			More than 25	Sufficient
Manganese	ppm		Less than 26	Deficiency of manganese
			26-75	Hidden hunger, need for use of Mn containing compound
			76-174	Sufficient
			175-223	More than required
			More than 300	Very high
Zinc	ppm		Less than 13	Deficiency of zinc
			13-52	Hidden hunger, need for use of Zn
			53-132	Sufficient, need for foliar spray of zinc sulphate
			More than 200	Very high
Copper	ppm		Less than 2.0	Deficiency of copper
			2.0-4.9	Hidden hunger, however, there is no need to apply Cu
			5.0-10.0	Sufficient
			10.0-100.00	More than required
			More than 100	Very high

Contd...

Annexure-III–*Contd...*

Parameter	*Unit*	*Value*	*Normal Range*	*Remark*
Sodium	per cent			
Water soluble K	per cent		Less than 0.60	Deficiency of potassium
			0.60-1.13	Hidden hunger, need for more N containing fertilizers and manure
			1.14-2.20	Sufficient use as per recommended
			2.21-2.73	More than required, Still use of K fertilizer is necessary
			More than 3.00	Very high, use limited quantity of K

Annexure-IV
List of Chemicals Required for Compulsory Testing for Export (Mg/Kg)

Sl.No.	Chemical	European Union	UK	Dutch
	Organo Chlorine Chemical			
1.	Aldrin	Banned	0.05	0.05
2.	Chlordane	Banned	0.02	0.02
3.	Chloro thalonil	1.00	1.00	1.00
4.	DDT	Banned	0.05	0.05
5.	Dichloflunid	10.00	15.00	10.00
6.	Dicofol	2.00	2.00	2.00
7.	Dildrin	Banned	0.05	0.05
8.	Endosulfan	0.50	0.50	0.50
9	Endrine	0.01	0.01	0.01
10.	Lindane	0.50	0.01	0.50
11.	SCH	#	#	0.01
12.	Heptachlore	0.01	0.01	0.01
	Organophosphorus			
13.	Accephate	0.02	0.02	0.02
14.	Ageenphos methyl	1.0	2.0	1.00
15.	Chloroenivphos	0.05	0.05	0.05
16.	Chlorpyriphos	0.50	0.30	0.50
17.	Chlorophyriphosmethyl	0.20	0.20	0.20
18.	Diazinon	0.02	0.02	0.02
19.	Dichlorovos	0.10	01.10	0.10
20.	Dimethoate	1.0	0.02	0.02
21.	Ethion	0.50	0.50	0.50
22.	Etrimphos	#	#	0.05
23.	Fenchlorophos	0.01	0.01	0.01
24.	Fenitrothion	0.50	0.50	0.50
25.	Malathion	0.50	0.50	0.50
26.	Profenophos	0.50	0.50	0.50
27.	2-Chlorophenol	–	–	–
28.	Methamidaphos	0.01	0.01	0.01
29.	Methedethion	0.50	0.50	0.50
30.	Mevinphos	0.10	0.10	0.10
31.	Parathionethyl	Banned	#	0.05
32.	Phoshomidon	0.15	#	0.15

Contd...

Annexure-IV–*Contd...*

Sl.No.	*Chemical*	*European Union*	*UK*	*Dutch*
33.	Fosmite	Banned	#	0.05
34.	Parathion methyl	0.20	0.20	0.20
35.	Quinolphos	0.05	0.05	0.05
36.	Triazophos Cynthate parathroid	0.02	0.02	0.02
37.	Gyelothrin	0.30	0.30	0.30
38.	Cyhalothrin lamda	Banned	#	0.02
39.	Cypermethrin	0.50	0.50	0.50
40.	Deltamethrin	0.10	0.10	0.10
41.	Fenvalarate	0.02	0.02	0.02
42.	Parmithrin	1.00	0.05	1.00
	Trizins			
43.	Altrazin Asayalannins	0.10	0.10	0.10
44.	Banlexil	0.20	0.20	0.20
45.	Metalaxil	2.0	2.0	2.0
	Carbamate			
46.	Carbofuron	0.10	0.10	0.10
47.	Methomyl	0.05	0.05	0.05
	Paramidins			
58.	Fenarimal	0.05	0.05	0.05
	Trizols			
59.	Bitertenol	0.05	0.05	0.05
60.	Fusiloczol	Banned	Banned	Banned
61.	Hexaconozol	0.02	0.10	0.10
62.	Propeconizol	0.50	0.50	0.50
	Imidozoles			
63.	Imizalil	0.02	0.02	0.02
64.	Captofol	0.02	0.02	0.02
65.	Captan	3.00	3.00	3.00
	Dicarboximes			
66.	Imrodine	10.00	10.00	10.00
67.	Prosamidon	5.0	5.0	5.0
68.	Vinchlozolin	5.0	5.0	5.0
	Benzimizol			
69.	Carbandizim	2.0	2.0	2.0
70.	Thiophenate methyl	1.0	2.0	2.0

Contd...

Annexure-IV–*Contd...*

Sl.No.	*Chemical*	*European Union*	*UK*	*Dutch*
	Thiocarbamate Fungicide			
71.	Mancozeb	0.01	2.00	2.00
72.	Maneb	2.00	2.00	2.00
73.	Thiram	0.01	#	3.00
74.	Propinab	0.01	#	0.02
75.	Ethylene Thiourea	0.01	#	0.02
	Others			
76.	Thiomethaxam	@#	@#	@#
	Oxydematon methyl	0.02	0.02	0.02
77.	Fostylal	#	–	0.20
78.	Copper oxychloride	2.00	20.00	20.00
79.	Tridemefon	–	2.00	2.00
	Penaconozol	#	2.00	2.00
80.	Sulphur	50.00	#	50.00
81.	Microbutanil	1.00	#	1.00
	Denocap	#	#	0.05
	Cymaczonil	0.01	3	0.05
	Carbaryl	3.00	3.00	3.00

As per Netherland MRL can be used @ # Syngenta Co. MRL 0.1 mg/kg, but further reduced to 002 mg/kg.

[*Courtesy* : Dept. of Horticulture, Maharashtra State Pune-411005].

Index

Thompson Seedless

Sonaka Seedless

Sharad Seedless

Flame Seedless

Red Globe

Superior Seedless

Fruit Bearing on Y-Trellis

Drip Irrigation with Diffusor

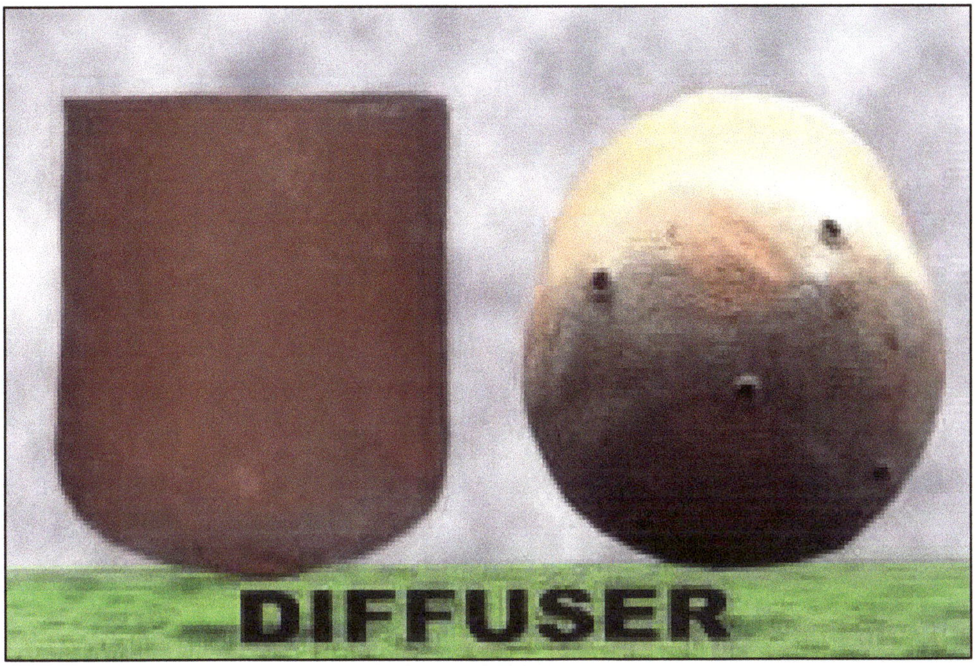

Diffuser (Drip irrigation system)

Covering with Paper to Avoid Sun Burn

Training System Y - Trellis

Mealy Bug
Courtesy - N.D. Sunita

Flea Beetle Damage
Courtesy - Dr. N.D. Sunita

Flea Beetle Damage
Courtesy - Dr. N.D. Sunita

Thrips Damage on Berries
Courtesy - N.D. Sunita

Mealy Bug on Stem
Courtesy - N.D. Sunita

Mealy Bug on Stem
Courtesy - N.D. Sunita

Sooty Mould on Leaf Due to Mealy Bug
***Courtesy* - N.D. Sunita**

Stem Borer Excreta
***Courtesy* - Dr. S.B. Jagginavar**

Mealy Bug in Bunch
Courtesy - N.D. Sunita

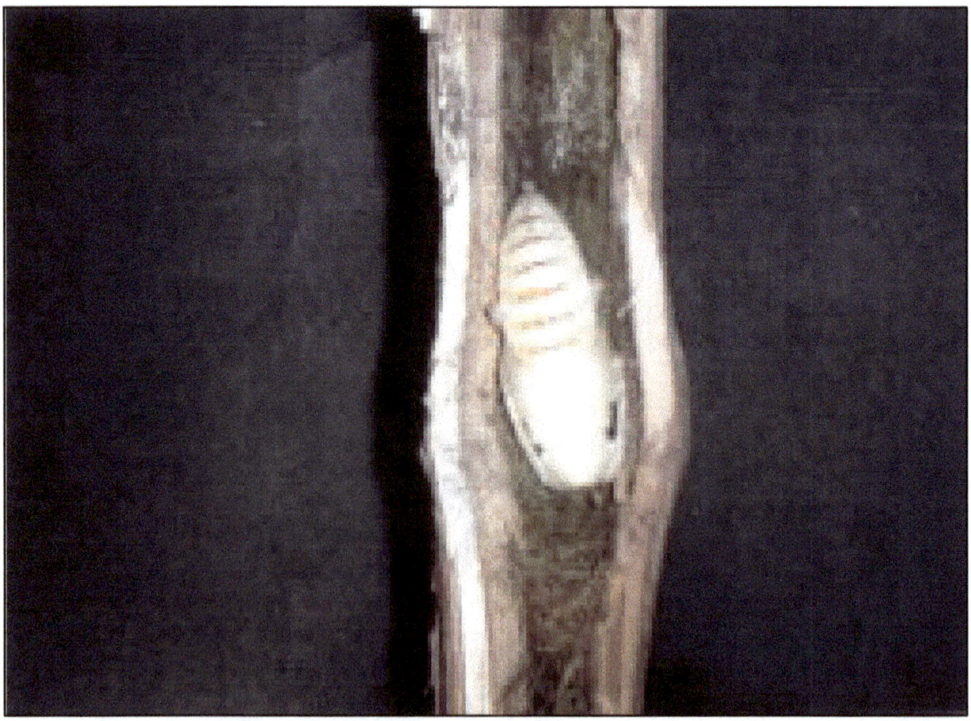

Stem Borer Pupae
Courtesy - Dr. S.B. Jagginavar

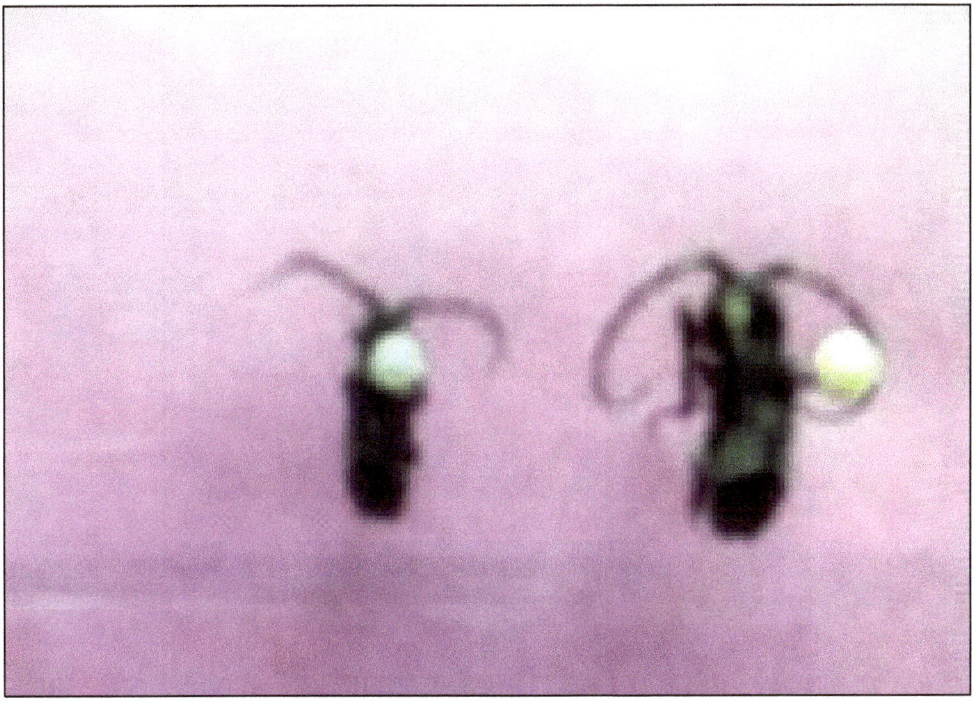

Stem Grinder
Courtesy - Dr. S.B. Jagginavar

Stem Borer Larvae
Courtesy - Dr. S.B. Jagginavar

Thrips

Courtesy - N.D. Sunita

Downy Mildew

Powdery mildew

Powdery mildew

Downy Mildew

Anthracnose Spots on Leaves and Shoots of Thompson Seedless

Grape Drying

Export Grape Box

Raisin Grading

Grading Machine

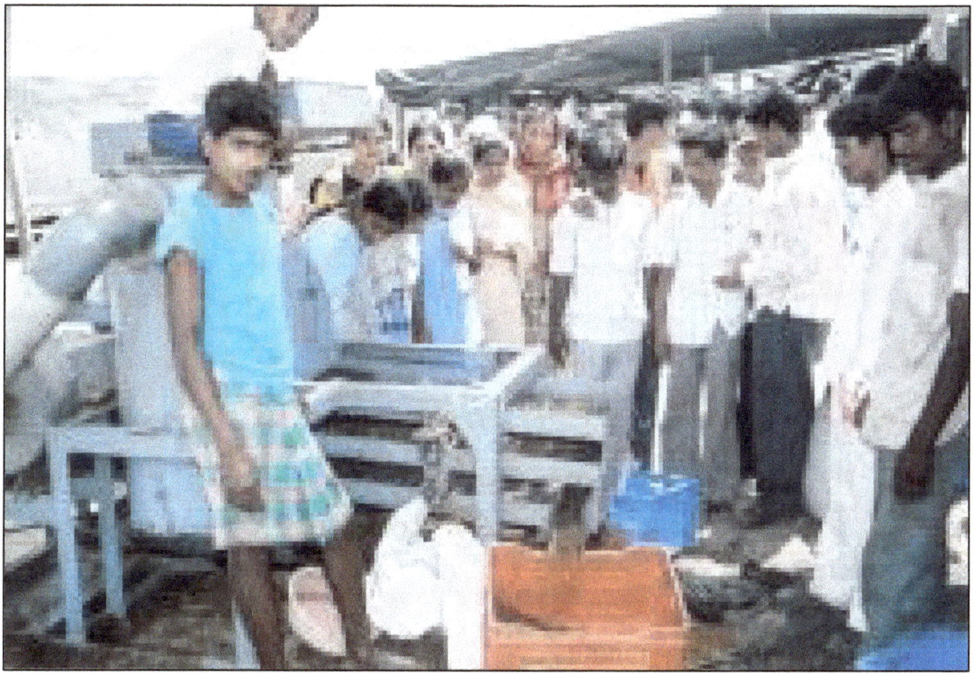

Grading Machine

Raisin in Cold Storage

Raisin Shed

Pre-cooling Chamber

Refrigerated Van

Cold Storage

Palletization of Grape

Pouch in Box

Export Grape Packing

Grape in Pouch

Stem Browning

Blemish

Wind Rub-stems and Berries

Shriveling

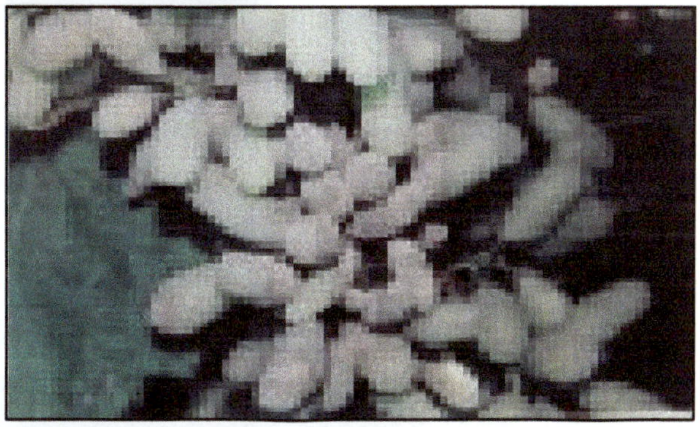

Powdery Mildew

Spray Damage

Surface Contamination

Circular Cracking

Splitting

80% Cap-Fall

100% Cap-Fall

Fruit Set

4 mm Berry size

6-8 mm Berry size

Girdling Stage

Bunch Closure

Woolly Bud

Bud Emergence

Three Leaf Stage

Bunch Emergence

Seven Leaf Stage

Pre-bloom Stage

Flowering Initiation

www.ingramcontent.com/pod-product-compliance
Ingram Content Group UK Ltd.
Pitfield, Milton Keynes, MK11 3LW, UK
UKHW022214051026
15624UKWH00001B/158

9 789351 301240